Air Pollution Reviews – Vol. 5

Urban Pollution and Changes to Materials and Building Surfaces

Editor

Peter Brimblecombe

City University of Hong Kong, Hong Kong

ICP

Imperial College Press

Published by

Imperial College Press
57 Shelton Street
Covent Garden
London WC2H 9HE

Distributed by

World Scientific Publishing Co. Pte. Ltd.
5 Toh Tuck Link, Singapore 596224
USA office: 27 Warren Street, Suite 401-402, Hackensack, NJ 07601
UK office: 57 Shelton Street, Covent Garden, London WC2H 9HE

Library of Congress Cataloging-in-Publication Data
Urban pollution and changes to materials and building surfaces / editor, Peter Brimblecombe, City University of Hong Kong, Hong Kong.
pages cm. -- (Air pollution reviews ; vol. 5)
Includes bibliographical references and index.
ISBN 978-1-78326-885-6 (hardcover : alk. paper) -- ISBN 1-78326-885-9 (hardcover : alk. paper)
1. Weathering of buildings. 2. Urban pollution. I. Brimblecombe, Peter, 1949– editor.
TH9039.U73 2016
624.1028'8--dc23
2015029899

British Library Cataloguing-in-Publication Data
A catalogue record for this book is available from the British Library.

Cover photo credits: Zhi Ning (background) and Carlota Grossi (inset)

Contents

CHAPTER 1

Past, Present and Future Damage to Materials and Building Surfaces in the Polluted Urban Environment

Peter Brimblecombe

School of Energy and Environment, City University of Hong Kong, Hong Kong

1.1 Introduction

Our cultural property is meant to last. We build in stone and durable metals. The memorials of our heritage are constructed to survive many centuries, thus damage processes have to be considered on such long timescales. It has become common to talk of our crumbling heritage as the victim of air pollution, sea-level rise, urban development, poor taste, etc. At such times, it is easy to see these processes as recent and to forget that the buildings we are so anxious to preserve have been confronted by continual environmental and social change. Buildings are records both of the forces that created them and those which have altered them. The patina they accrete is in itself a historic record (Sgobbi *et al.*, 2010).

Thus it is important to see buildings in their historical context. This is true of their artistic and social history, but here the primary concern is their environmental history. As with social history, it is partly to help us interpret the current form of the building, but also to aid with an understanding of the physical changes that have taken place. It further offers the opportunity to gain insight into the long term mechanisms of damage that can stretch for many hundred years. This story of long term damage to the built environment develops from an earlier one written some years ago (Brimblecombe and Camuffo, 2003).

1.2 Early Observations of Damage

Even in classical times the accretion of environmental damage to buildings was recognised. In his *Account of Egypt* Herodotus (~484–425 BC) described salt damage to the pyramids: "and an efflorescence of salt forms upon the surface, so that even the pyramids are being eaten away by it ...." Wind-blown sand buried the Great Sphinx, but recently there has been a wider concern over the accumulation of sand against the walls of monuments of Saharan Africa, especially where the margins of the desert are shifting. In ancient Rome, the poet Horace complained of smoke blackening to the temples. In religious buildings, smoke encrustations might be seen as desecration; but in the case of secular buildings, concern required an appreciation of architecture as an art form, before it received wide comment. This was reflected in the 17th century by discussions of smoke damage to English buildings (e.g. Evelyn, 1661). Even though there was a predominance of interest in the damage to religious buildings, this should not be taken to mean that it was primarily a religious concern.

1.2.1 *Early modern period*

Architects have often seen the environment as relevant to the long-term survival of buildings. In the early modern period, they began to believe that damage was caused by time, weather and smoke. Sir Christopher Wren and his colleagues recognised the interaction of urban sulfur dioxide with stone as seen from as thick corrosive crusts that are derived from sulfation processes, which had become dominant in 17th century London where large quantities of coal was burnt (Brimblecombe, 1992).

It was widely accepted that coal smoke was a most important cause of damage. Keepe (1683) wrote that at Westminster Abbey you see more: "the skeleton of a church than any great comeliness in her appearance, being so shrivelled and parcht by the continual blasts of the northern winds, to which she stands exposed, as also the continual smoke of the sea-coal which are of a corroding and fretting quality."

Two centuries later the cities of the Victorian period were so badly polluted that building stones had to be carefully chosen for new structures. Major construction activities often began with the formation of a select committee of experts — architects, chemists and stonemasons, who would deliberate over the right choice of stone to resist the polluted urban atmosphere. The Palace of Westminster (1837–1870), even though it was constructed after much thought on the choice of an ideal stone, began to manifest signs of stone decay in 1847. This showed that expert opinion on such matters did not always guarantee that buildings would not succumb to the enhanced weathering (Scott, 1861). At Westminster Abbey, a committee of architects and chemists reported on the best means of preserving the stonework (Carpenter, 1966). They used a scientific approach and concluded that disintegration resulted from acid vapours, particularly sulfuric acid, although they also saw hydrochloric acid as additionally responsible.

Writers of the Victorian period had many theories to explain stone decay so prevalent in their cities. The erratic classical scholar Paley (1878) argued that the blackening of St. Paul's cathedral was really due to the presence of lichens and should be treated with the biocide copper sulfate. While at York, the vulcanologist, Tempest Anderson, wrote that stone decayed from both air pollution and biological attack (Anderson, 1911). However, smoke abatement was increasingly seen as the most appropriate solution, and was strongly advocated by ecclesiastical officials and architects alike.

Pittaluga (1994) and Cesarino and Pittaluga (1995) studied past pollution by means of a stratigraphic analysis of the gases which have been absorbed and the particles are present within the multiple layers of paint found on the historical buildings in Genoa. Knowing the date of each paint layer, it was possible to establish the intervals of time across which each deposit of pollutants was formed. Ausset *et al.* (1998) have studied the history of air pollution and building damage through a direct examination of old crusts on stone. Ancient crusts of buildings have revealed unburnt wood fragments alongside micro-spherules from fossil fuel combustion (Bonazza *et al.*, 2009; Sgobbi, 2010).

1.2.2 *Early restoration*

Major restoration work has a very long history. There were many attempts to repair the damage to ancient monuments in Egypt and Herodotus wrote in classical times that Amasis II (570–526 BC) "...dedicated great colossal statues and man-headed sphinxes very large, and for restoration he caused [material] to be brought from the stone-quarries." Rome's Forum was in a state of continuous restoration, reuse and rebuilding for most of its millennium. By the Renaissance, there was considerable interest in the restoration of antiquities and despite a desire for restoring sculptures to completion, treatises on sculpture, such as Orfeo Boselli's *Observations Concerning Ancient Sculpture*, stress a conservatism and careful study of ancient materials.

In Britain, the 1700s saw great interest in the restoration of ecclesiastical buildings in response to pollution damage. At York Minster, many houses whose smoke damaged the fabric were demolished under the guidance of the Dean. Deposits of smoke and scurff (an archaic word for a saline or sulfurous encrustation) were removed from the stonework during cleaning operations (Drake, 1736). The aim of the repair work was to "stop up all the Cracks, Flaws and Perishing of the Stones, with excellent Cement and Mortar, that ... this Fabrick might yet bid Defiance to Time and Weather for many succeeding Generations". In a similar vein, Nicholas Hawksmoor concluded his repairs at Westminster "will stand 1000 years" (Westlake, 1863). Architects involved in such large scale restorations failed to appreciate the impermanence of their repairs. Interest gradually spread to dwellings, and in London of the 18th century, the rate of darkening to the paintwork of some houses was so rapid that repainting needed to be frequent (Malcolm, 1770).

By the nineteenth century, the Italian archaeologist Giacomo Boni was opposing restoration that led to the creation of historic fakes, while at the same time acknowledging the need to stop decay (Sgobbi *et al.*, 2010). Animated debates created an interest in the principles of conservation and restoration. This was not universal though, and despite Ruskin's sympathy with his colleague Boni, restoration, both in England and Italy, was often very aggressive, with an enthusiasm for induratating decaying stonework with protective solutions. Materials such as linseed oil, beeswax, paraffin, gums and resins were all tried, but these techniques were ultimately discredited. Inorganic materials were also applied to the stone. Baryta-water (barium hydroxide) was thought to prevent damage (Jackson, 1901; Church, 1901). Desch (1921) also recommended "magnesium silicofluoride", and for the more porous sections of the stonework, "double silicofluoride which contains zinc" to prevent the growth of vegetation. These techniques also failed and the early 20th century applications of silicates had produced grey patchy discolouration pitted with small spalls. On hindsight, it is now accepted that those who indurated stone during the period 1850s–1920s had often wreaked more havoc than centuries of decay and weathering (Ashurst, 1985).

1.2.3 *Responses in design*

The Victorians wrote that coal smoke was imparting a "new colour" to towns, but aesthetic opinions about this were divided. Some felt time and nature effected a "softening" appearance on urban buildings and accepted the effects of smoke, while others claimed that all that was required to bring colour back to London was to regularly wash buildings with water. Boni argued that cleaned buildings should not be "statue di sale" (i.e. made of salt or 'white') and suggested that pigmentation be incorporated into restored façades such that they should not appear too dirty or too clean. Today there is a reaction against the over-cleaning of buildings that seem to have been scraped clean of their history (Grossi and Brimblecombe, 2004a) and that our most significant and appreciated buildings have lost their patina. Grossi and Brimblecombe (2008) conclude that the rapid changes in the tone of facades is not always welcome, even where they are the result of buildings being cleaner.

By the late Victorian times, the issue of air pollution encompassed many aspects of the design of urban buildings, not just degradation or colour change. Architects recognised that within cities there were problems of overcrowding, access to light, air pollution, stone discolouration indoor pollution and inappropriate building styles. The urban environment needed to be an aspect of their designs. At the end of the 19th century, the range of responses included (Bowler and Brimblecombe, 2000a):

(1) abandonment of Gothic and promotion of less detailed classical styles not so susceptible to air pollutants,
(2) careful choice of pollution-resistant stone and glazed materials,
(3) extensive provision of clear window glass to let in more light,
(4) adoption of electric light and air filtration to lower pollution indoors, and
(5) development of dust excluding interior fittings to protect items in storage or on display, such as books.

Today, it is possible that modern architects are better trained in the issue of contemporary air pollution. They might recognise that in the past buildings were discoloured by coal smoke or corroded by sulfur dioxide, but it remains to be seen whether their buildings will respond well to the changing urban atmosphere and climate with more ozone and diesel soot along with rising temperature and changing precipitation patterns.

1.3 Long Term Changes in Air Pollution

1.3.1 *Past air pollution*

Urban air pollution was known from classical times. Imperial Rome, with a population in excess of one million, consumed enormous amounts of wood, such that temples were blackened by soot and the health of sensitive individuals were affected. Citizens had some redress under the Roman law if their property rights suffered interference and legal decisions against the production of smoke are known

(Brimblecombe, 1987a). Patterns of fuel use and more particularly, changing fuel types seemed to trigger concern about urban air pollution in the past. Especially noticeable was the transition to coal as a fuel. There are important differences in the historical development of fuel use, control policies and the climate of various European cities. It is possible to estimate the changes in pollutant concentration occurring within a city over the centuries by modelling the emissions that result from the use of fuels. This has been done for London (Brimblecombe, 1977), York (Hipkins and Watts, 1996) or American cities such as Boston (Sherwood and Bambaru, 1991) or Pittsburgh (Davidson, 1979). Alternatively, there have been studies of the exposure of individual monuments such as York Minster (Bowler and Brimblecombe, 1991) or Trajan's Column (Camuffo, 1993).

Studies of the long term change in air pollution may be based on the history of fuel use and be determined from civic and legal records, travellers' descriptions and medical writings. In some cities, scientific measurements of air pollutants began in the late 19th century. London has an interesting air pollution history because the shift from wood to coal as a fuel began so early. While the transition there began in the 13th century, it was as late as the early 20th century in some parts of Europe. When the fuel is changed, it may be accompanied by a noticeable change in the quality of the smoke. The unfamiliar smell of coal smoke led to early fears about the health risk, through the belief that disease was carried in malodorous air (miasmas). Coal in 13th century London was mostly used for industrial processes such as lime-mortar production and it was not until the late 16th century, with the widespread construction of household chimneys, that the fuel began to be used domestically (Brimblecombe, 1987). At this period it is possible to find early observations of damage. John Thornbrough, the Dean of York Minster, suggested that coal should be desulfurised to make it less offensive. In the early years of the 17th century, King James I complained of the terrible damage done to the fabric of St. Paul's Cathedral and Archbishop Laud fined the brewers upwind for damaging the same building (Brimblecombe, 1987). The diarist John Evelyn (1661) described the effects of smoke in his book *Fumifugium,* which illustrates the way in which air pollution made churches and palaces look old, clothes and furnishings foul, and paintings yellow.

The late 18th and 19th centuries saw an enormous amount of industrialisation across Europe. The development of the steam engine and the intense urbanisation created very special problems. In England, civic administrators soon became aware that their old legislative frameworks were inadequate to cope with pollution on this much-enlarged scale (see Manchester's problems in Bowler and Brimblecombe, 2000b). The problems continued throughout the 19th century with the development of much sanitary legislation concerned with the health of urban populations. These beginnings to modern pollution environmental legislation reflect a desire for cleaner air, but it was often thwarted by weak laws that were erratically applied and poor pollution abatement technology. The regulations usually failed to achieve substantial abatement of smoke in the 19th century. However, the framework initiated pressure on smoke emissions from industry; either their workers had to be trained or they had to be replaced by automated stoking. Gradually, the pall of smoke over cities lessened.

1.3.2 *Late 20th century*

The late 20th century has seen a gradual decline in the use of coal in cities of Europe and North America. This has been replaced by oil, gas or electricity (often generated outside the urban area). These changes and the inreasing importance of vehicles as a source of air pollutants have led to the emergence of an entirely new kind of air pollution — photochemical smog. This arises from reactions in the atmosphere, when volatile organic compounds (primarily from liquid fuels), sunlight and nitrogen oxides interact. The key damaging products from this change have been oxidants such as ozone. The late 20th century has also seen an increasing development of new air quality regulations. This has placed increasing pressure on emissions from solid fuel combustion initially and then emissions from motor vehicles. A shift to electricity has led to a sharp decline, in many countries, of the concentrations of sulfur dioxide and coarse soot from coal burning.

As a result, for the decades in the last half of the 20th century, many parts of Europe witnessed a pronounced decrease in the rate of damage to and discolouration of building facades in cities. This sensitised the public to any future discolouration. In particular, the increasing concentrations of diesel soot in the atmosphere were a special concern, with vehicular sources increasingly dominant (Montana *et al.*, 2012). London has shifted from a coal-burning to petroleum-burning phase; crusts are thinner and more brown than black — less gypsum cementation so less continuous (La Russa *et al.*, 2013). Particulate carbon in recent years has become associated with significant fractions of organic matter. The popularity of diesel engines has increased the emissions of fine particles and organic content of deposited soot that blacken buildings (Bonazza *et al.* 2007).

Ozone is the most characteristic compound found in the modern photochemical smog. Smogs were first noticed in Los Angeles during the Second World War and were actively studied in the 1950s when the mechanism of their formation began to be understood. At first, it was assumed that this type of smog would be restricted to warm sunny climates, but soon it began to be found in most modern cities. It profoundly influenced the nature of modern urban air, but also the way in which it affects materials, through its profoundly oxidising nature.

1.3.3 *Air pollution in the 21st century*

The modern urban atmosphere, while free of coal smoke, has been increasingly polluted with fine particles from diesel vehicles. This means that the blackening of facades has become a important contemporary issue. It has additionally suggested the potential for elemental carbon (often diesel soot) to be a key measure of aesthetic damage to the lighter coloured urban fabric. Public concern, once heightened by the attack on buildings by acid rain has been replaced with worries about the discolouration of buildings by soot. Soot may cause less physical damage than sulfation, but it means that understanding the perception of building facades is important. This will often need to go beyond blackening as in some modern urban environments oxidation of the soot, deposited organic material or minerals can impart a warmer tone to buildings (Grossi and Brimblecombe 2008). Rainfall can mobilise soot deposits and lead to an ugly streaking that further adds to the disfiguring effects of deposited soot. Over time, rain can clean building surfaces, but the intermediate appearance can be far from pleasing (Grossi and Brimblecombe, 2004).

The future will doubtlessly see continued pressures in the European Union and elsewhere to improve air quality. The regulatory framework pays scant attention to the preservation of buildings, although suggested guidelines tend to be footnoted in technical documents that relate to regulatory proposals or as secondary or optional regulations. The vehicle fleet is still increasing in many parts of the world, so improved technologies are definitely needed to maintain air quality. Emission standards for vehicle engines have tightened, although there may still be a mismatch between the aspiration on engine manufacturers and the on-road performance of the vehicles.

Climate change also creates the potential for shifts in urban air quality. Warmer winters decrease the need for heating in cool climates, while higher summer temperatures can lead to a wider adoption of air conditioning and therefore increased air pollution from energy generation. Higher temperatures also result in greater evaporation of solvents and fuels. However, solvents are increasing well-regulated in many countries, while improved design of fuel lines and petrol tanks mean less evaporative loss of fuels. These organic compounds represent the VOCs (volatile organic compounds) that enter reaction sequences that lead to ozone formation in photochemical smog, under potentially longer sunshine hours of the future. It might well be that regulations keep the VOCs at levels where the precursors to ozone creation remain stable. However, regulation has less control over the regional release of biological volatile organic compounds (BVOCs) that can also take part in the formation of ozone. Higher temperatures seem set to increase the release of BVOCs from forests.

The changes in air pollutants that have led to the contemporary urban atmosphere have been largely beneficial to the recession of stone facades, most notably through a reduction in the sulfur dioxide concentrations. The changes may not be equally beneficial to all materials because for some it is other air pollutants, such as ozone, that have become more important. The atmosphere has become more oxidising as a result of photochemical processes that are driven by sunlight and volatile organic compounds from automotive fuels. These smog components, most especially ozone, can attack some metals, perhaps copper and aluminium, and more particularly degrade polymers that are now an important component of buildings particularly in joints and as coatings (Brimblecombe and Grossi 2010).

1.4 Long-Term Recession Rates to Stone

There are long-term observations of recession rates of stone especially at St. Paul's Cathedral, but also from Parisian buildings along with changes in the composition of crusts. Brimblecombe and Grossi (2008) used historical data on the historic and projected climate and air pollution for London along with modelling to estimate the changing recession of limestone for the period 1100–2100 CE. Modelled output using the Lipfert function suggests that karst weathering was dominant across the millennium, while contributions from sulfur deposition predominated across a much shorter period 1700–2000 CE. These predictions were compared with observations from the late seventeenth century which suggest London architects witnessed a notable increase in the recession rate and attributed this to pollution derived from coal burning (Fig. 1.1).

The recession rates (per annum, a^{-1}) measured in the late twentieth century lend some support to the estimates that are made with the Lipfert model:

$$-\mathrm{d}x/\mathrm{d}t = Lv \cdot Rn + 0.016[\mathrm{H}^+]Rn + 0.18(V_{dS}\,\mathrm{SO}_{2(\mathrm{g})} + V_{dN}\,\mathrm{HNO}_{3(\mathrm{g})}) \tag{1.1}$$

where $-\mathrm{d}x/\mathrm{d}t$ is the rate of recession as μm a^{-1}; Rn is rainfall as m a^{-1}; V_{dS} and V_{dN} are deposition velocities, taken as 0.38 cm s^{-1} and 0.32 cm s^{-1} (Sabbioni, 2003); $SO_{2(g)}$ and $HNO_{3(g)}$ are concentrations as μg m^{-3} and $[H^+]$ is the concentration as μmol l^{-1}. The term Lv is the Lipfert value, which is typically set at 18.8 μm (recession) per m (precipitation), but may need to be tuned to higher future carbon

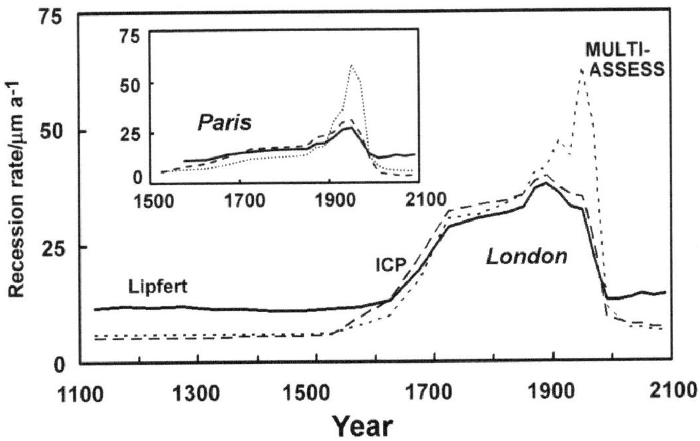

Fig. 1.1 Recession rate estimated from the Lipfert function and the ICP and MULTI-ASSESS programmes (for details see Brimblecombe and Grossi, 2008). The inset shows a similar plot for Paris.

dioxide concentration. Recession rates can also arise from two other functions, ICP and MULTI-ASSESS (Kucera, 2005):

$$\Delta x = 2.7\mathrm{SO}_{2(g)}^{0.48} \exp(-0.018T)\, t^{0.96} + 0.019Rn[\mathbf{H}^+]t^{0.96} \quad (1.2)$$

$$\Delta x = 3.1 + (0.85 + 0.0059\mathrm{SO}_{2(g)}\mathrm{RH}_{60} + 0.054Rn[\mathrm{H}^+] + 0.078\mathrm{HNO}_{3(g)}\mathrm{RH}_{60} + 0.0258\mathrm{PM}_{10})\,t \quad (1.3)$$

where T is temperature (°C), t is the length of exposure in years, RH_{60} is $RH_{60}, = RH - 60$ where $RH > 60$ and zero elsewhere and PM_{10} is the concentration of particulate matter expressed as μg m^{-3}.

Reduction in pollution has led to a notable reduction in recession rates in the 20th century as predicted by the models, which agree with historical and contemporary observations (e.g. Brimblecombe and Grossi, 2008; Inkpen *et al.*, 2012) In the future, recession looks to increase only slightly because of increases in carbon dioxide concentrations and a decrease in the amount of precipitation that arrives as snow. Some unconstrained calculations for Paris are shown in the inset in Fig. 1.1 and give a rather similar picture. This work has used pollution and climate data from Ionescu *et al.* (2012), and has begun as part of a research project, which will hopefully improve these estimates. However, the initial results suggest that in the past, pollution dominates in Paris very much as it did in London, although the atmosphere never appears quite as aggressive and the damaging period is of shortened duration.

1.5 Long-Term Corrosion Rates for Iron and Copper

The estimates of long-term corrosion rates for metals can be made in rather the same way as the recession of stone. Here, iron has used the functions of Kucera (2005) for the time dependent mass loss Δm_t of mild steel

$$\Delta m_t = 29.1 + (21.7 + 1.39\mathrm{SO}_{2(\mathrm{g})}^{0.6} RH_{60} \exp(f_{\mathrm{Fe}}) + 1.29 Rn[\mathrm{H}^+] + 0.593\mathrm{PM}_{10})\, t^{0.6} \tag{1.4}$$

where f_{Fe} is $0.15(T - 10)$ when $T < 10°\mathrm{C}$ and $-0.061(T - 10)$ at $T > 10°\mathrm{C}$. The equation of the National Materials Exposure Programme (NMEP) is simpler (Butlin *et al.*, 1995), but care is needed with the units. Here the equation is taken to represent a four-year loss which is reduced to an average annual loss $(-\mathrm{d}m/\mathrm{d}t)$, adjusted using a specific gravity of 7.86.

$$-\frac{\mathrm{d}m}{\mathrm{d}t} = \frac{7.86(41 + 1.9\mathrm{SO}_{2(\mathrm{g})} + 12Rn)}{4} \tag{1.5}$$

Copper has long been used in London and here the ICP and MULTI-ASSESS damage function are adopted from Kucera *et al.*, 2007:

$$\Delta m_t = 0.0027\mathrm{SO}_{2(\mathrm{g})}^{0.32}\mathrm{O}_{3(\mathrm{g})}^{0.79} RH \exp(f_{\mathrm{Cu}})\, t^{0.78} + 0.05 Rn[\mathrm{H}^+] t^{0.89} \tag{1.6}$$

$$\Delta m_t = 3.12 + (1.09 + 0.002\mathrm{SO}_{2(\mathrm{g})}^{0.4}\mathrm{O}_{3(\mathrm{g})} RH_{60} \exp(f_{\mathrm{Cu}}) + 0.0878 Rn[\mathrm{H}^+])\, t \tag{1.7}$$

where f_{Cu} is $0.083(T - 10)$ when $T < 10°\mathrm{C}$ and $-0.061(T - 10)$ at $T > 10°\mathrm{C}$

Estimates of long-term potential corrosion rates for iron and copper (Fig. 1.2(a)–(b)) suggest a rather similar picture to that for stone, with a significant peak in corrosion rates across the two-three centuries when coal burning dominated.

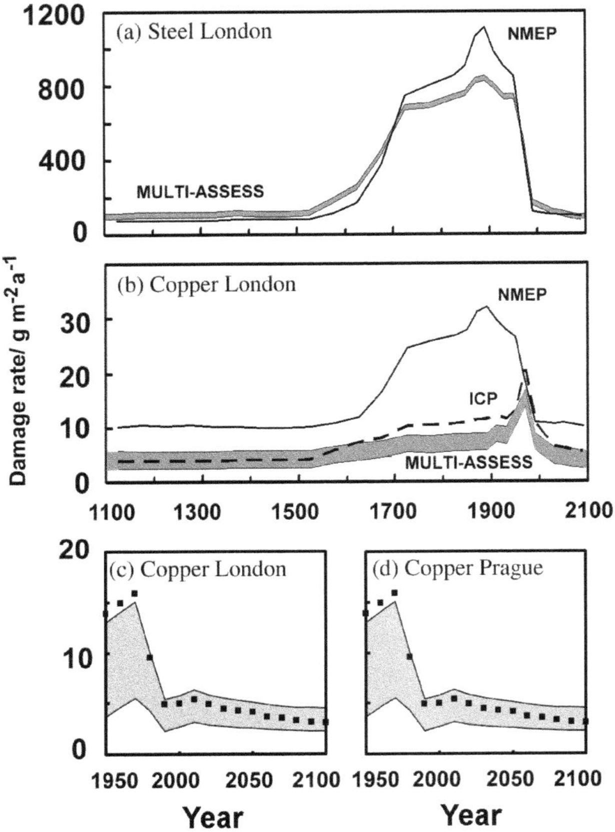

Fig. 1.2 (a) The initial rate of damage to mild steel using the MULTI-ASSESS and NMEP functions. The upper part of the shaded area shows the effect of adding the time independent rate of 29.1 g $m^{-2}a^{-1}$. (b) The initial rate of damage to copper using the ICP, MULTI-ASSESS and NMEP functions. The upper part of the shaded area shows the effect of adding the time independent rate of 3.12 g $m^{-2}a^{-1}$. Initial yearly rate of mass loss for copper in London (c) and Prague (d). The shaded area marks the lower bound (sheltered) and the upper bound (the exposed) metal from the ICP equation. The points mark the values from the MULTI-ASSESS equation and the dotted line is from the NMEP function. Details are to be found in Brimblecombe and Grossi (2008, 2010).

Figure 1.2 shows the corrosion of copper across the recent past and through the 21st century for London and Prague. These sites represent two idealised climates, one for a maritime city (London) and one for Prague, a city with a colder continental winter climate and potentially higher ozone concentrations in the future given its more central European location. The overall pollution trend for London over the years 1960–2000 is available for sulfur dioxide, nitrogen dioxide and smoke (Brimblecombe, 2006). The city has seen annual mean ozone increases of about 50% over the period 1996–2006, but this sharp increase is not reflected very strongly in the time trends for central London and in Prague where ozone is likely to be higher in future, but it seems to have limited effect on the rates of corrosion of copper.

1.6 Future Damage to Polymers from Ozone

Ozone is harmful to organic materials, especially those with double bonds. However, the oxidising atmosphere of a photochemical smog

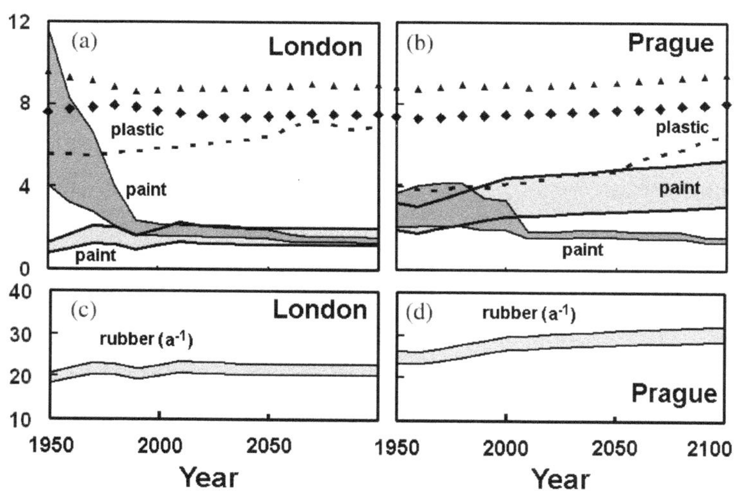

Fig. 1.3 Rate of deterioration of plastic in London (a) and Prague (b) in terms of changes in IR absorption in polyethylene (dotted line), damage depth in polyurethane (diamonds), and colour change in polyester (triangles). Damage to paints is shown as ICP estimates for damage to steel panels with alkyd and coil coated galvanised steel with alkyd melamine, as the upper and lower bounds of the dark shaded area and the % increase in erosion rate of acrylic coil coating (upper bound) and oil-based house paint (lower bound) of the light shaded area. Rate of damage to rubber in London (c) and Prague (d) where the lower bound is the middle of the rubber and the upper bound the edges (for details see Brimblecombe and Grossi, 2010).

is also able to transform the nitrogen oxides to nitric acid, which can damage metals and stone. Ozone may also enhance sulfate formation on building surfaces.

Although increases in ozone can enhance the rate of damage to copper and aluminium, it is paints and polymers that are likely to suffer most (Brimblecombe and Grossi, 2010). This is shown in Fig. 1.3, where panes (a) and (b) represent damage to plastics and paints in London and Prague respectively through the 20th century. The relative damage is given in terms of lines that denote changes in IR absorption in polyethylene (dotted line), damage depth in polyurethane (diamonds), and colour change in polyester (triangles). The damage to most of these alters only slightly into the future, except in the case of the IR absorption in polyethylene in Prague, where increases in ozone prove a enhanced threat.

Paint is also affected as shown by the shaded parts of Fig. 1.3(a) and (b) representing the steel panels painted with alkyd and coil coated galvanised steel with alkyd melamine (shown as the upper and lower bounds of the dark shaded area). Not surprisingly, rubber, which is very sensitive to ozone shows a more dramatic change in Prague than London with its lower future ozone (Fig. 1.3(c) and 1.3(d)).

1.7 Conclusion

Over the last few centuries buildings have suffered badly from the aggressive atmosphere of cities, although traditional materials, such as stone, are affected less in the contemporary urban atmosphere. However, we should also reflect on the suitability of modern architecture to the evolving urban environment, where oxidants are likely to be present and attack polymers and paint. It is also probable the climate change, as reflected in rising temperatures and sunlight hours, would further increase damage to these materials. Light-coloured surfaces that are much in favour will soil rapidly from deposited diesel soot.

References

Anderson, T. (1911). The decay of stone antiquities, *Museums J.*, 10, pp. 100–106.

Ashurst, J. (1985). Options to Bodging, *Architectural Journal Special Conservation Issue*, pp. 40–43.

Ausset, P., Bannery, F., Del Monte, M. and Lefevre, R.A. (1998). Recording of pre-industrial atmospheric environment by ancient crusts on stone monuments, *Atmos. Environ.*, 32, pp. 2859–2863.

Bonazza, A., Brimblecombe, P., Grossi, C.M. and Sabbioni, C. (2007). Carbon in black layers at the Tower of London, *Environ. Sci. Technol.*, 41, pp. 4199–4204.

Bowler, C. and Brimblecombe, P. (1991). Long term stone decay at York Minster, *Europ. Cult. Herit. Newslett. Res.*, 5, pp. 47–57.

Bowler, C. and Brimblecombe, P. (2000a). Environmental pressures on the design of Manchester's John Rylands Library, *J. Des. Hist.*, 13, pp. 175–191.

Bowler, C. and Brimblecombe, P. (2000b). Control of air pollution in Manchester prior to the Public Health Act, 1875, *Environment and History*, 6, pp. 71–98.

Brimblecombe, P. (1977). London air pollution 1500–1900, *Atmos. Environ.*, 11, pp. 1157–1162.

Brimblecombe, P. (1987a). The antiquity of 'smokeless zones', *Atmos. Environ.*, 21, p. 2485.

Brimblecombe, P. (1987b). *The Big Smoke*, Methuen, London.

Brimblecombe, P. (2006). The Clean Air Act after Fifty Years, *Weather*, 61, pp. 311–314.

Brimblecombe, P. and Camuffo, D. (2003). 'Long term damage to the built environment', in P. Brimblecombe (Ed.), *The Effects of Air Pollution on the Built Environment*, Imperial College Press, London, pp. 1–30.

Brimblecombe, P. and Grossi, C.M. (2008). Millennium-long recession of limestone facades in London, *Environ. Geol.*, 56, pp. 463–471.

Brimblecombe, P. and Grossi, C.M. (2010). Potential damage to modern building materials from 21st century air pollution, *Scientific World J.*, 10, pp. 116–125.

Butlin, R.N., Yates, T.J.S., Murray, M., and Ashall, G. (1995). The United Kingdom national materials exposure programme, *Water Air Soil Pollut.*, 85, pp. 2655–2660.

Camuffo, D. (1993). Reconstructing the climate and the air-pollution of Rome during the life of the Trajan Column, *Sci. Total Environ.*, 128, pp. 205–226.

Cesarino, A. and Pittaluga, D. (1995). Superfici intonacate: inquinamento, degrado e pulitura nei secoli XVI-XIX, *Scienza e Beni Culturali*, 11, pp. 127–138.

Church, A.H. (1901). *Chapter House Westminster, Prof Church's Report on the Treatment of Decayed Stone*, Westminster Abbey Library, City of Westminster, 28 May 1901.

Davidson, C.I. (1970). Air pollution in Pittsburgh: a historical perspective, *J. Air Pollut. Control Assoc.*, 29, pp. 1035–1041.

Desche, C. (1901). *Note on the Preservation of Exterior Stonework of Westminster Abbey*, Westminster Abbey Library, City of Westminster, 25 July 1921.

Drake, F. (1736) *Eboracum*, London.

Evelyn, J. (1661) *Fumifugium*, London.

Grossi, C.M. and Brimblecombe, P. (2004). Aesthetics of simulated soiling patterns on architecture, *Environ. Sci. Technol.*, 38, pp. 3971–3976.

Grossi, C.M. and Brimblecombe, P. (2008). Past and future colouring patterns of historic stone buildings, *Revista Materiales de Construcción*, 58, pp. 143–160.

Hipkins, S. and Watts, S.F. (1996). Estimates of air pollution in York, 1381–1891, *Environment and History*, 2, pp. 337–345.

Inkpen, R.J., Viles, H.A., Moses, C., Baily, B., Collier, P., Trudgill, S.T. and Cooke, R.U. (2012). Thirty years of erosion and declining atmospheric pollution at St Paul's Cathedral, London, *Atmos. Environ.*, 62, pp. 521–529.

Ionescu, A., Lefèvre, R.A., Brimblecombe, P. and Grossi, C.M. (2012). Long-term damage to glass in Paris in a changing environment, *Sci. Total Environ.*, 431, pp. 151–156.

Jackson, H. (1901). *Mr Jackson's Report Upon the Examination of the Stone Work of Westminster Abbey*, Westminster Abbey Library, City of Westminster.

Keepe, H. (1683). *Monumenta Westmonasteriensa*, London.

Kucera, V. (2005). *Model for Multipollutant Impact and Assessment of Threshold Levels for Cultural Heritage*, Swedish Corrosion Institute, Stockholm.

Kucera, V., Tidblad, J., Kreislova, K., Knotkova, D., Faller, M., and Reiss, D., (2007). UN/ECE ICP materials dose–response functions for the multi-pollutant situation, *Water Air Soil Pollut.: Focus*, pp. 249–58.

La Russa, M.F., Belfiore, C.M., Comite, V., Barca, D., Bonazza, A., Ruffolo, S.A., Crisci, G.M. and Pezzino, A. (2013). Geochemical study of black crusts as a diagnostic tool in cultural heritage, *Appl. Phys. A*, 113, pp. 1151–1162.

Malcolm, J.R. (1770). *Anecdotes of the Manners and Customs of London during the 18th Century*, London.

Montana, G., Randazzo, L. and Mazzoleni, P. (2012). Natural and anthropogenic sources of total suspended particulate and their contribution to the formation of black crusts on building stone materials of Catania (Sicily), *Environ. Earth Sci.*, 67, pp. 1097–1110.

Paley, F.A. (1878). *Science Gossip*, 14, pp. 170–171.

Pittaluga, D. (1994). Nuovi sviluppi interdisciplinari nell'analisi stratigrafica delle superfici intoinacate: esperienze e prospettive, *Scienza e Beni Culturali*, 10, pp. 123–133.

Sabbioni, C. (2003). 'Mechanisms of air pollution damage to stone', in P. Brimblecombe (Ed.), *The Effects of Air Pollution on the Built Environment*, Imperial College Press, London, pp. 63–88.

Sgobbi, M., Brimblecombe, P., Grossi, C.M., Biscontin, G. and Zendri, E. (2010). Surface stratigraphy on limestone of Venetian palaces, *Journal of Architectural Conservation*, 16, pp. 52–70.

Sherwood, S.I. and Bumbaru, D. (1991). Historical urban SO_2 levels, *APT Bulletin*, 23, p. 72.

Westlake, H.F. (1863). *Westminster Abbey*, Bell and Hyman, London.

CHAPTER 2

Weathering of Building Materials

Dario Camuffo

Institute of Atmospheric Sciences and Climate, National Research Council (ISAC-CNR), Italy

2.1 Introduction

Outdoor monuments and building envelopes are built with materials resistant to weather injuries, at least for reasonably long periods. However, their vulnerability is exacerbated by specific atmospheric factors that may accelerate deterioration, causing characteristic forms of decay. Weathering is due to a number of atmospheric agents, often in synergism with atmospheric pollutants. Weather and pollution act on various scales, depending on their nature, and their action may be enhanced or attenuated in the proximity of building surfaces. Especially, the microscale interaction that occurs at the interface between the material surface and the free air is crucial. It explains the various deterioration patterns that may be visible on different places of the same building.

Over the years, and especially in the most recent period, an abundant literature including a number of comprehensive books (e.g. Prikryl and Viles, 2002; Fort *et al.*, 2006; Smith *et al.*, 2008; Dohene and Price, 2010; Siegesmund and Snethlage, 2011; Lichtfouse *et al.*, 2012; Camuffo, 2013; Forsyth, 2014) have been produced to elucidate various forms of weathering and decay of built cultural heritage. The aim of this chapter is to shortly explain how weather and air pollution, separately or in synergism, concur to affect historic building materials.

A number of weathering mechanisms exist that may considerably shorten the life of each specific material, depending on characteristics of the material itself, and the intensity and frequency of occurrence of some adverse situations.

2.2 Main Building Materials and Their Vulnerability

A list of the most common materials used in historic buildings is made, with a short mention of their key deterioration mechanisms.

2.2.1 *Wood*

Biological attack determined by insects or various species of fungi, rodents or birds is the chief degradation mechanism of wood (Simpson and TenWolde, 1999; Gelbrich *et al.*, 2008; Caneva *et al.*, 2009). Wood is very resistant in dry-cold conditions. However, in temperate to tropical climates, in the presence of dampness (i.e. when the moisture content (MC) reaches or exceeds the fibre saturation point, i.e. with MC around 20–30%), it is easily destroyed by the attack of xylofagus insects (e.g. beetles, termites) and moulds.

A further problem for wood is yield and the formation of cracks for shrinkage-swelling cycles in response to changes in the equilibrium MC (Bratasz, 2010).

Freeze-thaw cycles may cause cell structures to break with a subsequent loss of strength. Some wooden species used for structural timber may be subject to delignification if located in coastal areas. The breakdown of the wooden cells affected by marine salts may only constitute a visual deterioration. However, in some extreme cases, it may cause significant structural loss of timber strength (Simpson and TenWolde, 1999).

2.2.2 *Plaster, mortar, brick and stone*

Water either supplied by precipitation, percolation or capillary rise, is the main agent for the deterioration of masonry materials. In polluted areas, chemical reactions with atmospheric pollutants are responsible for dissolution and severe surface recession especially driven by percolating rainwater. Salts can damage building materials by chemical reactions or crystallisation, especially on the front of capillary rise. Capillary action may pull soluble salts that result in efflorescence from the ground into the masonry, such as chlorides from salting of streets and sidewalks in winter and nitrates from fertilisers. The key deterioration mechanism is the cumulative effect of salt crystallisation-dissolution cycles related to the pore size and combination (Zehnder and Arnold, 1989; Arnold and Zehnder, 1991; Camuffo, 1995; Flatt, 2002; Rijniers *et al.*, 2005; Fort 2006; Goudie and H. Viles, 2008; Smith *et al.*, 2008; Dohene and Price, 2010; Steiger *et al.*, 2011; Sanjurjo-Sánchez and Alves, 2012; Camuffo, 2013).

Biodeterioration is typical in damp environments. Bioorganisms may cause mechanical damage with their hyphae, and chemical damage for the organic acids they produce, which can eventually dissolve or transform the stone (Warscheid and Braams, 2000; Sterflinger, 2000; Piñar and Sterflinger, 2009; Steiger *et al.*, 2011; Macedo *et al.*, 2011).

Severe mechanical deterioration may follow freeze-thaw cycles or thermal expansion cycles especially in the parts overheated by direct solar radiation (Siegesmund *et al.*, 2000).

2.2.3 *Terracotta and ceramic roof tiles*

These building elements have varying salt and frost tolerance, depending on how they have been constructed and with the type of used clay. Acidic rain or dry deposition of atmospheric pollutants may be responsible for chemical deterioration of tiles (Cultrone *et al.*, 2000; Sanjurjo-Sánchez and Alves, 2012). In coastal areas, or in cold climates, tiles are subject to varying degrees to salt/frost deterioration. For their specific location, tiles are particularly vulnerable to hailstone crashes.

Exposure to light, large temperature ranges and frequent exposure to rain will favour bacteria, fungi, algae, lichens and plant colonisation. Biodeterioration is more common on stone than on ceramic or glass. However, biologic communities may develop, damaging tiles or covering them with black or green biofilm patinas (Radeka *et al.*, 2007a,b; Coutinho *et al.*, 2013).

2.2.4 *Glass*

Rainfall and condensation supply water to glass panes and historic stained glass windows. Water contains natural ions (i.e. H^+ and/or H_3O^+) or pollutants that may replace alkali ions (e.g. Na^+ or K^+ that are typical constituents of mediaeval windows and are less strongly bound than Ca^{2+}), and may hydrate the silica lattice. Deterioration begins with leaching, i.e. an ion exchange between penetrating ions from the moisture and an alkali ion that is removed from the glass, forming a brittle, deteriorated surface layer. Modern glass made with Ca^{2+} is more resistant to leaching. Pitting and heavy corrosion may occur when the silica constituent is affected. In dry condition, or under the action of solar radiation, the temperature gradients and the loss of the absorbed water cause dehydration and mechanical stress on the surface layer (Munier *et al.*, 2002; Leissner, 2003; Lefèvre *et al.*, 2005; Schreiner, 2005; Cailleteau *et al.*, 2008; Kunicki-Goldfinger, 2008; Lombardo *et al.*, 2010; Schreiner, this book).

2.2.5 *Metals*

Although corrosion is the key deterioration mechanism of metals, all materials, including mortar, stone and wood may, to a certain extent, deteriorate when they are exposed to certain combinations of solar radiation (UV light), high temperature, water, oxygen and pollutants.

Metal corrosion can be subdivided into general, localised, galvanic, stress cracking, and caustic-agent corrosion. Particularly common examples of general corrosion are rusting and flaking of iron; of localised corrosion is pitting; galvanic corrosion occurs when two metals with different electrochemical charges are linked via a conductive path; stress cracking may occur in the presence of strong tensile stress; caustic-agent corrosion is activated when water makes a solution of atmospheric pollutants, especially those already deposited in the dry phase on the metal surface (Leygraf and Graedel, 2000; Roberge 2000; Roberge *et al.*, 2002; Tidblad and Kucera, 2003).

In general, water acts as an electrolyte and its aggressive power is determined by the presence of pollutants dissolved in it.

The temperature will contribute to increasing the energy above the activation threshold. This means that in dry, cold climates, metals suffer little corrosion, as opposed to what happens in humid, warm and polluted environments.

The most effective and common procedure for preventing or slowing down corrosion is to prevent contact with water, maybe with coatings or envelopes.

2.3 Atmospheric Factors and Key Deterioration Mechanisms

In order to avoid misunderstanding for a number of terms and definitions that are in popular use, in the following we will make reference to the World Weather Organization meteorological vocabulary (WMO, 1966), cloud atlas (WMO, 1987) and hydrological guides (WMO, 2008a,b).

2.3.1 *Solar radiation*

The solar energy impinging on a building surface depends not only on the surface exposure (i.e. surface orientation and tilting angle) and albedo, but also on a number of factors depending on the Sun, the atmosphere and the surrounding environment, i.e. the actual position of the sun (i.e. solar height over the horizon and azimuth angle), the atmospheric absorption (i.e. cloud cover, thickness of crossed atmospheric layer and atmospheric turbidity), and the presence of nearby obstacles or reflections.

For instance, the solar declination $\delta_\odot$ ranges between $\pm 23°27'$. At 60° Lat North, at the Summer Solstice (when $\delta_\odot = 23°27'$) the Sun reaches the same height over the horizon as that at the Equinoxes (when $\delta_\odot = 0°$) at the $60° - 23°27' = 36°33'$ Lat North (Fig. 2.1). In this example, the highest irradiation that on June 21st reaches Oslo (Norway, 50° 57′ Lat N), Helsinki (Finland, 60° 10′ Lat N), Stockholm (Sweden, 59° 17′ Lat N) and Tallinn (Estonia, 59° 26′ Lat N), closely equals the levels that Seville (Spain, 37° 23′ Lat N), Granada (Spain, 37° 10′ Lat N), Alger (Algeria, 36° 49′ Lat N), Catania (Italy, 37° 10′ Lat N), Siracusa (Italy, 37° 5′ Lat N) and Athens (Greece, 37° 59′ Lat N) receive on 21st March at the beginning of spring, or

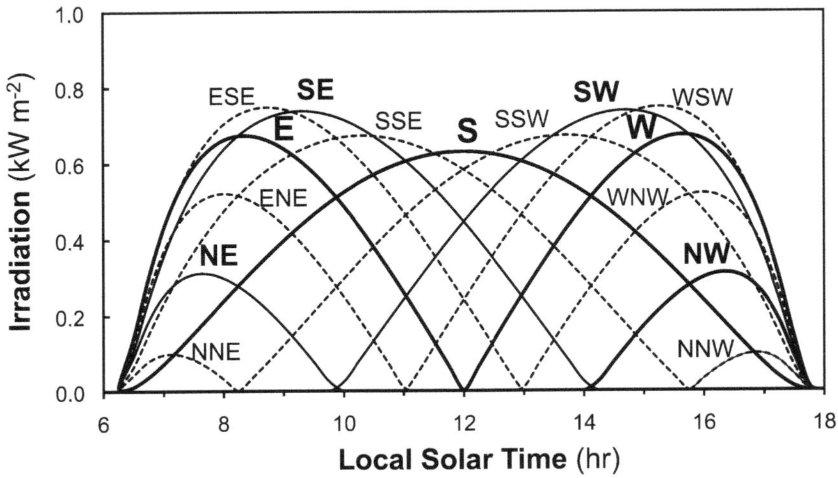

Fig. 2.1 Hourly flux density of solar irradiation impinging on walls with various orientations, at the Summer Solstice at 60° Latitude North which equals the corresponding fluxes at the Equinoxes at 37° Lat North. Astronomic Local Solar Time, i.e. solar culmination at noon. Legend: N for North, E for East, S for South and W for West.

on 21st September, at the beginning of autumn. The same holds for vertical walls. This example gives a clear idea of how much the latitude is relevant to the aims of solar radiation and building envelope overheating.

Solar radiation is a typical long-scale phenomenon, with regular cycles over the day and the calendar year, but continually perturbed by weather variability. Its main effect is to warm surfaces depending on their albedo and heat capacity, and to create internal temperature gradients while the heat travels across the building structure. Temperature levels and imbalances depend on both solar radiation and material characteristics. They may affect the mechanical properties of materials (e.g. thermal expansion), chemical reaction rates (e.g. the kinetics in the Arrhenius equation), change the physical state of water (e.g. condensation-evaporation cycles, freeze-thaw cycles), and other effects as we will see later (Siegesmund *et al.*, 2000; Gómez-Heras, 2006; Camuffo, 2013).

Another important deterioration mechanism related to solar radiation, and especially the highly energetic UV component, is discoloration (Wypych, 1995; Bacci *et al.*, 2004).

2.3.2 *Wind*

Wind is a weather variable that develops on a wide scale but that is strongly affected by the local urban context. Fresh to strong winds are responsible for a number of adverse effects: unplanned transverse structural loads; turbulence and aerodynamic drag removing tiles, or breaking windows; driving hailstones, sand or raindrops; accumulating snow; transporting and diffusing marine salts, volcanic acids and pollutants (Dohene and Price, 2010). In particular environments wind erosion is generated by windborne dust (Camuffo, 1993; 2013; Hussein and El-Shishiny, 2009). Tornadoes may develop terrific forces with enormous destructive impact on the built environment.

Even under normal conditions wind may cause some direct or indirect effects, and undergo local accelerations. For instance, when wind is channelled through a passage that narrows at the end, it creates a 'jet effect', which increases the wind velocity (i.e. Venturi effect) with a strong impact on heat and moisture transport. When wind is blowing, the non-aerodynamic shape of buildings and especially the presence of edges generate turbulence, surface drag and dispersion of airborne effluents.

The capture efficiency of airborne particles on the building surface is a complex function of the particle mass, size and shape, velocity and direction of the wind, turbulence intensity, concentration of airborne particles and physical conditions of the surface (e.g. damp or dry, roughness, location and exposure).

2.3.3 *Atmospheric precipitation*

Precipitation occurs in a wide variety of configurations determined by local thermodynamic situations (e.g. warm-air uprising cells, stratification, slope uplift, urban heat island), environmental wind shear, spatial variability of humidity, and proximity to triggering mechanisms (e.g. topography, built environment, fronts, dry lines) that may generate convective or stratiform clouds. Precipitation is composed of falling water drops or ice aggregates. The impact of raindrops and hailstones on buildings is determined by their size, which is expressed in terms of their diameter (D).

From the meteorological point of view, precipitation is classified as follows.

Drizzle is composed of small sized droplet, i.e. $0.5 < D < 1$ mm, and typically falls from stratus clouds in association with calm air or light wind. The falling velocity of a droplet of 1 mm D is 3 m s^{-1} in still air, and less in the presence of turbulence. Such small droplets do not exert mechanical effects on building surfaces.

Drizzle wets, dissolves and mobilises particulate matter already deposited in the dry phase on the building surfaces, but is unable to remove and take them out. On white stone it forms small rivulets similar to zebra stains (Fig. 2.2). These rivulets are particularly visible after a statue has been cleaned and some black particles have accumulated on the monument.

Fig. 2.2 This recently cleaned statue in Vienna has accumulated black particles on arms, shoulders and club. Light rain and drizzle have formed a black solution that flows in small zebra rivulets along the white stone surface. The amount of water is not sufficient to remove dirt and wash out the monument.

Rain drops are in the size range $1 < D < 5$ mm, typically from nimbostratus or cumulus clouds. Rain may occur under a wide spectrum of wind forces. The falling velocity of raindrops increases with size and ranges between 3 to 9 m s^{-1}, where the upper limit is determined by the equilibrium between two forces: gravity (i.e. the raindrop weight, downwards) and the aerodynamic drag (upwards) generated by the turbulent wake behind the falling drop. Drops break up in the falling velocity range 8 to 9 m s^{-1}.

The amount of water is generally sufficient to remove dirt and wash monuments or walls exposed to driven rain.

Showers are characterised by large drops with $5 < D < 7$ mm, typically from cumulonimbus clouds, which in addition are associated with strong winds and stormy weather. Falling drops travel at 9 m s^{-1}. Drop size cannot exceed 7 mm because the friction with the air and the instability generated by the turbulent shear behind it overcome the strength of the surface tension, with the effect of breaking up a giant drop into a number of smaller droplets.

Showers supply large amounts of water with energetic drops splashing at their impact. They are very efficient in removing dirt.

Hail may be classified in terms of soft hail ($1 < D < 5$ mm) and hailstones ($5 < D < 100$ mm). Soft hail may include snow and is softer (i.e. less violent impact) than the more energetic hailstones made of compact ice. Hail typically occurs with cumulonimbus clouds, and falls with strong winds and stormy weather.

The density of ice is not much different from water, and the falling velocity of water drops or soft hail with the same diameter is almost the same. However, the situation differs for very large hailstones, with falling velocity given by $1.4\, D^{0.8}$, e.g. a hailstone with $D = 8$ cm falls at a rate of 48 m s^{-1}. The energy associated with falling hailstones may break tiles, roofs, statues and decorations. Windborne hailstone may break panes and stained glass windows.

Snow. Snowflakes are released from stratiform clouds (typical snow) and more rarely from cumuliform clouds (snow showers). Snowflakes of any size falls at about 1 m s^{-1}.

Snow may accumulate on roofs and terraces, causing high (temporary) static loads. When an exceptional load exceeds the planned threshold, or persists for a long time, or finds a degraded timber structure, it may cause the roof to collapse.

Some snow may melt during the daytime, and the percolated water may freeze again during the night time, exerting large freezing pressures and causing severe mechanical damage.

2.3.3.1 *Precipitation roses*

In addition to the falling speed along the vertical, we should consider the dragging velocity of blowing wind along the horizontal axis of the wind direction. This may be responsible for violent impact on vertical surfaces, possibly with mechanical damage or drive water deep into porous materials. Typically, medium and large drops or hail hit building surfaces facing the wind.

Rose diagrams composed of circular frequency histograms related to the occurrence of wind-driven precipitation (Fig. 2.3) are useful to pinpoint the building surfaces that most frequently intercept precipitation, with either positive or negative consequences. Rain and showers may bring remarkable amounts of water, which washes the monuments surface in the parts where rainwater flows, removing deposited particles (positive factor) and possibly eroding the underlying material (negative factor).

The situation is different for small droplets or snowflakes, because of their small inertia and large wind drag. They are airborne and practically remain entrapped in the airstream, flowing around the obstacle. The local turbulence may cause deposition, and the

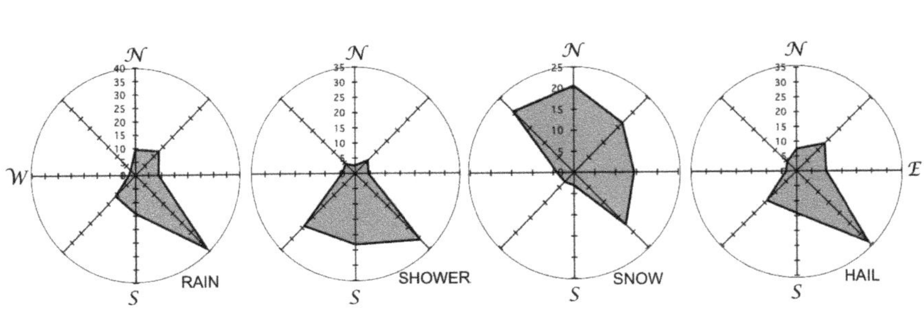

Fig. 2.3 Wind-driven precipitation roses for Rain, Shower, Snow, Hail in Rome, Italy (After Camuffo, 2013).

probability increases when the airstream is tangent to the building surface, especially in the presence of increased surface roughness.

Drizzle causes modest dampness, except when it lasts for days. Hail and snow may wet only after melting.

Precipitation is a phenomenon generated on the wide or regional scale but with marked local departures triggered by urban or land tography or related to convective cells and cloud size. However, the spatial distribution of wind-driven drops or ice is strongly affected by the wind field and turbulence on the very local scale. This means that in proximity of a building, the wind intensity and direction will differ very much from the record of weather stations located in the free countryside or in airports.

2.3.3.2 *Occult precipitation*

Fog, mist, dew, hoar frost and rime are hydrometeors affecting the water balance and the physics/chemistry of building surfaces. However, they cannot be included in the "*precipitation*" class for the simple reason that they do not fall, but are directly formed in the air, or upon coming in contact with a cold surface. For this reason they are sometimes called "*occult precipitation*".

All of these phenomena need specific weather conditions, e.g. high relative humidity levels in the atmosphere, which are determined on the regional scale. However, the amount of water that can be transferred from the atmosphere to monuments strongly depends on the monument surface temperature and very local conditions.

Fog is composed of small droplets (typically $1 < D < 300\,\mu m$) suspended in air that reduce visibility. They fall so slowly that normal turbulence and air currents keep them aloft almost indefinitely, until the fog disappears. Fog may disappear for two reasons. The first is evaporation by heat supplied by solar radiation, or the urban heat island, or by mixing with drier air. The second is precipitation. When the humid environment persists for days, the smaller droplets evaporate and condense on the coarser ones feeding them (i.e. coalescence). This mechanism reduces the population of the finest droplets and increases the coarser ones, which grow in size until they reach 1 mm and their falling velocity becomes relevant. At this point the fog is transformed into drizzle.

Fog is unable to supply a substantial amount of water to building materials for the simple reason that the droplets are very small and hardly deposit on the surface. The key fact is that fog is associated with high relative humidity (*RH*) levels, i.e. $RH > 95\%$, and especially in the case that the building surface is colder than the air, condensation may occur on the surface.

The difference in the equilibrium pressure over a spherical droplet and the damp building surface favours the transfer of vapour from the droplets to the surface. The amount of transferred water is modest. The problem is that dampness may activate pollutants already deposited in the dry phase, or that high concentrations of pollutants may reach and dissolve in the fog droplets (Fisak *et al.*, 2002).

Dew is water condensed on a cold surface, whose temperature falls below the dew point (*DP*). On porous materials (e.g. plaster, brick, sandstone, wood) this water may be adsorbed into pores, increasing the material water content without any external appearance. On less adsorbent materials (e.g. marble, metals), condensed water may form a film covering the surface (especially in the case of hydrophilic surfaces) with a uniform thickness layer for the strong adhesion forces of water in contact with the surface. If the surface is water-repellent (e.g. glass, or a surface treated with oil, wax, siloxane, fluoropolymers or other water-repellent products), the high surface tension of condensed water may form hemispherical drops to minimise the free liquid surface.

The amount of water deposited by condensation (or be absorbed) on a building material depends on a number of factors: the moisture content (*MC*) in air, the spread between the surface temperature (T_S) and the *DP*, also called dew-point spread ($DPS = T_S - DP$), the time duration τ while the *DPS* is $DPS < 0$. The last variable is the wind speed.

The amount of moisture in air at saturation (i.e. when the water vapour becomes saturated reaching the highest possible concentration at that temperature) is exponentially related to the air temperature, and may be expressed in terms of humidity mixing ratio (*MR*, expressed as the number of grams of water vapour dispersed in 1 kg of dry air), or absolute humidity (*AH*, expressed as the number of grams of water vapour dispersed in 1 cubic metre of dry air), or dew point (*DP*, expressed as the temperature at which condensation starts to occur at the same mixing ratio and atmospheric pressure).

The local ventilation is responsible for bringing new moisture to feed the deposition mechanism while the air in proximity of the surface is impoverished of the water vapour that has changed phase. In other words, the water deposited by dew (*DW*) is related to the product

$$DW = MR \times DPS \times \tau \times c(V) \tag{2.1}$$

where $c(V)$ is a dynamic coefficient increasing with the wind speed (V) (Camuffo and Giorio, 2003).

At mid latitudes the amount of water that may deposit in a single summer night on a marble surface by nocturnal condensation with clear skies (Fig. 2.4) is of the order of 10–30 mg cm^{-2} if the surface is horizontal, facing the sky, which implies the largest loss of radiant heat. Condensed water is a very small amount (e.g. $DW < 5$ mg cm^{-2}) if the surface is vertical, which implies a very modest loss of radiant

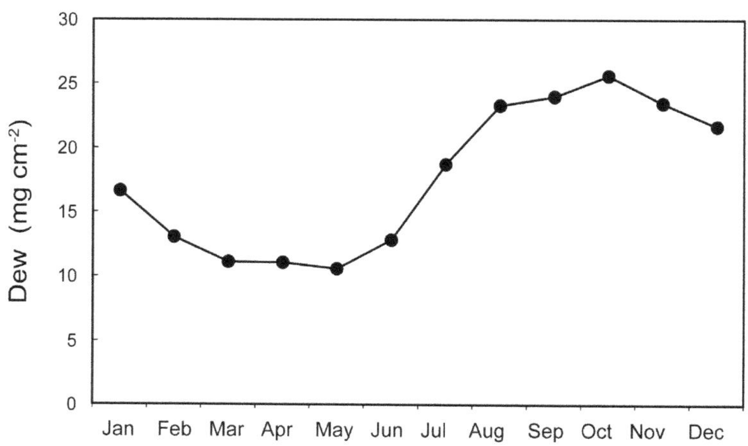

Fig. 2.4 Amount of water that deposits by nocturnal condensation (dew) under a clear sky, on a horizontal marble surface at Padua, Italy.

heat. There is no condensed water if the surface is horizontal, facing the soil, which implies almost no loss of radiant heat.

The above figure shows that in spring the dew formed in a single clear night reaches the minimum value, half that of autumn. This apparently surprising behaviour is explained by the lag of phase between two families of key variables. The first is the pair composed of solar radiation and night duration: both follow their astronomical cycles starting with the Winter Solstice, close to the calendar year. The second pair is the air temperature *T* and the saturation *MR* that is exponentially related to *T*. Both follow the seasonal cycle delayed by about two-three months with reference to the astronomic cycle. The theory suggests a three-month lag, but the weather variability reduces this delay. The combination of all the above factors, the related variables and regional features explain the observed readings that will vary from site to site.

Each time dew, or fog, occurs, the amount of water transferred by the atmosphere to the building is modest, and may damp the very surface layer, e.g. making a solution of salt efflorescence, deposited pollutants, or allowing chemical reactions. If the next day is sunny and windy, the little water deposited by occult precipitation is easily removed.

The repetition of wet-dry cycles induces dissolution-crystallisation cycles of soluble salts on the surface (efflorescence) or the subsurface (sub-efflorescence). If the dew frequency is elevated, dampness may overcome the threshold to form a favourable habitat for moulds.

The most effective dewing mechanism to supply water to building materials is when moist air is blown by wind and the monument is cold, below the dew point. This typically occurs when warm winds, e.g. the Sirocco from Africa, blow northwards, crossing colder regions and meeting cold monuments. A critical season is spring, when the large thermal inertia of buildings is in contrast with the warm winds, which hold large concentrations of water vapour. In this period the temperature of building envelopes is low, being conditioned by the winter months. When moist air reaches a cold building envelope, condensation starts at a fast rate and may continue for days, weeks or months, depending on the atmospheric circulation, transferring relevant amounts of water (Camuffo, 2013).

Ice Crystals. Hoar frost is composed of white ice crystals directly formed on a cold surface at sub-freezing temperature. *Rime ice* is originated from supercooled water drops that have changed state in contact with cold surfaces. The determinant factor is the surface temperature. Like dew, hoar frost and rime ice mobilise a small amount of water and constitute a limited risk for conservation.

The main risk factor is when an abundant amount of water, e.g. supplied by rainfall or melting snow, passes from the liquid to the solid state when the building material drops below 0°C, i.e. ice from frozen water. The thermal expansion of ice is responsible for mechanical damage, especially when the freeze-thaw cycles are frequent. The frost susceptibility of plasters, mortars, bricks, stones and tiles are largely controlled by the material porosity and pore size distribution (Grossi *et al.*, 2007; Camuffo, 2013).

It may be useful to underline that the material temperature is a key factor: it may differ by several degrees °C from the air, for the solar radiation, the radiant heat loss, the wind convective exchanges, the rain, the related changes of phase of water, and the thermal inertia of the envelope. Freeze-thaw cycles become more frequent when the envelope temperature approaches the 0°C transition threshold.

Tests on freeze-thaw cycles made under controlled conditions show that rocks remain unweathered during the most part of the frost cycle test, until some cracks start to appear and rapidly propagate until complete destruction of the sample. An open porous structurte and connectivity between pores controls the effectiveness of ice crystallisation. The most porous materials, e.g. bricks and rocks, as a result are vulnerable to frost crystallisation (Martínez-Martínez *et al.*, 2013).

2.3.3.3 *Water from the soil*

Capillary rise is the mechanism by which adsorbed water mobilises soluble salts from the soil, bricks and mortar, and transports them upwards inside the masonry. A common source of wall dampness is rainwater or melting snow locally collected by the soil after a precipitation. Compacted clay soils, asphalt, concrete and paving make good catchment surfaces for collecting rainwater and melting snow. Groundwater also comes from precipitation that percolates into the soil and may eventually reach building foundations. The origin of the above forms of soil water is precipitation, but the chemical composition of the soil tends to stabilise (buffer) the pH of the water. Depending on the type of soil, the pH may range from 6 in acidic soils (e.g. volcanic origin) to 8.5 in alkaline soils, especially carbonate soil (including $CaCO_3$) (Rodriguez-Navarro and Doehne, 1999; Charola, 2000).

Capillary water raises salts to a level determined by the material porosity. The wall surface characterised by rising water evaporates, increasing the concentration of salts in the aqueous solution. When this solution reaches the top level of the capillary rise (i.e. the front) it brings salts (continuous input) but there is no further mechanism to remove salts that will necessarily accumulate around the front. The dissolution-crystallisation cycles that the soluble salts undergo on the front will cause severe damage disaggregating plaster, bricks and stones with consistent loss of material.

Capillary rise is the main source of dampness in buildings, followed by condensation and percolation. Whatever the origin, the combination of dampness, especially if enriched in mineral salts, and mild temperature in mortar, stone or wood constitutes the favourable habitat for bacteria, algae or mould colonisation (Steiger *et al.*, 2011). This situation may eventually lead to microbiological sick building syndrome (SBS), not much different from the chemical SBS for newly renovated or built offices that release volatile organic compounds (VOC).

2.4 Atmospheric Factors and Acidic Deposition

In order to interpret the decay observed on historical buildings, we should keep in mind that atmospheric pollution and acid rain of natural and anthropogenic origin are not a novelty of the most recent times or the Industrial Revolution. They were frequent even in the Middle Age and the antiquity, as it has been documented in previous works (Brimblecombe, 1987; Camuffo, 1992; Camuffo and Enzi, 1995; Camuffo *et al.*, 2000; Camuffo and Sturaro, 2001; Brimblecombe and Camuffo, 2003) and today see their cumulative effect over time. In the Trajan's Column, Rome, dating back to 105 A.D., the Carrara marble of which is made has undergone some 3 mm surface retreat, compared with the parts transformed in insoluble oxalate patina, called *scialbatura*. This means 1.5 μm/year surface retreat with yearly precipitation in the order of 781.7 cm (1862–2006 average), i.e. 0.002 μm per 1 cm of precipitated water. However, in less resistant stones, either for composition, solubility or porosity, e.g. sandstone, the effect is much greater and visible in a few decades. In this book we will leave aside the history, and we will concentrate on the deposition and deterioration mechanisms. Acidic deposition may occur either in the wet or the dry phase, i.e. in the presence or absence of precipitation, as follows.

2.4.1 *Wet deposition*

Wet deposition is dominant in the northern European countries, where precipitation is frequent in all seasons. It is responsible for several forms of decay and has been intensively studied since the early 1970s (Seinfeld and Pandis, 1997; Fallemberg, 2000; Charola and Ware, 2002; Dohene and Price, 2010). Precipitation originates in clouds in the presence of condensation nuclei. Acidic precipitation is generated when pollutants in the gaseous or aqueous phase are absorbed onto cloud drops or ice crystals. The process in cloud is called *rainout*. When drops, snowflakes or hail fall, they scavenge atmospheric pollutants suspended in the lower atmosphere, increasing their acidity. This mechanism is called *washout*. Rainout and washout constitute the two basic mechanisms of acidic deposition in the wet phase. Rainout occurs aloft and may include pollutants originated in the long distance; washout is due to airborne pollutants collected in the air column connecting the soil with the cloud, and is more sensitive to the local pollution.

The pH of uncontaminated water is 7. However, unpolluted rainwater in equilibrium with clean air reaches a pH of 5.6 at 15 °C for the equilibrium reached with the atmospheric carbon dioxide (CO_2, average: 400 ppm). This pH level may further decrease in the presence of NO (average: 0.01 ppm) for electric discharges (i.e. lightning on thunderstorms produced by cumulonimbus clouds, usually generating heavy rain and sometimes hail) or the presence of natural SO_2 (average: 0-0.01 ppm) for volcanic emissions. Carbon dioxide, produced in the decomposition of organic materials, wild fires, ocean release and volcanic emissions, is the primary source of acidity in unpolluted rainwater. Rain with low pH levels is efficient in dissolving carbonate stones. In areas where rainwater is relatively free from pollutants, the dissolution of most common building stones is not a serious problem.

2.4.2 *Dry deposition of suspended particles*

In the dry period between two precipitation episodes, aerosols may deposit and accumulate on building surfaces via a number of deposition mechanisms, dependent on the particle diameter D, atmospheric temperature and moisture gradients, turbulence and surface characteristics (Sehmel, 1980; Seinfeld, 1986; Camuffo, 2013), as follows.

Electrostatic capture is generated by attractive electric forces in the presence of charged surfaces or particles. This mechanism is favoured in the presence of dryness, i.e. low *RH* levels, *UV* radiation and wind friction. The deposition efficiency is independent of the size of the suspended particles.

This deposition is frequent on the surfaces most exposed to solar radiation and more easily overheated, e.g. wood, tiles.

Brownian deposition is due to the energy transferred by the air molecules to the finest aerosol particles (e.g. smoke, $D \ll 0.1\ \mu m$). When a particle is very small, at each instant it receives a limited number of simultaneous impacts from the surrounding air molecules. The net result of all of these impacts is not zero; the smaller the particle surface, the higher the unbalance. As a consequence, the particle will gain energy and momentum, a vector changeable in intensity and direction at each instant. This will cause a continual series of jumps forming the so-called drunk walk.

The deposition efficiency of this mechanism increases with the kinetic energy E_C of the air molecules that is proportional to air temperature T. It decreases with increasing D because when the particle surface increases, the random molecular impacts tend to be compensated between them and the net result tends to vanish. In addition, the particle diffusivity decreases with increasing size. This random deposition is irrespective of the building surface orientation.

Thermophoresis is the transport of suspended particles towards a cold building surface that generates a temperature gradient in the air next to it. Let us suppose that the gradient of air temperature corresponds to $T+\Delta T$, T, and T-ΔT respectively on the warmer side, the middle and the colder side of the suspended particle. On the same positions, the kinetic energy of the air will be $E_C+\Delta E_C$, E_C, E_C-ΔE_C. The kinetic energy of the air molecules is higher at higher temperature levels, and the molecular collisions are more violent on the "warm" side of suspended particles. As a consequence, a suspended particle will be eventually moved in the direction of the weaker impacts, i.e. the direction of the lower temperature, until it reaches and sticks on the cold surface. The opposite occurs when a surface is warmer than the air.

The highest deposition efficiency is in the fine and medium-sized particle range (i.e. $0.1 < D < 1\,\mu m$) because the increase in size reduces the particle diffusivity. The deposition occurs when the building surface temperature is lower than the air temperature for radiative cooling, evaporation, or in the presence of warm air (e.g. daily temperature cycle, southern winds).

Thermophoresis is per se independent of the building surface orientation. However, it is more frequent where there is some reason for a surface to be colder, e.g. on the northern side of buildings. The parts of the building exposed to solar radiation tend to be protected from this specific mechanism while they are warmer. However, solar overheating increases Brownian deposition and triggers inertial impaction (see later), with a net result increasing soiling.

Diffusiophoresis and *Stefan flow* determine the transport of particles towards a surface on which condensation or absorption of vapour is occurring. In practice, it happens when water vapour molecules apparently disappear from the gaseous system, changing phase on contact with a building surface.

This complex mechanism is due to the uneven distribution of the concentration of water vapour and the same for the dry air molecules next to a condensing surface. These determine two opposed diffusive transports (i.e. *diffusiophoresis*) but with different efficiency for the different molecular masses of dry air (dry air mass: 29, as a weighted average of N_2 (mass: 28) and O_2 (mass: 32)) and water vapour (H_2O mass: 18). The equilibrium between the two diffusive concentration gradients (i.e. water vapour and dry air) determines a micro hydrodynamic flow (called *Stefan flow*) that generates a net transport and deposition of all the suspended particles that are next to the condensing surface.

The deposition rate is almost independent of the particle size and surface exposure. This deposition mechanism is very efficient and occurs on the surfaces where condensation is frequent: especially those exposed to north, or facing a clear nocturnal sky.

Inertial impaction, also called *aerodynamic deposition* occurs when an airstream with airborne particles flows next to a building surface. The origin of the airstream motion may be either dynamic (e.g. wind) or thermal (e.g. buoyancy). When an airstream is blowing particles in close proximity of a surface, the surface roughness or the air turbulence induces sharp changes in the airflow direction on the microscale. An airborne particle is subjected to two forces: a deflecting viscous drag for the change in air direction, and the inertial force of the particle tending to keep the previous motion. The balance tends to favour drag for the smaller particles, and inertial motion for the larger ones. The finest particles tend to remain entrapped within the airstream; the coarser ones tend to be scattered out. The highest efficiency is in with medium-sized and coarse particle range, i.e. $D > 1\ \mu m$).

The key factor is the turbulence generated in the proximity of architectural features or decorations, or microturbulence generated by friction with rough surfaces. This mechanism is particularly enhanced above surfaces that may be overheated by the solar radiation (e.g. dark surfaces, wooden doors or window shutters), generating uprising flows. If the surface is rough, it enhances turbulence and increases soiling. The most popular example of this physical mechanism is the wall blackening above a thermo-convective radiator.

Gravitational settling is the natural settling, in calm air, of medium and large particles that fall under their own weight onto horizontal surfaces. The falling velocity reaches equilibrium between the weight that tends to accelerate the particle, and the viscous drag of the air that opposes to the particle motion. Therefore, the falling velocity is constant throughout the whole falling trajectory, but increases with the particle diameter. It sharply decreases with increasing turbulence that tends to stop deposition and even re-suspend all particles. The highest efficiency is with still air and medium and large-sized particle range ($D > 1\,\mu m$).

Only horizontal or protruding surfaces are exposed to this deposition mechanism. However, this mechanism is very efficient and may collect large amounts of settled material, including dust, debris and dirt of various types.

2.4.3 *From deposition to decay*

All of the above mechanisms contribute to transfer pollutants from the air to the building surface. Acid rain is composed of water and pollutants and constitutes an aggressive solution. Dry deposition occurs in the absence of water in the liquid phase, and this makes the difference. In the latter case, not necessarily all deposited pollutants will remain on the surface and will react with the underlying material.

Especially, particulate matter deposited in dry phase may be removed by wind or rain, or with regular maintenance cleaning. In any case, no chemical reactions will occur while the building surface is dry, because most of the chemical reactions affecting stone or masonry occur in the aqueous phase. Therefore, dry deposition is the first step for various forms of decay to occur, but the presence of water is needed, and this may be found in the material, or is supplied by the air, but not necessarily with direct delivery. Except for precipitation, this transfer is mediated by the action of a viscous layer as we will see later.

Surface recession of building stone (Grossi *et al.*, 2008) is due to the synergistic contribution of clean rain, acid rain and dry deposition of pollutants that may be removed or react in the presence of water from precipitation, or supplied by other weather phenomena, or in the presence of high humidity levels. The surface temperature will drive the reaction rate, or will stop it if is so high as to evaporate all the present water.

The various damage functions give different results, but this is largely dependent on their controlling assumption and completeness, and especially on what is happening on the microscale at the interface between air and building materials.

2.5 The Boundary Layer Next to Building Materials

In fluid mechanics, the *'boundary layer'* (BL) is a layer of a fluid in the immediate vicinity of a bounding solid surface where the fluid flows in a different physical regime. The bounding surface, e.g. the surface of a wall, constitutes a no-slip boundary that exerts a strong resistance to the hydrodynamic motion next to it that is locally affected. Both the flow velocity and the turbulence intensity will decrease from the 'free air' towards the no-slip boundary, changing their profiles. In practice, the BL constitutes a bulk transition layer between the unperturbed free air motions and the wall surface, with a progressive reduction of momentum that vanishes at the wall surface. The air molecules in contact with the wall have null average, i.e. molecular motions but no ordered transport.

The thickness of the BL is not exactly defined, but is generally assumed as the distance from the wall to a point where the flow velocity has essentially reached a given percentage (e.g. 90%) of the free air stream (FAS) velocity. This thickness varies with the wind speed, surface roughness and difference between the thermal level of the wall and the free air, reaching centimetres, decimetres or even metres.

The BL is composed of two sub-layers (Fig. 2.5). The first is a very thin viscous sub-layer (VSL) adhering to the surface, where turbulent

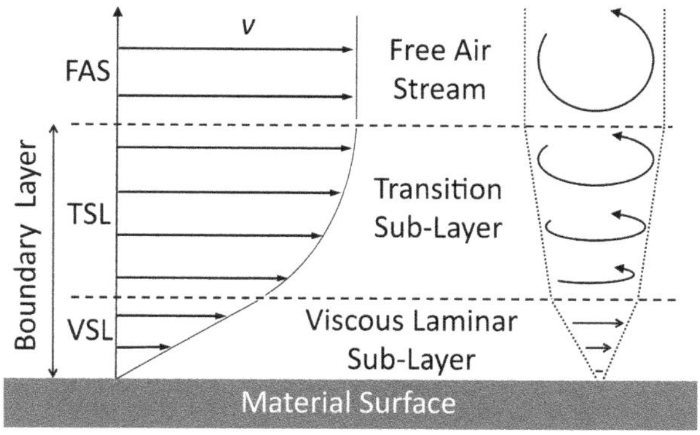

Fig. 2.5 The boundary layer (BL) over a material surface. It is composed of a viscous sub-layer (VSL) with laminar motions and a transition sub-layer (TSL). Above them, the undisturbed free air stream (FAS). Left side: the profile of the velocity V vectors in the above sub-layers. Right side: the passage from the 2D laminar motions to the 3D air turbulence, related to the above sub-layers and the distance from the material bounding surface.

fluctuations are damped. The second is a thicker transition sub-layer (TSL) where motions pass from laminar to turbulent. In the TSL the friction resistance is attenuated, and the air motions gain freedom on the third direction, gradually passing from two-dimensional (2D, at the top of the VSL) to three-dimensional (3D) turbulence, typical of FAS. Briefly, in the TSL the air velocity and turbulence increase with the distance from the top of the VSL until the undisturbed FAS is reached.

The BL may be of dynamic origin, i.e. the layer generated by the friction with the building surfaces. In this case it is called *dynamic boundary layer* (DBL). The DBL is triggered and starts on the upwind edge of building and develops, growing until it reaches the downwind edge. It covers the whole wall but its thickness grows while the airflow progresses over the rough surface. The DBL has a horizontal development, being more homogeneous on the vertical direction.

The BL may be of thermal origin, i.e. the layer generated by the contact with a warm, or cold building surface. Sunshine, thermal radiation loss, evaporation and other forms of heat and mass balance may substantially affect the surface temperature of building materials, which may depart of some degrees, or tens of degrees °C from the free air temperature. The air next to the surface gains buoyancy and rises along the surface (e.g. a wall) if this is overheated, or becomes denser and sinks if the wall is colder than the free air, forming a *thermal boundary layer* (TBL). The TBL has a vertical development, being more homogeneous on the horizontal direction.

The thickness of TBL increases while it flows along the wall surface, gaining further heat, or increasing the number of colder air molecules. Briefly, after a few metres of path along a wall, the BL reaches an appreciable thickness and (positive or negative) buoyancy, thus enveloping the whole building facade or side walls. The warm, or cold, air of the TBL eventually meets fissures or openings, and penetrates inside by leakage or through open windows. In the countries where sunshine is powerful, when a window is opened to ventilate a room, one should consider that the air that will enter the room might be considerably warmer than the outdoor free air, perturbing the indoor temperature and RH.

The BL, either of dynamic or thermal origin, is characterised by an airstream bearing particulate matter, flowing along the building surface. The micro-turbulence and other deposition mechanisms will contribute to deposit part of the entrapped particles, thus increasing soiling and surface blackening. The main deposition mechanism is inertial impaction.

2.6 The Viscous Sub-Layer at the Interface Between Air and Building Materials

The VSL plays a key role in governing a number of exchanges between building materials and their environment, as follows. Next to a wall surface, the free air motions and the 3D turbulence of the wind are strongly affected (and damped) by the presence of the wall surface. The surface constitutes a limiting solid boundary and a no-slip constraint. At a very close distance from the boundary surface, the air velocity component normal to the surface is naturally zero, as mass cannot penetrate the surface. The only possible motions are tangent to the surface, i.e. 2D laminar motions, like sheets of paper slipping above each other. The surface roughness generates a friction force that opposes and tends to stop any air motion, and the velocity of the very first air layer in contact with the surface is null. The molecular viscosity transmits this passive force to the air in close proximity, i.e. to the next layer, but with intensity decreasing with increasing distance from the boundary. Briefly, next to a wall surface a thin VSL is formed where viscosity makes motions slow, laminar and parallel to the surface, determining a shear.

The thickness of the VSL depends on the velocity of the undisturbed flow and the surface roughness. It may be in the order of micrometres to millimetres, or even more. For a smooth surface (Fig. 2.6a) characterised by a low roughness with typical height h, the thickness δ of the VSL is greater than h. The roughness is completely immersed in the viscous sub-layer, and it can neither affect the velocity profile nor the friction resistance. For a rough surface (Fig. 2.6b), characterised by high roughness, δ is smaller than h. The top of the rough peaks exceeds the VSL thickness, with the result of changing the velocity profile and causing an increase in the friction resistance to the upper airflow, as well as the thickness of film of air entrapped next to the surface.

The VSL acts as a filter controlling the exchanges between the material surface and the free air. However, it is a special air pocket that sometimes reaches specific characteristics that may substantially alter synergisms and interactions between the ambient atmosphere and buildings.

2.6.1 *Aerodynamic filter of particulate matter*

The VSL may constitute an obstacle for airborne particles to reach the surface. Airborne particles are carried by wind, but when the wind approaches the building structure, it locally deviates and/or increases the turbulence. The kinetic energy gained by the particles pushes them along their inertial motion, thus leaving the swirling air stream. Depending on the balance between the kinetic energy of the particle and the viscous friction in proximity of the wall, an energetic particle may reach the surface, and a less energetic particle may be stopped by viscosity before reaching it.

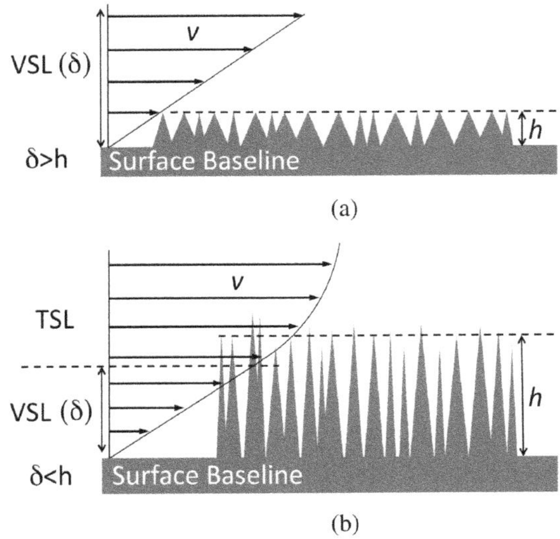

Fig. 2.6 The viscous sub-layer (VSL) thickness δ compared with the typical height h of the surface roughness considered above the surface baseline of the building material. (a) The surface is smooth and the small roughness stays within the VSL, i.e. $\delta > h$. (b) The surface is dynamically rough (i.e. coarse roughness) and the peaks of the roughness exceed the VSL, i.e. $\delta < h$. In both cases the VSL is topped by the TSL, although the TSL is indicated only in (b) to show that the exceedingly high roughness favours the onset of turbulence and is almost coincidental with the base of the TSL.

The VSL makes a natural aerodynamic selection of the types of particles that may reach the surface under different environmental conditions.

2.6.2 *Synergism with gaseous pollutants*

The VSL is composed of air, water vapour and pollutants that are eventually renewed with some external free air. The exchange rate is low, and varies with the diffusivity of the molecular species or aggregates (e.g. H_2O, air, gaseous and particulate pollutants). This may lead to remarkable differences with the concentration in the free air at some distance from the wall.

The pollutants that will react with the materials are necessarily those that can reside in the VSL, especially those sticking to the material surface, and the reaction is possibly limited to the time of simultaneous presence with a film of liquid water. Once a reactive pollutant has been removed from the VSL, or has reacted with the surface, is out of the system. Records of environmental air pollution are representative of the air quality, but are only a part of the information needed to know the chemical reactions that will occur on the building surfaces.

2.6.3 *Temperature and relative humidity*

The small slip factor of H_2O molecules and their high velocity lead to their diffusivity in air large. In contrast, larger air molecules are characterised by smaller diffusivity. This means that in the VSL where molecular diffusivity dominates, moisture exchange is faster than heat exchange. The temperature T of the almost stagnant air in close proximity of the surface will be determined by the building material. In contrast, the humidity mixing ratio (*MR*, i.e. the ratio between the mass of water vapour and the mass of dry air in the same air volume) will follow the *MR* of the free air. In the VSL, the *MR* may be, to a few percent, affected by possible condensation on the wall (i.e. subtraction of water molecules from VSL) or evaporation from it (i.e. addition of water molecules to VSL), but in a first approximation we can neglect this very small departure.

Therefore, in the VSL the relative humidity (*RH*) will be substantially determined by the temperature of the wall and the *MR* of the free air. An example is shown in Fig. 2.7 where buildings with the same exposure on the same street show different T and *RH* levels depending on the individual response of the various building materials. The wall plaster is some 10–15°C warmer than the stone of the two balconies and the parts in the shadow. The *RH* drops accordingly, with crystallisation of soluble salts. The largest temperature difference is found with the wooden door that is soon overheated because of its low thermal diffusivity. It may be useful to remember that the *thermal diffusivity* is the ratio between the thermal conductivity and heat capacity of a material. Physically, it expresses the speed of propagation of a thermal wave into a

Fig. 2.7 Two balconies viewed as infrared surface temperature image (thermogram, *T*) and *RH* calculated at the interface between the air and building materials. The upper balcony has glass panes that appear cold for the reflected sky; the lower has a wooden door. Wood is overheated by solar radiation. Images taken in Padua, Italy, 9 August 2014, 10:30 a.m., with variable cloud cover.

medium, which is proportional to the capability of transmitting heat in the presence of the unit temperature gradient (*thermal conductivity*) and inversely proportional to the capability of storing heat (*heat capacity*).

2.6.4 *Crystallisation cycles*

Although the role of hygroscopic salts in causing deterioration of porous stones and masonry is known and extensive overviews exist (Charola, 2000; Dohene and Price, 2010), some notes may be useful to clarify mechanisms. We will start from the definition. It is often claimed that: "a hygroscopic substance *attracts* water out of air or materials". Water molecules are not *attracted*, but are simply captured and held when they reach, or impact on a hygroscopic salt. A hygroscopic salt is typically composed of a crystal lattice including ions of opposite signs, e.g. sodium chloride: Na^+ Cl^-, bound by electric Coulomb's forces. The water molecule is also polar, exerting forces from the slight positive electrical charge on the H and negative on the O, although the bulk H_2O molecule is electrically neutral, and for this reason it is not attracted by the salt. However, when a water molecule impacts with a $H^{\delta+}$ (or a $O^{\delta-}$) on a sodium chloride crystal, hitting a Cl^- (or a Na^+) ion constituting the crystal lattice, it remains bond in that position.[1] However, the original crystalline structure had a regular alternation of positive and negative ions, but the newly included H_2O molecule contains other charges that do not exactly fit with the regular electric structure of the lattice, forming a local discontinuity. The salt may continue to take in more water, depending on the affinity of the ions for water molecules. The uptake increases with increasing *RH* levels. The water uptake weakens the crystal. This process will continue until a solution is formed. The original salt is the solute, and the water is the solvent. This process is favoured because some heat energy (enthalpy) of the system is released when the salt becomes hydrated.

In the case of certain salts, the water uptake does not occur at a regular rate, but increases very much when a certain *RH* threshold is reached, called deliquescence relative humidity (*DRH*). At this *DRH* threshold, the salt becomes *deliquescent*, which means that it absorbs a large enough amount of water into which it will dissolve,

[1]The charges on Na+ in the crystal lattice and H+ in a water molecule are very different. The hydrogen never has full unit charge in the water molecule. The same holds for oxygen. For this reason chemists prefer to write the above charges $\delta+$ and $\delta-$.

Table 2.1 Deliquescence Relative Humidity (*DRH*) of selected salts at 20°C.

Salt	DRH(%)	Salt	DRH (%)
LiCl	11.3	$MgCl_2.6H_2O$	32
$CaCl_2.6H_2O$	32	$Zn(NO_3)_2.6H_2O$	42
$K_2CO_3.2H_2O$	44	$Mg(NO_3)_2.6H_2O$	56
$NaBr.2H_2O$	58	NaCl	76
KBr	84	$ZnS0_4.7H_2O$	90
$Na_2CO_3.10H_2O$	92	KNO_3	95

passing from the crystalline to the solute (liquid) phase. Each deliquescent salt has a particular *DRH* threshold (as shown in Table 2.1) that slightly changes with temperature, either increasing or decreasing. Of course, both *RH* and temperature should be referred to the micro-environment around the salt crystal.

Once a salt has dissolved, the ambient *RH* should be lowered again for crystallisation to occur. When *RH* decreases, the solvent (i.e. pure water) will evaporate and the solution will concentrate until it reaches saturation and precipitates salt crystals. However, the process may be affected by some hysteresis and after the *DRH* threshold has been reached, the solution may become supersaturated and a further decrease in *RH* may be necessary to reach the nucleation point and crystallisation, i.e. *CRH* (crystallisation relative humidity), as shown in Fig. 2.8. At *CRH* all the solvent will evaporate and the whole solute will pass to the crystalline form (Hinds, 1982).

The hysteresis is typical of aerosols (Randles *et al.*, 2004) for the effect of the curved meniscus that increases the equilibrium partial pressure of water vapour, i.e. Kelvin effect for very small positive radii of curvature, i.e. convex meniscus (Fig. 2.9) (Camuffo, 2013). As a consequence, the equilibrium *RH* over a fine water droplet is affected by the increase in partial pressure over a curved meniscus and may differ from the level of the electrolyte solution with plane free surface. Similarly, for the curvature effect, a fine droplet cannot freeze and becomes supercooled. In clouds, the freezing point depression may reach −40°C. The difference $\Delta RH = DRH - CRH \geq 0$ depends on the chemical composition of the solution and the physical and

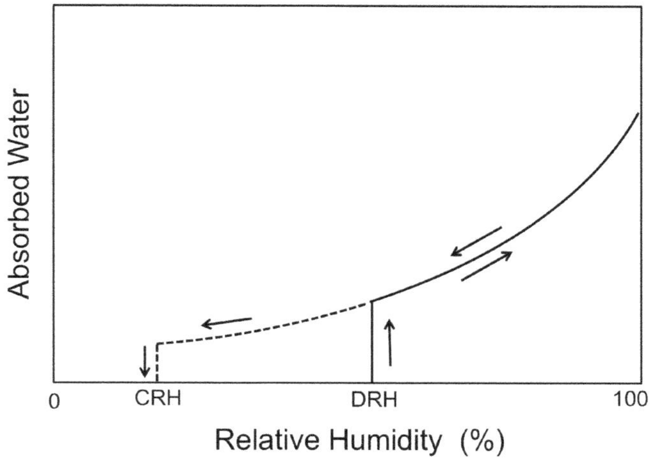

Fig. 2.8 Diagram showing the amount of absorbed water versus relative humidity for a deliquescent salt with hysteresis. *CRH* is the crystallisation relative humidity; *DRH* the deliquescence relative humidity. Arrows show the direction of the thermodynamic transformation.

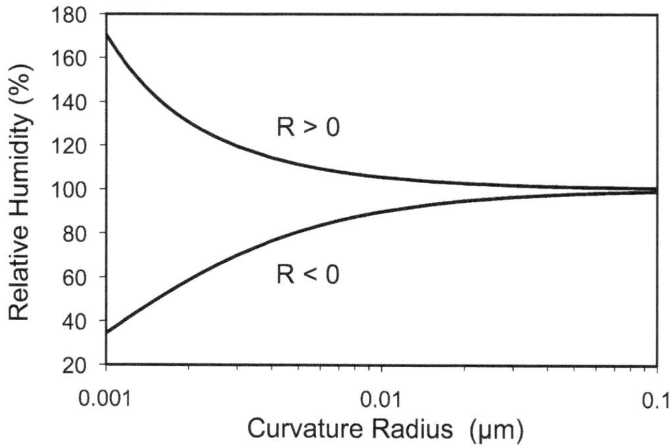

Fig. 2.9 Effect of the curvature of a pure water meniscus in increasing or decreasing the equilibrium partial pressure of water vapour and related relative humidity. Droplets have convex meniscus with positive radius R, i.e. $R > 0$, upper line; pores have concave meniscus with negative radius, i.e. $R < 0$, lower line.

chemical properties of the material surface. When $\Delta RH = 0$ the thermodynamic system is reversible. When $\Delta RH > 0$ the thermodynamic system is characterised by hysteresis.

Hysteresis is especially relevant for aerosol particles, and this explains the persistence of fog in relatively unsaturated atmosphere, or the radiative impact of aerosols in atmosphere (Sportisse, 2009; Anaf *et al.*, 2015). For instance, NaCl is characterised by $DRH = 75\%$ and $CRH = 47\%$ for an aerosol of NaCl solution, (Andrews and Larson, 1993; Tang, 1997; Spurny, 1999; 2000; Randles *et al.*, 2004). *CRH* of NaCl in complex salt mixtures occurs at significantly lower *RH* than implied by the deliquescence humidity of the pure salt (Steiger *et al.* 2011). In addition, the *CRH* of NaCl in complex salt mixtures occurs at significantly lower *RH* than implied by the deliquescence humidity of the pure salt (Steiger *et al.*, 2011).

Things are different on a material surface, or inside pores, because the larger size of pores makes the curvature effect negligible. In addition, the effect would be opposite, because the equilibrium partial pressure of water vapour is lowered when the radius of curvature R is negative, i.e. concave meniscus, and condensation is favoured.

The second factor to be considered is the presence of crystallisation nuclei that trigger the self-organisation of the crystal lattice. Nucleation is very sensitive to impurities in the system. On airborne droplets or aerosols, the spontaneous formation of nuclei within the liquid (i.e. homogeneous nucleation) is difficult and may generate some hysteresis. On a building material, the physical and chemical properties of the surface in contact with the solution may act as crystallisation nuclei (i.e. heterogeneous nucleation) and catalyse the formation of a regular lattice. The system will have small or no hysteresis.

In conclusion, the *CRH* may be much lower than *DRH* in fine aerosols with $R < 0.1\ \mu m$; on building materials, either on the surface or inside pores, this difference should be disregarded.

Not only *RH* is the problem, but also temperature, presence of other salts and/or a combination of these factors. Let us consider a building material soaked with marine water and the temperature dropping below the freezing point. When the temperature drops below 1.9°C, the NaCl is no more soluble in water and precipitates, while the solvent (i.e. pure water) will form large ice crystals. As the amount of solvent is dominant in comparison with the solute, the damage is likely due to the ice crystals.

When salts crystallise on a building surface, the crystals are termed efflorescence; when below the surface, sub-efflorescence. The behaviour of a salt efflorescence is determined by the *T* and *RH* levels reached in the VSL. For instance, suppose that the free air temperature *T* is 20°C and the *RH* level is 80%, which implies 11.6 g kg^{-1} *MR*. Suppose that a wall is hit by solar radiation and its temperature is 30°C. Combining this temperature with the above *MR*, one obtains $RH = 44\%$. Practically, the VSL at the interface is at $T = 30°C$ and $RH = 44\%$, in contrast with the free air which is characterised by $T = 20°C$ and $RH = 80\%$. A NaCl salt efflorescence that is located on the wall surface (i.e. at the interface with the VBL) is in crystal form, while a traditional environmental monitoring indicates that NaCl has exceeded $DRH = 75\%$ and has formed a solution with moisture absorbed from the air. This is a very misleading conclusion especially if one is concerned with the crystallisation-dissolution cycles and their frequency of occurrence.

From the conservation point of view, the presence of hygroscopic salts constitutes a negative factor, especially when crystallisation dissolution cycles occur. They exert a cumulative effect over time that is exacerbated in the case micro-climate will swing around *DRH-CRH*. When *RH* falls below the *CRH*, crystals exert pressure inside pores and damage the hosting material with disaggregation, exfoliation, delamination, scaling and flaking. When *RH* rises above the *DRH*, the aqueous solution that will form may dissolve and corrode the stone, or feed halophilic organisms, or at very low temperatures may generate disruptive ice crystals.

It may be useful to consider the relationship between the *MC* of a porous material and the *RH* in equilibrium with it. Let us start with an uncontaminated material in which the *MC* is only due to adsorption and Kelvin condensation in pores (Brunauer, 1945; Gregg and Sing, 1967; Mikhail and Robens, 1983; Camuffo, 2013). In general, when *RH* increases, the *MC* also increases, as shown in Fig. 2.10a. The plot can be divided into three parts. In the first part, when *RH* passes from

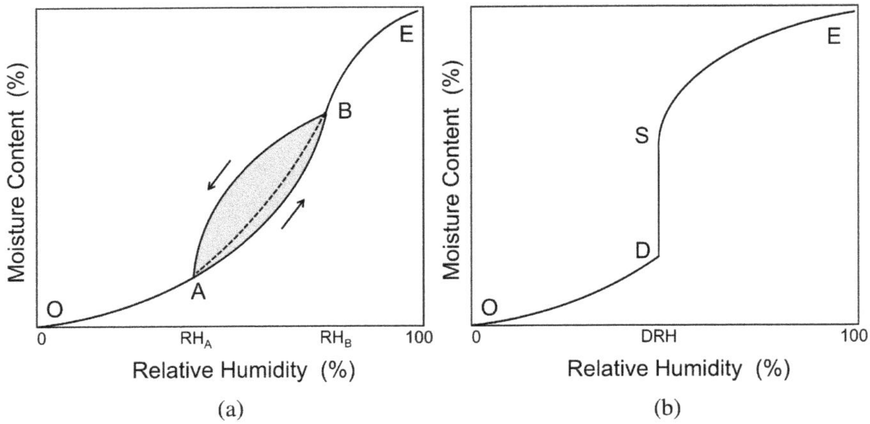

Fig. 2.10 (a) Diagram showing the moisture content versus relative humidity for a porous material with hysteresis loop (grey area) from A to B. (b) The same diagram but for a material impregnated with a deliquescent salt (only for increasing *RH*). O to D adsorption, D to S deliquescence absorption; S to E mainly dilution absorption.

0 to RH_A, the adsorbed water increases the *MC*, passing from the origin O to A. This adsorption branch is reversible. The second part, i.e. the *RH* interval from RH_A to RH_B, is characterised by a hysteresis loop (grey area). The lower line is for increasing *RH*, the upper for decreasing *RH*. The former is due to water condensing in pores, the latter to water evaporating from the mouths of pores. The difference between the dimension of the radii of pores and the radii of mouths justifies the loop. If the system were without hysteresis, passing from point A to B along the dashed line, then one well-determined *MC* value would be related to each *RH* level, and vice-versa. However, the hysteresis in the grey area makes obscure any relationship between *RH* and *MC*, the uncertainty band depending on the loop thickness. Finally, the third part from B to the ending point E is also characterised by reversible adsorption and one-to-one correspondence between *RH* and *MC*.

Let us consider what happens in the case where the material is contaminated with a deliquescent salt (Fig. 2.10b). Starting from a completely dry material, an increase in *RH* from 0 to *DRH* will cause in reversible adsorption and will displace the plot from the origin O to D, thus increasing the *MC*, as already discussed. However, as the system will reach the deliquescence point D, the system will absorb water to dilute the super-saturated solution until saturation is reached in S. This part of the process occurs at the same *RH* level, i.e. *DRH*, to which a certain *MC* interval is corresponding. The distance from D to S, and the corresponding amount of water absorbed in the material, depends on the amount of soluble salt present in the material. Once saturation has been reached, an increase in *RH* will continue to dilute the solution by further absorption (if the empty space in pores is large enough), thus increasing the *MC* of the material. The return pattern for decreasing *RH* levels may not be the same, for the already discussed hysteresis mechanisms.

The saline solution (either formed for water uptake or for water supplied by dew, rain, percolation, or capillary raise) may migrate inside the building material. Two driving forces should be considered. The main mechanism from inside to outside transport is evaporation. When a wall is soaked, the water evaporating from the wall surface draws further water from the interior across the system of pores and capillaries acting as siphons and communicating vessels. On the evaporating surface, only water vapour may leave the system, while salts will increase their concentration in the solution until they precipitate, forming efflorescence or sub-efflorescence.

The transport from the external surface to inside a wall is governed by the Fick's law of diffusion, driven by differences in the solute concentration χ. The diffusive flux J is proportional to the concentration gradient $\partial\chi/\partial x$, i.e.

$$J = -\mathscr{D}\frac{\partial\chi}{\partial x}$$

where x is the spatial coordinate along the gradient (i.e. from the external surface to the wall inside) and $\mathscr{D}$ is the diffusion coefficient of the solute. The solute will move from a region of high concentration to a region of low concentration.

The aqueous solution may evaporate on the surface or inside pores, depending on what level pores and capillaries are filled with water, allowing (or not allowing) internal diffusion and migration of water molecules in the vapour phase (Fig. 2.11).

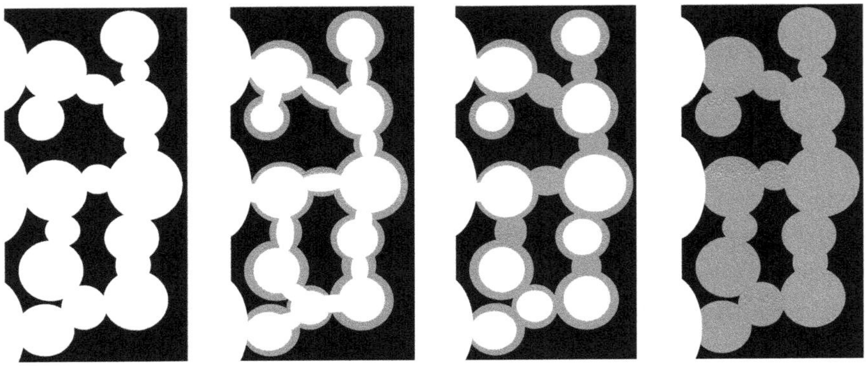

(a) (b) (c) (d)

Fig. 2.11 (a) A cross-section of a wall with pores completely dry (white dots). Free air is on the left side. (b) Inside pores some water remains adhering to the material (grey), and the connections between pores are not obstructed by water. Internal evaporation may occur. (c) As in (b) but with pore connections obstructed by condensed water. No internal evaporation is possible. (d) The same cross-section, but with pores completely full of water in the liquid phase (grey). Evaporation is possible from the wall surface, not from the interior.

When some water remains adherent to the material inside pores, but the connection between pores is open (Fig. 2.11b), evaporation may occur in the internal pores, and the released vapour molecules may diffuse through the network of free connecting spaces, until they reach the free air.

When the condensed water obstructs the entrance of pores (Fig. 2.11c), each pore remains isolated. Each cavity contains an air pocket, like an air bubble, separated from the others and the external environment. In such a case, internal pore condensation is stopped for the presence of the air bubbles that cannot be removed. Internal evaporation cannot start because the internal bubbles are always saturated being surrounded by liquid water, and there is no way to remove water vapour. Evaporation may only occur starting from the exterior, on the wall surface, and will slowly propagate inside, like a front, after the water obstructing the pore entrances has evaporated. However, if the wall temperature increases, the air bubbles will expand and will move outwards the water from each pore entrance to the interior of the next pore, passing to the situation depicted in Fig. 2.11b. If temperature decreases, the bubbles will shrink, and will move inwards the water from the entrance, also returning to Fig. 2.11b.

When pores are completely full of water (Fig. 2.11d), evaporation is only possible from the wall surface. In general, condensation cannot completely fill the internal pores with water because of the entrapped air that will stop the process. Pores may be full of water after rain, water percolation or capillary rise if the network of free spaces allows the displacement of the internal air.

In the case of internal pore evaporation, the dissolved salts will increase the concentration, reach saturation and then form a crystalline precipitate. The thermal expansion coefficients of the salt crystals and the building material are in general different. When a crystal fills the whole internal volume of a pore, if the crystal expands more than the pore cavity, or the cavity shrinks more than the salt, the pressure exerted on the cavity borders may damage the stone, or the plaster, with loss of the surface and sub-surface layers.

Some minerals may undergo a crystalline transition from hydrated to dehydrated forms under specific environmental conditions. In coastal areas, some combinations of temperature and NaCl may trigger the transition of calcium sulfate from gypsum ($CaSO_4 \cdot 2H_2O$) to anhydrite ($CaSO_4$), or vice-versa (Fig. 2.12) (McDonald, 1953; Price and Brimblecombe, 1994). Gypsum passes to anhydrite either for an increase in temperature, or an increase in NaCl concentration, or a combination of the two. When the amount of sea spray deposited on the wall surface increases, or the evaporation of the water solvent increases the solute concentration, the transition temperature decreases. When the solution contains 3 mol of NaCl, the transition temperature is 30°C; at 4 mol, 25°C; at 5 mol, 21°C; and at saturation (i.e. 6.15 mol) the transition occurs at 14°C. When gypsum changes to anhydrite, from the hydrated to the anhydrous form, the volume shrinks and some crystallisation water is released. In contrast, the hydration implies an increase in volume that may damage the material on which gypsum crystals are present.

Another example is given by thenardite (Na_2SO_4), a mineral impurity frequently found in cement-based mortars and concrete

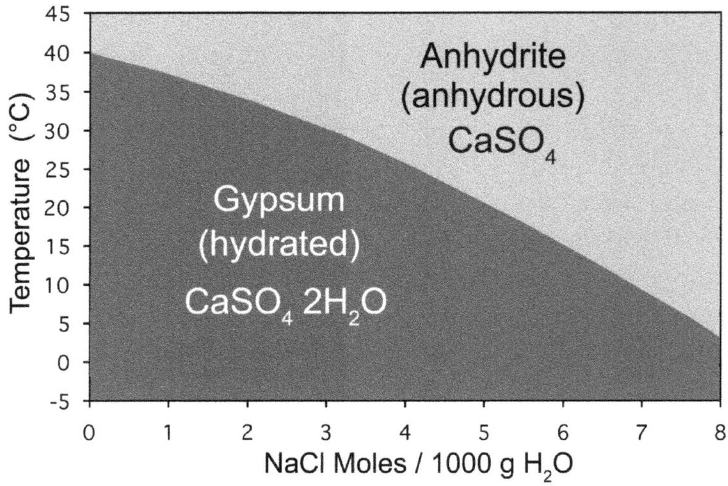

Fig. 2.12 Diagram of the gypsum-anhydrite mineral transition for temperature and/or NaCl concentration changes.

(Espinosa Marzal and Scherer, 2008; Haynes *et al.*, 2008). Under certain combinations of temperature and *RH* (shown in Fig. 2.13), the anhydrous thenardite crystals undergo a transition to the mirabilite ($Na_2SO_4 \cdot 10H_2O$) hydrated form, which includes 10 molecules of water. In the graph, one can see that the transition to the hydrated phase may occur for a decrease in temperature (horizontal displacement from the grey to the black area), an increase in *RH* (vertical displacement) or a combination of the two. The enormous increase in crystal volume exerts pressures, damaging the hosting material. In contrast, the transformation to the anhydrous form will reduce the crystal size without exerting mechanical forces, but the release of liquid water will dampen the concrete and will rust reinforcement iron bars with further increase in volume and related damage. Repeated cycles may cause severe weathering.

2.6.5 *Chemical reactions*

A chemical reaction between a building material and a pollutant is governed by the presence of dampness inside the material, or on its surface, or the *RH* level at the interface, i.e., the VSL. Of course,

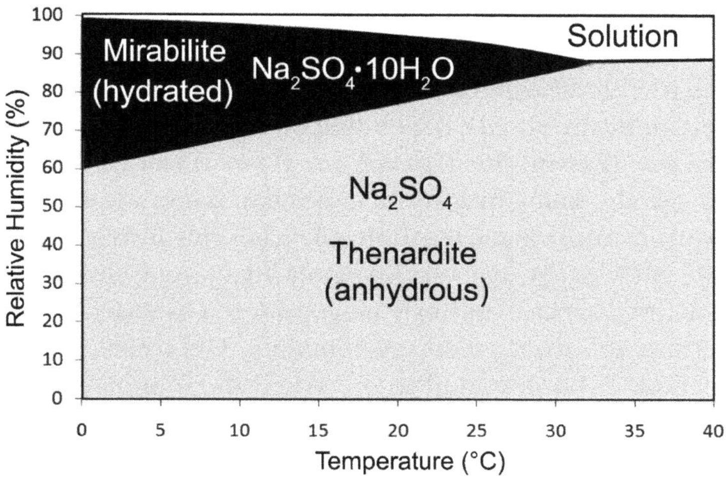

Fig. 2.13 Diagram of the thenardite-mirabilite mineral transition for temperature and/or *RH* changes. Thenardite is found in the grey area, mirabilite in the black, and aqueous solution in the white one.

the kinetic rate is related to the material temperature (Arrhenius equation) and not to the free air temperature.

In the presence of sunshine, however, the elevated temperature reached by the material surface may force evaporation, leaving the outer layer of the material dry, thus preventing any chemical reaction for water in the liquid phase to form aqueous solutions.

2.7 The Physical Phases of Moisture in Building Materials

It is obvious that precipitation supplies water thus causing dissolution, surface recession, feeding bio-organisms and other forms of decay. Less clear is the interaction between materials and the atmospheric H_2O molecules in the vapour phase. It is hardly probable that a material may have physical or chemical reactions with the moisture of the free air stream. We know that the key variable for the deterioration mechanisms and the damage functions is *RH*, and in the previous sections we have seen that we should consider the *RH* in the VSL, not in the free air. The moisture that passes from the air to the material may either be absorbed into the volume, e.g. as a solvent for the efflorescence, or adsorbed on the surface with stronger or weaker bounding forces. The absorption may be related to a variable *RH* level, or to some critical thresholds, depending on the material's nature.

Similarly, the adsorption may be regulated by continuously changing levels, or by thresholds, depending on the micro-pore and micro-capillary size (Kelvin effect). Even on glossy surfaces, e.g. marble, glass or metals, water may form molecular layers whose thickness may vary from one to some ten molecules (Leygraf and Graedel, 2000; Camuffo, 2010; 2013). The very first molecular layer is strongly bound to the material surface and may be considered as water in the solid state because the attractive forces dominate. The second layer is also strongly bound, but less, and so on. When the monolayer films are only a few, they can be substantially considered as a solid coating, like a varnish. In contrast, when the *RH* exceeds 70% and the layer thickness is 6–7 monolayers, the H_2O molecules are only weakly bound, have gained mobility and start to behave as liquid water does, e.g. forming solutions with the deposited pollutants, allowing chemical reactions in the aqueous phase (e.g. oxidation, corrosion), feeding water to micro-organisms that may colonise and grow on the moist surface. For instance, gaseous pollutants, e.g. SO_2, reaching a stone or a metal, but in dry phase, are blown away by the wind leaving the material unaffected. Deposited aerosols or bacterial contamination remain dormant until liquid water will allow them to grow.

2.8 Conclusions

Weather is responsible for several deterioration mechanisms, but each weather variable acts in a particular way: in some cases directly, in others indirectly and with a certain delay to reach a dynamic equilibrium between the environment and building materials. For instance, solar radiation and heavy precipitation reach the building surface directly, while other variables, e.g. the action of wind, fog, gaseous and particulate pollutants are regulated or filtered by the presence of a BL.

We should keep in mind that building materials do not have direct heat and moisture exchanges and chemical reactions with the outdoor free flowing air. Most exchanges and reactions are only possible with moisture and substances entrapped into the VSL, or already present on the material surface for previous deposition. This makes a big difference.

Field monitoring for conservation diagnostics or restoration intervention purposes that consider the temperature and the relative humidity of the air, but neglect the building surface temperature, will provide incomplete or misleading information. In particular, the *RH* level observed in the free air may drastically differ from the RH reached in the VSL. In addition, a certain (longer or shorter) time period is needed to reach a dynamic equilibrium between the moisture content, or the filming moisture, and the atmosphere. This is crucial for the formation of efflorescence, or crystallisation-dissolution cycles.

The moisture dependent deterioration mechanisms proceed synergistically with the molecules of water that have been already absorbed (i.e. penetrated inside the volume) or adsorbed (i.e. filming on the surface) onto building materials. This water may be either directly supplied by precipitation, percolation, capillary rise etc. or derived from the atmosphere, but through the mediation of the VSL.

Not all the moisture inside a material is in the same physical state. Part of the H_2O molecules may be inert in the solid state, part may be gaseous inside the air pockets of pores and capillaries, and part may be in the liquid phase, constituting what is popularly considered as the dangerous dampness responsible for various forms of decay.

The puzzle emerging from this chapter is the first step to understand complex deterioration mechanisms affecting built heritage and to take appropriate conservation measures.

Acknowledgments

The Author is grateful to the colleague Dr Chiara Bertolin for useful discussions.

References

Anaf, W., Bencs, L., Van Grieken, R., Janssens, K, DeWael, K. (2015). Indoor particulate matter in four Belgian heritage sites: case studies on the deposition of dark-colored and hygroscopic particles, *Sci. Total Environ.*, 506–507, pp. 361–368.

Andrews, E., and Larson, S.M. (1993). Effect of surfactant layers on the size changes of aerosol-particles as a function of relative-humidity, *Environ. Sci. Technol.*, 27, pp. 857–865.

Arnold, A. and Zehnder, K. (1991). 'Monitoring wall paintings affected by soluble salts', in S. Cather (Ed.), *The Conservation of Wall Paintings*, Paul Getty Trust, Thien Wah Press, Singapore, pp. 103–135.

Bacci, M., Cucci, C., Mencaglia, A.A., Mignani, A.G., Porcinai, S. (2004). Calibration and use of photosensitive materials for light monitoring in museums: Blue Wool Standard 1 as a case study, *Stud. Conserv.*, 49, pp. 85–98.

Bratasz, L. (2010). 'Acceptable and non-acceptable microclimate variability: the case of wood' in: Camuffo, D., Fassina, V., Havermans, J. (Eds.), *Basic Environmental Mechanisms Affecting Cultural Heritage — Understanding Deterioration Mechanisms for Conservation Purposes*, COST Action D42 "Enviart", Nardini, Florence, Italy.

Brimblecombe, P. (1987). *The Big Smoke*, Methuen, London.

Brimblecombe, P. and Camuffo, D. (2003). 'Long term damage to the built environment' in: Brimblecombe, P. (Ed.), *The Effects of Air Pollution on the Built Environment*, Imperial College Press, London, pp. 1–30.

Brunauer, S. (1945). *The Adsorption of Gases and Vapors*, Princeton University Press, Princeton.

Cailleteau, C., Angeli, F., Devreux, F., Gin, S., Jestin, J., Jollivet, P. and Spalla, O. (2008). Insight into silicate-glass corrosion mechanisms, *Nat. Mat.*, 7, pp. 978–983.

Camuffo, D. (1992). Acid rain and deterioration of monuments: how old is the phenomenon? *Atmos. Environ.*, 26B, pp. 241–247.

Camuffo, D. (1993). Controlling the aeolian erosion of the Great Sphinx, *Stud. Conserv.*, 38, pp. 198–205.

Camuffo, D. (1995). Physical weathering of stones, *Sci. Total Environ.*, 167, pp. 1–14.

Camuffo, D. (2010). 'The role of temperature and moisture', in Camuffo, D., Fassina, V. and Havermans, J. (Eds.), *Basic Environmental Mechanisms Affecting Cultural Heritage — Understanding Deterioration Mechanisms for Conservation Purposes*, COST Action D42 "Enviart", Nardini, Florence, Italy, pp. 9–30.

Camuffo, D. (2013). *Microclimate for Cultural Heritage — Conservation, Restoration and Maintenance of Indoor and Outdoor Monuments*, 2nd Ed., Elsevier, New York.

Camuffo, D. and Enzi, S. (1995). Impact of clouds of volcanic aerosols in Italy in the past centuries, *Nat. Hazards*, 11, pp. 135–161.

Camuffo, D. and Sturaro, G. (2001). The climate of Rome and its action on monument decay, *Clim. Res.* 16, pp. 145–155.

Camuffo, D. and Giorio, R. (2003). Quantitative evaluation of water deposited by dew on monuments, *Bound-Lay. Meteorol.*, 107, pp. 665–672.

Camuffo, D., Daffara, C. and Sghedoni, M. (2000). Archaeometry of air pollution: urban emission in Italy during the 17th century, *J. Archaeol. Sci.*, 27, pp. 685–690.

Caneva, G., Nugari, M.P. and Salvadori, O., (2009). *Plant Biology for Cultural Heritage — Biodeterioration and Conservation*, The Getty Conservation Institute, Los Angeles.
Charola, A.E. (2000). Salts in the deterioration of porous materials: an overview, *J. Am. Inst. Conserv.*, 39, pp. 327–343.
Charola, A.E. and Ware, R. (2002). 'Acid deposition and the deterioration of stone: a brief review of a broad topic', in Siegesmund, G.S., Vollbrecht, A., Weiss, T. (Eds.) *Natural Stone, Weathering Phenomena, Conservation Strategies and Case Studies*, Special publication 205, Geological Society, London, pp. 393–406.
Coutinho, M.L., Miller, A.Z., Gutierrez-Patricio, S., Hernandez-Marine, M., Gomez-Bolea, A., Rogerio-Candelera, M.A., Philips, A.J.L., Jurado, V., Saiz-Jimenez, C. and Macedo, M.F. (2013). Microbial communities on deteriorated artistic tiles from Pena National Palace (Sintra, Portugal), *Int. Biodeter. Biodegrad.*, 84, pp. 322–332.
Cultrone, G., De-la-Torre, M., Sebastian, J.E.M. and Cazalla, O. (2000). Behavior of brick samples in aggressive environments, *Water Air Soil Pollut.*, 119, pp. 191–207.
Dohene, E., and Price, C. (2010). Stone conservation — an overview of current research, The Getty Conservation Institute, Los Angeles.
Espinosa Marzal, R.M., and Scherer, G.W. (2008). Crystallization of sodium sulfate salts in limestone, *Environ. Geol.*, 56, pp. 605–621.
Fellenberg, G. (2000). *The Chemistry of Pollution*, Wiley, Chichester.
Fisak, J., Tesar, M., Rezacova, D., Elias, V., Weignerova, V. and Fottova, D. (2002). Pollutant concentrations in fog and low cloud water at selected sites of the Czech Republic, *Atmos. Res.*, 64, pp. 75–87.
Flatt, R.J. (2002). Salt damage in porous materials: how high supersaturations are generated, *J. Cryst. Growth*, 242, pp. 435–454.
Forsyth, M. (Ed.) (2014). *Structures and Construction in Historic Building Conservation*, Wiley-Blackwell, New York.
Fort, R., Alvarez de Buergo, M., Gomez-Heras, M. and Vazquez-Calvo, C. (Eds). (2006). *Heritage Weathering and Conservation*, Balkema, Rotterdam.
Gelbrich, J., Mai, C. and Militz, H. (2008). Chemical changes in wood degraded by bacteria, *Int. Biodeter. Biodegrad.*, 61, pp. 24–32.
Gómez-Heras, M., Smith, B.J. and Fort, R. (2006). Surface temperature differences between minerals in crystalline rocks: implications for granular disaggregation of granites through thermal fatigue, *Geomorphology*, 78, pp. 236–249.
Goudie, A. and Viles, H. (2008). 'Weathering processes and forms' in Burt, T.P., Chorley R.J., Brunsden, D., Cox, N.J., and Goudie, A.S. (Eds). *The History of the Study of Landforms or the Development of Geomorphology*, Geological Society of London, London, pp. 129–164.
Gregg, S.J., Sing, K.S.W. (1967). *Adsorption, Surface Area and Porosity*, Academic Press, London.
Grossi, C.M., Bonazza, A., Brimblecombe, P., Harris, I. and Sabbioni, C. (2008). Predicting twenty-first century recession of architectural limestone in European cities, *Environ. Geol.*, 56, pp. 455–461.

Grossi, C.M., Brimblecombe, P. and Harris, I. (2007). Predicting long-term freeze–thaw risks on Europe built heritage and archaeological sites in a changing climate, *Sci. Total Environ.*, 377, pp. 273–281.

Haynes, H., O'Neill, R., Neff, M., Mehta, P.K. (2008). Salt weathering distress on concrete exposed to sodium sulfate environment, *ACI Mater. J.*, 105, pp. 35–43.

Hinds, W.C. (1982). *Aerosol Technology: Properties, Behavior, and Measurement of Airborne Particles*, John Wiley & Sons, New York.

Hussein, A.S., and El-Shishiny, H. (2009). Influences of wind flow over heritage sites: A case study of the wind environment over the Giza Plateau in Egypt, *Environ. Model. Softw.*, 24, pp. 389–410.

Jacobson, M.Z. (2000). 'Reversible chemical reactions in aerosols', in Spurny, R.K. (Ed.) *Aerosol Chemical Processes in the Environment*, CRC Press LLC, Boca Raton, Florida, pp. 135–156.

Kunicki-Goldfinger, J. (2008). Unstable historic glass: symptoms, causes, mechanisms and conservation, *Rev. Conserv.*, 9, pp. 47–60.

Lefèvre, R.A., Chabas, A., Munier, I. and Lombardo, T. (2005). 'Modern and ancient glass in the polluted atmosphere. What is the prevailing phenomenon? Leaching, corrosion, soiling, encrusting…?' in: van Grieken, R. and Janssens, K. (Eds.), *Cultural Heritage Conservation and Environmental Impact Assessment by Non-Destructive Testing and Micro-Analysis*, Taylor & Francis Group plc, London, pp. 77–90.

Leissner, J. (2003). 'The effect of air pollution on glass', in Brimblecombe, P. (Ed.), *The Effects of Air Pollution on the Built Environment*, Imperial College Press, London, pp. 249–265.

Leygraf, C. and Graedel, T.E. (2000). *Atmospheric Corrosion*, Wiley-Interscience, New York.

Lombardo, T., Ionescu, A., Chabas, A., Lefèvre, R.A., Ausset, P. and Candau, Y. (2010). Dose-response function for the soiling of silica-soda-lime glass due to dry deposition, *Sci. Total Environ.*, 408, pp. 976–984.

MacDonald, G.J.F. (1953). Anhydrite-gypsum equilibrium relations, *Am. J. Sci.*, 251, pp. 884–898.

Macedo, M.F., Miller, A.Z., Dionísio, A. and Saiz-Jimenez, C. (2009). Biodiversity of cyanobacteria and green algae on monuments in the Mediterranean Basin: an overview, *Microbiology*, 155, pp. 3476–3490.

Martínez-Martínez, J., Benavente, D., Gomez-Heras, M., Marco-Castaño, L. and García-del-Cura, M.A. (2013). Non-linear decay of building stones during freeze–thaw weathering processes, *Constr. Build. Mater.*, 38, pp. 443–454.

Mikhail, R.S., Robens, E. (1983). *Microstructure and Thermal Analysis of Solid Surfaces*, Wiley, New York.

Munier, I., Lefèvre, R.A. and Losno, R. (2002). Atmospheric factors influencing the formation of neocrystallisations on low-durability glass exposed to urban atmosphere, *Glass Technol.*, 43C, pp. 114–124.

Piñar, G., and Sterflinger, K. (2009). 'Microbes and building materials', in Cornejo, D.N., and Haro, J.L. (Eds.), *New Building Materials: Properties, Performance and Applications,* Nova Science Publishers, York, pp. 163–188.

Price, C. and Brimblecombe, P. (1994). Preventing salt damage in porous materials, *Stud. Conserv.*, pp. 90–93.

Prikryl, R. and Viles, H.A. (Eds.) (2002). *Understanding and Managing Stone Decay,* The Karolinum Press, Prague.

Radeka, M., Ranogajec, J., Kiurski, J., Markov, S. and Marinkovic-Neducin, R. (2007a). Influence of lichen biocorrosion on the quality of ceramic roofing tiles, *J. Eur. Ceram. Soc.*, 27, pp. 1763–1766.

Radeka, M., Kiurski, J., Markov, S., Marinkovic-Neducin, R., and Ranogajec, J. (2007b). 'Microbial deterioration of clay roofing tiles', in Brebbia, C.A. (Ed.), *Studies, Repairs and Maintenance of Heritage Architecture X,* WIT Press, Southampton, pp. 567–575.

Randles, C.A., Russell, L.M, and Ramaswamy, V. (2004). Hygroscopic and optical properties of organic sea salt aerosol and consequences for climate forcing. *Geophys. Res. Lett.*, 31, L16108.

Rijniers, L.A., Pel, L., Huinink, H.P. and Kopinga, K. (2005). Salt crystallization as damage mechanism in porous building materials — a nuclear magnetic resonance study, *Magn. Reson. Imaging,* 23, pp. 273–276.

Roberge, P.R. (2000). *Handbook of Corrosion Engineering,* McGraw-Hill, New York.

Roberge, P.R., Klassen, R.D. and Haberecht, P.W. (2002). Atmospheric corrosivity modeling — a review, *Mater. Design,* 23, pp. 321–330.

Rodriguez-Navarro, C. and Doehne, E. (1999). Salt weathering: influence of evaporation rate, supersaturation and crystallization pattern, *Earth Surf. Proc. Land.*, 24, pp. 191–209.

Sanjurjo-Sánchez, J. and Alves, C. (2012). 'Pollutant-induced decay of building materials', in Lichtfouse, E., Schwarzbauer, J. and Robert, D. (Eds.), *Environmental Chemistry for a Sustainable World — Volume 2: Remediation of Air and Water Pollution,* Springer Verlag, Berlin, pp. 47–120.

Schreiner, M. (2005). 'Corrosion of historic glass and enamels', in: Janssens, K., Van Grieken, R. (Eds.), *Non-destructive Micro Analysis of Cultural Heritage Materials,* Elsevier, Amsterdam, pp. 713–754.

Sehmel, G. (1980). Particle and gas deposition: a review, *Atmos. Environ.*, 14, pp. 983–1011.

Seinfeld, J.H. (1986). *Atmospheric Chemistry and Physics of Air Pollution,* Wiley, New York.

Seinfeld, J.H. and Pandis, S.N. (1997). *Atmospheric Chemistry and Physics,* Wiley, New York.

Siegesmund, S., Ullemeyer, K., Weiss, T. and Tschegg, E.K. (2000). Physical weathering of marbles caused by anisotropic thermal expansion, *Int. J. Earth Sci.*, 89, pp. 170–182.

Siegesmund, S. and Snethlage, R., (2011). *Stone in Architecture — Properties, Durability*, 4th Ed., Springer Verlag, Heidelberg.

Simpson, W. and TenWolde, A. (1999). *Physical Properties and Moisture Relations of Wood. Wood Handbook: Wood as an Engineering Material*, General technical report FPL, GTR-113, USDA Forest Service, Forest Products Laboratory 1999, Madison, pp 3.1–3.24.

Smith, B.J., Gomez-Heras, M. and McCabe, S. (2008). Understanding the decay of stone-built cultural heritage, *Prog. Phys. Geog.*, 32, pp. 361–439.

Sportisse, B. (2009). *Fundamentals in Air Pollution: From Processes to Modelling*, Springer Verlag, Heidelberg.

Spurny, K.R. (1999). *Analytical Chemistry of Aerosols: Science and Technology*, CRC Press, Lewis Publishers, New York.

Spurny, K.R. (2000). *Aerosol Chemical Processes in the Environment*, CRC Press, Lewis Publishers, New York.

Steiger, M., Charola, A.E. and Sterflinger, K. (2011). 'Weathering and deterioration', in Siegemund, S., and Snethlage, R. (Eds.), *Stone in Architecture*, Springer Verlag, Heidelberg.

Sterflinger, K. (2000). Fungi as geologic agents, *Geomicrobiol. J.*, 17, pp. 97–124.

Tang, I.N. (1997). Thermodynamics and optical properties of mixed-salt aerosols of atmospheric importance, *J. Geophys. Res.*, 102, pp. 1883–1893.

Tidblad, J. and Kucera, V. (2003). 'Air pollution damage to metals', in Brimblecombe, P. (Ed.), *Air Pollution Damage to Metals*, Imperial College Press, London, pp. 227–247.

Warscheid, Th. and Braams, J. (2000). Biodeterioration of stone: a review, *Int. Biodeterior. Biodegrad.*, 46, pp. 343–368.

WMO (1966). *International Meteorological Vocabulary*, WMO No. 182, World Meteorological Organization, Geneva.

WMO (1987). *International Cloud Atlas*, World Meteorological Organization, Geneva.

WMO (2008a). *Guide to Meteorological Instruments and Methods of Observation*, WMO Technical Publication No. 8, World Meteorological Organization, Geneva.

WMO (2008b). 'Guide to hydrological practices', in *Hydrology — from Measurement to Hydrological Information*, 6th Ed., Vol. 1, WMO Technical Publication No. 168, World Meteorological Organization, Geneva.

Wypych, G. (1995). *Handbook of Material Weathering*, ChemTec Publishing, Toronto.

Zehnder, K. and Arnold, A. (1989). Crystal growth in salt efflorescence, *J. Cryst. Growth*, 97, pp. 513–521.

CHAPTER 3

Air Pollution Damage to Stone

Michael Steiger

Department of Chemistry, University of Hamburg, Germany

3.1 Introduction

The damage mechanisms of porous building stones are usually classified into chemical and physical (mechanical) processes and may also include biochemical and biophysical mechanisms. Chemical damage processes refer to the dissolution and alteration of the mineral constituents of a stone material by chemical reactions. Mechanical damage results if stone is subjected to load or stress exceeding its mechanical strength. There are a number of different processes that can induce stress in a stone fabric. While thermal cycling and hygric swelling induce expansion of the matrix constituents themselves, the formation of new phases in the void space of a fabric can generate substantial stress by confined crystal growth.

Though chemical weathering and salt weathering are considered as different weathering mechanisms of rock breakdown, they generally operate together as the products of chemical dissolution reactions are more soluble than the parent minerals and are subsequently involved in dissolution–recrystallisation cycles, then act as the agents in (mechanical) salt damage processes.

Natural and acidic rain and acidic gases are the major agents of chemical processes leading to stone deterioration. Today, it is commonly accepted that air pollution is a major cause of damage in old buildings made of natural stone. Several authors presented archival photographs of sculptures and other ornamental stone to demonstrate the rapid loss of details during the 20th century (Winkler, 1973; Amoroso and Fassina, 1983). Winkler in his classical book presented photographs of a sculpture at Herten Castle (Recklinghausen, Germany), one taken in 1908, the other one in 1968. The sculpture dates from 1702 and shows only moderate damage in 1908. In the following 60 years until 1969, it suffered from a more or less complete loss of details. A similar series of photographs taken at the façade of St Mark's Basilica in Venice is shown in the famous book of Amoroso and Fassina. The authors concluded that 'the stone suffered more damage from weathering during 40 years of the present century than during the first four to five centuries of its existence' (Amoroso and Fassina, 1983). Similarly, Winkler drew a semi-quantitative diagram of the visual rate of damage vs. time and concluded that there is an exponential increase in the rate of damage due to the action of air pollution (Winkler 1973, 1987).

However, at the time of publication of Winkler's book there was no consensus among conservation scientists about the role of air pollution. In contrast to the view of Winkler, some researchers categorically denied that there is a contribution of air pollution to stone decay. For example, Josef Riederer who later became the head of the Rathgen Research Laboratory in Berlin (one of the oldest museums and conservation laboratories in the world) published a paper in 1973 entitled 'Die Wirkungslosigkeit von Luftverunreinigungen beim Steinzerfall' ('The *ineffectiveness* of air pollution in stone degradation') (Riederer, 1973). He concluded that the degradation of stone monuments is essentially the result of natural weathering processes and is largely due to a lack of maintenance. Similarly, Künzel (1987) argued that an exponential increase of damage rates is also expected for natural weathering processes. Of course, those ideas did not remain unchallenged. Only 6 months after the publication of Riederer's paper Luckat, one of pioneers of field exposure experiments with stone, published a paper in the same journal and with the very similar title 'Die Wirkung von Luftverunreinigungen beim Steinzerfall' ('The *effectiveness* of air pollution in stone degradation') (Luckat, 1973).

Since the days of that discussion, nearly 40 years have passed; today, there is no doubt that air pollution is an important cause of damage in building materials. Nonetheless, the quantification of the contribution of past and future air pollution to the degradation of stone monuments using damage functions is still an issue. Also, the concentrations of the major air pollutants showed significant changes since the aforementioned discussion in the seventies of the last century. While there is a significant reduction in sulfur dioxide concentrations in Europe and the United States, there is a rapid increase in threshold countries (Smith *et al.*, 2011). In contrast, nitrogen oxide concentrations do not decline as dramatically and carbon concentrations increase. In some parts of the world stone monuments are past the peak of acidic air pollution, at other places they face a period of increasing or ongoing air pollution. In order to evaluate the actual state of damage, to determine rates of damage in the past and to estimate future damage, the mode of action of air pollutants on building stones needs to be understood. This chapter tries to briefly review some of the relevant processes.

3.2 Air Pollutants

As will be discussed in more detail below, mineral dissolution reactions in building stones involve the attack of water and its involved acidity. Deposition of air pollutants is the major source of H^+ in building materials and significantly increases the natural rates of mineral dissolution reactions. Here, acid deposition refers to both wet deposition, i.e. acid precipitation, and the dry deposition of gaseous and particulate pollutants into water films on mineral surfaces.

Atmospheric acidity is closely related to the atmospheric chemistry of sulfur dioxide (SO_2) and the nitrogen oxides (NO, NO_2) which become oxidised to sulfate and nitrate through gas and aqueous phase processes (Seinfeld and Pandis, 1998). The atmospheric chemistry of sulfur and nitrogen compounds is only briefly summarised here.

3.2.1 *Sulfur dioxide*

The major source of sulfur dioxide emissions is the combustion of fossil fuels that contain sulfur as an impurity. Major pathways of

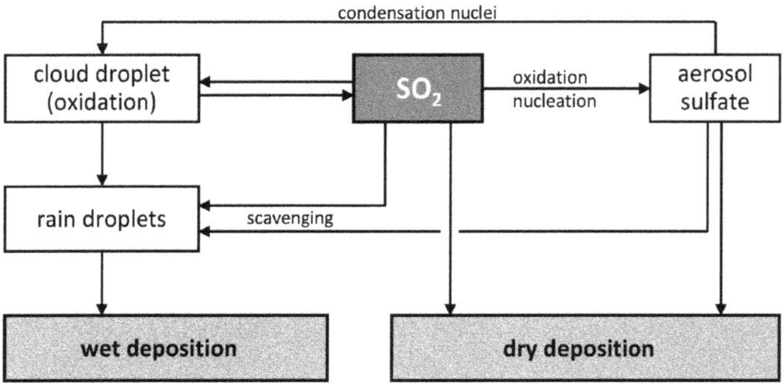

Fig. 3.1 Major pathways of SO_2 transformation and deposition.

transformation and acid deposition associated with sulfur dioxide are summarised in Fig. 3.1. The gas phase oxidation of SO_2 leads to the formation of sulfur trioxide (SO_3) which is converted rapidly into sulfuric acid (H_2SO_4) in the presence of water vapor and a third molecule, denoted M (which is usually an O_2 or N_2 molecule):

$$SO_2 \xrightarrow{[O]} SO_3 \tag{3.1}$$

$$SO_3 + H_2O + M \rightarrow H_2SO_4 + M \tag{3.2}$$

Though thermodynamically favorable, SO_2 is not directly oxidised by O_2 for kinetic reasons. In the gas phase, the oxidation is initiated by reaction with the OH radical and follows a chain reaction with the radicals $HOSO_2$ and HO_2 as reactive intermediates. The final reaction products are SO_3 and H_2SO_4. Sulfuric acid will rapidly nucleate to form aqueous sulfuric acid droplets, acting themselves as condensation nuclei for cloud droplets. Ammonia, the primary basic gas in the atmosphere, is readily absorbed by sulfuric acid droplets resulting in the formation of neutral or acidic ammonium sulfate which is a major constituent of the atmospheric aerosol.

Due to its high solubility in water, another important pathway of sulfur dioxide oxidation and sulfuric acid formation proceeds via absorption of SO_2 into hydrosols (cloud, fog or rain droplets and moist aerosol particles) and subsequent oxidation. There are a number of possible reactions that may be responsible for the oxidation of sulfur (IV) in aqueous solutions. Examples include the oxidation by dissolved ozone (O_3), hydrogen peroxide (H_2O_2), nitrogen dioxide

(NO_2) and the catalysed or uncatalysed oxidation with oxygen (Seinfeld and Pandis, 1998).

Both aerosol sulfate and gaseous SO_2 are subject to below cloud scavenging by rainfall. In effect, a large fraction of the atmospheric sulfur dioxide is removed by acid precipitation. In acid rain the pH is significantly lower than the background value of about 5.6, which is the pH of rainwater in equilibrium with an ambient CO_2 mixing ratio of 350 ppm. The dry removal of acidic sulfur compounds involves the deposition of gaseous SO_2 and acidic particulate sulfate. After dry deposition to surfaces, SO_2 is also oxidised to sulfuric acid; thus, both wet and dry deposition of sulfur compounds are important sources of acidity in building stones.

3.2.2 *Nitrogen oxides*

The atmospheric chemistry of the nitrogen oxides is considerably more complex than that of SO_2. The transformation and deposition pathways of nitrogen compounds are summarised in Fig. 3.2. Nitric oxide (NO) is the major oxide formed during the high temperature combustion of fuels. Nitrogen dioxide (NO_2) is formed in the atmosphere by oxidation of NO with ozone:

$$NO + O_3 \rightarrow NO_2 + O_2 \tag{3.3}$$

The final oxidation product of the nitrogen oxides (NO_x) is nitric acid (HNO_3), which is formed by the following reaction of NO_2 with

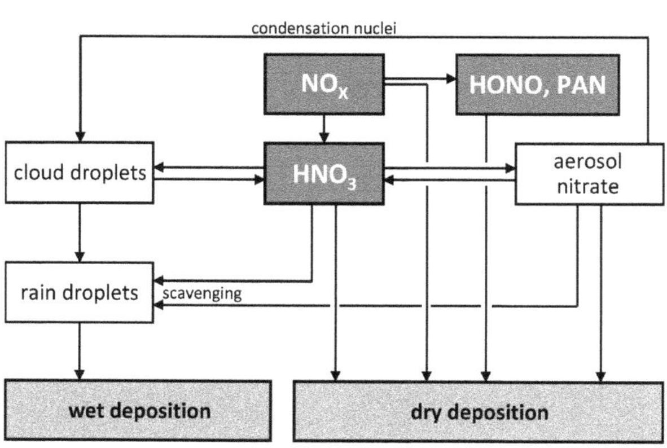

Fig. 3.2 Major pathways of NO_x transformation and deposition.

the OH radical:

$$NO_2 + OH \rightarrow HNO_3 \tag{3.4}$$

Nitric acid is more volatile than sulfuric acid and can exist in significant concentrations in the gas phase. However, it is also highly soluble and is readily absorbed by cloud droplets and moist aerosol particles.

Another important reaction of nitric acid is the neutralisation with NH_3 and the formation of ammonium nitrate aerosols (NH_4NO_3). Actually, there is a dynamic equilibrium between NH_3, HNO_3, and NH_4NO_3 and both NH_3 and HNO_3 can revolatilise from NH_4NO_3 aerosols according to

$$NH_4NO_3 \rightleftharpoons NH_3 + HNO_3 \tag{3.5}$$

Other reactions of NO_x with free radicals, including OH, RO, RO_2 and RCO_3, lead to a number of gaseous nitrogen compounds among which nitrous acid (HONO) and peroxyacetyl nitrate ($CH_3C(O)OONO_2$, PAN) are the most prominent species. Thus, the dry deposition of nitrate to surfaces includes several gaseous precursors and particulate nitrate. Table 3.1 lists typical concentration ranges of several gaseous nitrogen compounds for rural and moderately polluted atmospheres, compiled by Finlayson-Pitts and Pitts (1986).

The incorporation of nitrate into precipitation follows similar paths as that of sulfate including nucleation scavenging of nitrate-containing aerosols in cloud formation, absorption of HNO_3 by

Table 3.1 Typical concentrations of some gaseous nitrogen species in rural and moderately polluted atmospheres, units are ppb (data from Finlayson-Pitts and Pitts, 1986).

Pollutant	Rural	Moderately polluted
NO	0.05–20	20–1000
NO_2	1–20	20–200
HNO_3	0.1–4	1–10
HONO	0.03–0.8	0.8–2
PAN	2	2–20
NH_3	1–10	1–10

cloud droplets and precipitation scavenging of HNO_3 and nitrate-containing aerosols (washout). A major difference to the chemistry of sulfur dioxide is the fact that aqueous phase oxidation of NO_x is of very limited importance in the atmosphere. This is largely due to solubility limitations — nitric oxide is practically insoluble in water, and nitrogen dioxide is only slightly soluble and not very reactive in solution at typical ambient concentrations (Lee and Schwartz, 1981).

3.2.3 *Other pollutants*

The hydrogen halides have to be considered as powerful agents in stone deterioration, either, because they are strong acids (e.g. HCl), or, because they react with silicate minerals (HF). Important sources of hydrogen chloride (HCl) are the combustion of fuels containing chlorine, e.g. waste incineration, wood combustion, the combustion of high chlorine coal and the reaction of nitric acid with sea salt aerosols. Natural sources include volcanic emissions. The major sources of hydrogen fluoride (HF) are emissions from brick kilns and aluminium smelters.

Despite of their aggressiveness, the hydrogen halides are usually considered as being less important stone deterioration agents as their concentrations are significantly lower than concentrations of other acidic gases.

The principal role played by ammonia, NH_3, in the atmosphere is the neutralisation of acidic species such as sulfuric acid and nitric acid with the main reaction products ammonium sulfate, $(NH_4)_2SO_4$, and ammonium nitrate, NH_4NO_3. Thus, the ammonium ion is an important constituent of both cloud droplets and the submicron aerosol. It has been already mentioned that ammonia can revolatilise from ammonium nitrate aerosols such that ammonia and nitric acid can coexist in the gas phase.

The interest in the deposition of ammonium and ammonia on building materials and soil mainly arises from the fact that ammonium is biologically oxidised by the process of nitrification (van Breemen *et al.*, 1982). This two step process is given by the overall reaction:

$$NH_4^+ + 2O_2 \rightleftharpoons 2H^+ + NO_3^- + H_2O \quad (3.6)$$

Thus, the deposition of ammonia and ammonium to building materials may be regarded as an indirect pathway of acid deposition.

The major constituents of atmospheric aerosols are inorganic (soot) and organic carbon compounds, mineral particles, sea salt particles and secondary aerosols formed from gaseous precursors such as ammonium sulfate and ammonium nitrate aerosols. Sea salt and mineral dust are the major natural aerosol sources. However, re-suspension of particulate matter from streets and construction sites is also an important man-made source of mineral particles in the atmosphere. The contribution of atmospheric aerosols to acid deposition is negligible. The deposition of Ca-rich particles has to be considered as an important source of alkaline particles on silicate stone surfaces where they may form surface layers that can react with SO_2. The deposition of particulate matter to building surfaces is also an important source of soiling and discolouration (Davidson *et al.*, 2000; Bonazza *et al.*, 2007; Grossi *et al.*, 2007). This issue is discussed in more detail in Chapter 5.

3.3 Damage to Architectural Stone

Chemical damage processes refer to the dissolution or alteration of the mineral constituents of a stone material via chemical reactions. This section provides a brief summary of such reactions and will also discuss the resulting damage patterns in building stones and the changes in the appearance of building façades.

3.3.1 *Mineral dissolution reactions*

Many minerals present in rocks and building stones are thermodynamically unstable in the Earth's surface conditions. Therefore, these minerals are under-saturated if they are in contact with natural waters, e.g. rainwater. Mineral dissolution reactions in building stones involve the attack of water and its associated acidity. The H^+ ion attacks on the mineral components and causes their dissolution and the formation of deterioration products. Some examples of mineral dissolution reactions relevant to stone deterioration are:

$$CaCO_3 + 2H^+ \rightleftharpoons Ca^{2+} + CO_2 + H_2O \tag{3.7}$$

$$2KAlSi_3O_8 + 2H^+ + 9H_2O \rightleftharpoons 2K^+ + Al_2Si_2O_5(OH)_4 + 4Si(OH)_4 \tag{3.8}$$

$$Mg_5Al_2Si_3O_{10}(OH)_8 + 10H^+ \rightleftharpoons 5Mg^{2+} + Al_2Si_2O_5(OH)_4 + 4Si(OH)_4 + 5H_2O \tag{3.9}$$

$$Fe_2SiO_4 + 1/2O_2 + 3H_2O \rightleftharpoons 2FeOOH + Si(OH)_4 \tag{3.10}$$

Acid attack generally causes the mobilisation of metal cations and dissolved silica, $Si(OH)_4$, from the parent minerals. The weathering of silicates also leads to the formation of new mineral compounds, e.g. iron or aluminium oxyhydroxides and clays such as kaolinite, $Al_2Si_2O_5(OH)_4$. Compared to the parent materials, the mineral reaction products are depleted in metal cations relative to Al and Si. In addition, the product minerals are usually more hydrated. If the parent mineral contains Fe(II), the dissolution reaction usually also involves an oxidation step, as in, for example, Eq. (3.10). Table 3.2 lists the cations released from the weathering of some common mineral constituents of building materials. It can be seen that mineral

Table 3.2 Ions released from weathering of some common minerals in building stone.

Mineral name	Idealised formula	Ions released
carbonate minerals:		
– calcite	$CaCO_3$	Ca^{2+}
– dolomite	$CaMg(CO_3)_2$	Ca^{2+}, Mg^{2+}
feldspars:		
– plagioclase feldspar	$Na_xCa_{1-x}Al_{2-x}Si_{2+x}O_8$	Na^+, Ca^{2+}
– microcline (K-feldspar)	$KAlSi_3O_8$	K^+
clay minerals:		
– biotite	$K(Mg,Fe)_3AlSi_3O_{10}(OH)_2$	K^+, Mg^{2+}
– chlorite	$Mg_5Al_2Si_3O_{10}(OH)_8$	Mg^{2+}

dissolution reactions lead mainly to the formation of sodium, potassium, magnesium, and calcium salts, i.e. the major cations in natural waters.

The rates of mineral dissolution reactions depend on the amount and contact time of liquid water available at mineral surfaces, the solubility of the mineral in question and the pH of the solution in contact with the mineral surface. The carbonate minerals, i.e. calcite and dolomite, exhibit the highest solubilities, followed by feldspars, clay minerals, and quartz. As a rule of thumb, the solubility increases with decreasing silica content and with an increasing content of calcium and magnesium.

Carbon dioxide (CO_2) is a major source of acidity in natural waters, and it is the acid that is most responsible for natural rock weathering on a geological time scale. The solubility of carbonate materials in water is enhanced in the presence of CO_2 according to the reaction:

$$CaCO_3 + CO_2 + H_2O \rightleftharpoons Ca^{2+} + 2HCO_3^- \qquad (3.11)$$

The pH value of unpolluted rainwater in equilibrium with atmospheric CO_2 is about 5.6, i.e. slightly acidic. The equilibrium solubility of calcite in rainwater in equilibrium with atmospheric CO_2 ($5.5 \cdot 10^{-4}$ mol L^{-1}) is about four times higher than in pure water. Thus, even in the absence of other sources of acidity, there is a natural dissolution reaction of carbonate minerals which is commonly referred to as the karst effect. The dissolution rate of carbonates is significantly increased in calcareous building stones exposed to acidic rainfall.

Although the dissolution rates of silicate minerals are far slower than those of carbonate minerals, the former are also subject to chemical weathering. The same reactions are responsible for the dissolution of silicate minerals in building stones exposed to acid air pollution.

3.3.2 *Damage patterns*

The deterioration of building stone in natural and polluted atmospheres causes dramatic changes in the appearance of a building façade which often follow distinct spatial patterns clearly depending on the geometric configuration of the façade and the exposure to the local weather conditions. These patterns reflect the interplay between the dry deposition of pollutants and the action of rainfall. The dry deposition of atmospheric pollutants such as SO_2 or NO_y leads to the formation of reaction products (i.e. sulfates and nitrates) that are more soluble than the parent minerals. Though the rate of dry deposition is variable and depends on several parameters (see Section 3.5), dry deposition as a continuous process results in the continuous formation of reaction products on the stone surface. In contrast, rainfall is not only a discontinuous source of acidity leading to the dissolution of the mineral constituents of the stone. It is also a sink for the reaction products which are dissolved and removed from where they originally formed during preceding dry periods.

In effect, black and white areas can usually be clearly distinguished on limestone and marble façades (Camuffo *et al.*, 1982) corresponding to rain-sheltered and rain-exposed surfaces. In rain-sheltered areas, both the reaction products of acid deposition and deposited aerosol particles are enriched and form dense surface deposits and crusts (Fig. 3.3). In the rain-exposed areas, reaction products and dust particles can be removed by the cleansing effect of rainwater. This process is particularly efficient on limestone and marble surfaces due to the karst effect, i.e. the comparably large solubility of calcite in natural and acidic rainwater.

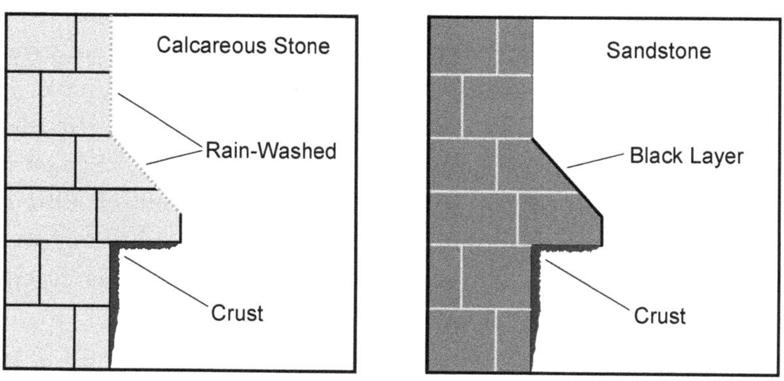

Fig. 3.3 Typical macroscopic appearance of exposed and sheltered surfaces of calcareous materials (limestone, marble) and sandstone.

Crusts

While most reaction products of mineral dissolution reactions are very soluble, calcium sulfate is an important exception and, therefore, tends to be enriched in building materials. Calcium sulfate in the form of gypsum, $CaSO_4 \cdot 2H_2O$, is the major reaction product of the dissolution of carbonate stone, i.e. limestone and marble, and is a ubiquitous compound found in building materials. The properties of gypsum and its role in the deterioration of building stones have been reviewed by Charola *et al.* (2007). Gypsum crusts are preferentially formed on rain-sheltered surfaces of carbonate building materials. Due to its low solubility however, gypsum can also be enriched in building stones that are exposed to rainwater.

Gypsum crusts on calcareous stones are the result of dry deposition of sulfur dioxide to the stone surface followed by reaction with calcite and subsequent precipitation of gypsum. Hence, the major damage mechanism is a chemical process. The replacement of calcite by gypsum and the growth mechanism of crusts have been studied in detail, both, for limestone (Schiavon, 1992) and marble (Vergès-Belmin, 1994). It was shown that gypsum crusts on limestone and marble grow inward which is in clear contradiction to the earlier growth model of Skoulikidis and Charalambous (1981) who proposed a solid state diffusion model according to which gypsum growth would proceed outward.

The formation of gypsum crusts on sandstone is less obvious. However, gypsum crusts are commonly found on calcite cemented sandstones (e.g. Zehnder, 1982) and nearly calcium-free materials such as quartz sandstone and granite (Whalley *et al.*, 1992; Neumann *et al.*, 1993; Smith *et al.*, 1994). In contrast to calcareous stone, the mechanism of crust formation on sandstone must be different, because both sulfate and calcium must originate from external sources. One obvious source is the mobilisation of calcium from mortar joints. Another source of calcium is the deposition of particulate matter from the atmosphere, either as gypsum or calcite.

A more detailed discussion on the composition and mineralogy of crusts on buildings stones and the blackening and other discolorations of stone surfaces is provided in Chapters 4 and 5 of this book.

Infiltration and run-off water

The situation becomes a lot more complicated if a stone surface is exposed to rainfall. If rainfall hits the surface of a porous material, it will be absorbed by capillary transport. As water is freely available at the surface of a porous material, the absorption of water obeys the equation

$$i = At^{1/2} \tag{3.12}$$

where i is the cumulative infiltration (mass per unit surface area), A is the water absorption coefficient and t is time. Equation (3.12) does not strictly apply to the case of driving rain penetrating into a porous material because the supply of water is limited by the rainfall rate. A more rigorous treatment of the problem based on unsaturated flow theory is provided by Hall and Kalimeris (1982). For the simple qualitative approach followed here, Eq. (3.12) is appropriate, however.

It directly follows that the rate of infiltration, $u = \mathrm{d}i/\mathrm{d}t$, decreases as $t^{-1/2}$, resulting in very large rates at small values of t. During the initial phase of absorption, the actual infiltration rate is therefore limited by the rainfall rate, u_0 as schematically depicted in Fig. 3.4a. As the infiltration rate decreases, a point will be reached at which the rainfall rate exceeds the infiltration rate. The porous material becomes saturated and the excess water will start to run off. The time to reach saturation (t_{sat}) depends on the rainfall rate and the water absorption coefficient A.

Hall and Kalimeris (1982, 1984) provide a simplified but useful model equation for the calculation of t_{sat} that is valid for many building materials:

$$t_{\mathrm{sat}} = \gamma \frac{S^2}{u_0^2} \tag{3.13}$$

where $S = A/\rho_{\mathrm{w}}$ is the sorptivity, ρ_{w} is the density of water and γ is a constant with values in the range 0.5–0.75. Hall and Kalimeris (1982) recommend $\gamma = 0.64$. Figure 3.4b depicts saturation times of typical building materials and their dependence on the rainfall rate, calculated using Eq. (3.13). For typical rates of driving rain in the range 1–3 mm h^{-1} (e.g. Blocken *et al.*, 2013), it can be seen that

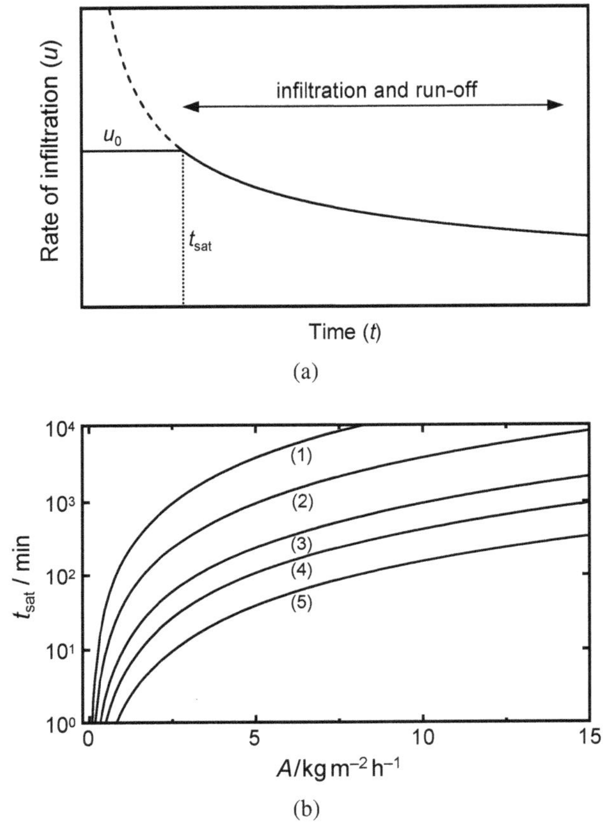

Fig. 3.4 Supply limited infiltration rate and formation of run-off water (a) and dependence of saturation time t_{sat} on water absorption coefficient A at various rates of driving rain: (1), 0.5 mm/h; (2), 1 mm/h; (3), 2 mm/h; (4), 3 mm/h; (5), 5 mm/h.

saturation times are in the order of minutes to several tens of hours depending on the water absorption coefficient. Once t_{sat} is reached at a given rainfall rate, a water film forms on the stone surface. From then on, the total flux of rainwater reaching the surface splits into two fractions — one fraction infiltrating the stone, and an excess fraction forming surface run-off water. With increasing duration of the rain event, the fraction of rainwater infiltrating the stone decreases and the amount of run-off water increases. Dissolved matter may be either transported into the porous material, or may be removed by running water. The significance of these processes depends on the intensity, duration and frequency of rainfall, its chemical composition and the sorptivity of the stone.

Different limiting cases may be discussed. The simplest situation that can be considered is a nearly impermeable material, i.e. a very dense material of low sorptivity such as marble. The saturation times of such materials are small and run-off water is the dominant process. This situation corresponds to the surfaces classified as white areas by Camuffo *et al.* (1982). The action of rainfall reduces to the dissolution of calcite and the removal of the reaction products (mainly gypsum) formed due to the dry deposition of acidic species.

In this case, the collection and analysis of stone run-off solutions is a useful experimental technique (e.g. Reddy, 1988; Cooper *et al.*, 1992). Any differences in the concentrations between run-off and incident rainfall must be due to interactions with the stone. For carbonate stones, the excess calcium concentrations in run-off solutions provides a direct measure of stone dissolution. Here, excess calcium concentration refers to a corrected concentration taking into account contributions from the dry deposition of particles and the calcium concentrations in the incident rainfall itself (Reddy, 1988). Data obtained from run-off water experiments were used to determine the relative contributions of the karst effect, the neutralisation of acid rain, and the dry deposition of sulfur dioxide (Livingstone, 1992; Baedecker *et al.*, 1992). It was found that in urban areas with significant SO_2 pollution, the dry deposition of local sulfur dioxide was the major source of calcium dissolution (Roekens and van Grieken, 1989; Livingstone, 1992; Steiger *et al.*, 1993; Cardell-Fernández *et al.*, 2002).

The mobilisation of metal cations as an indicator for the dissolution of silicate minerals, mainly clay minerals and feldspars, was also observed in run-off water from sandstone surfaces (Steiger and Dannecker, 1994; Halsey *et al.*, 1995). Mineral dissolution reactions in sandstone can also be detected by careful analysis of weathered stone samples. Mausfeld and Grassegger (1992, 1994) detected feldspar and clay mineral leaching by careful petrographic analysis. Schäfer and Steiger (2002) measured profiles of cation exchange capacities (CEC) in clay minerals containing sandstones. In all cases, they found distinct profiles with significantly lower CEC close to the exposed stone surfaces, indicating a significant loss of the mineral surface area available for ion exchange processes due to partial dissolution of clay minerals.

Another limiting case that might be considered is that of a stone with high sorptivity exposed to high intensities of driving rain which are usually observed in the upper part of a façade (Davidson *et al.*, 2000; Blocken *et al.*, 2013). In this case saturation times are very large and infiltration is dominant. Soluble material at the stone surface, i.e. the mineral constituents, the reaction products of deposition during dry periods and dust deposits, migrate into the porous stone. In fact, there are many buildings where the maximum gypsum enrichment is not found at the exposed stone surface, but rather at some distance in the interior of the stone. Zehnder (1982) provided several examples from different buildings in Switzerland with gypsum maxima at depths from a few millimeters up to several centimeters. The transport of gypsum to such great depths requires substantial penetration of water into the stone. In addition, as water evaporates during the drying of the stone, capillary transport back to the surface must not take place. These conditions are most likely to be met as a result of intense wetting and rapid drying. This is typically the case on the top of a building (especially near corners) where high wind speeds favor both, high driving rain intensities and fast drying.

The amount of gypsum that can be transported into the stone is limited because of its rather low solubility. In effect, gypsum accumulation proceeds slowly. Typical maximum gypsum concentrations do not exceed 2–3% by weight. Compared to the enormous enrichment of gypsum in crusts, these concentrations are much lower and are far from filling the available pore space. Nonetheless, the presence of subsurface gypsum enrichments is often associated with severe damage because of the formation of contour scales (Zehnder, 1982).

In contrast to gypsum, most other salts that are formed by acid deposition on limestone and sandstone are very soluble and are easily transported into porous stone by rainwater infiltration. An example of a typical depth profile of soluble salts is shown in Fig. 3.5 as analysed in a drill core taken at Schloss Schillingsfürst, which is located in a rural environment in Bavaria, South Eastern Germany, where the level

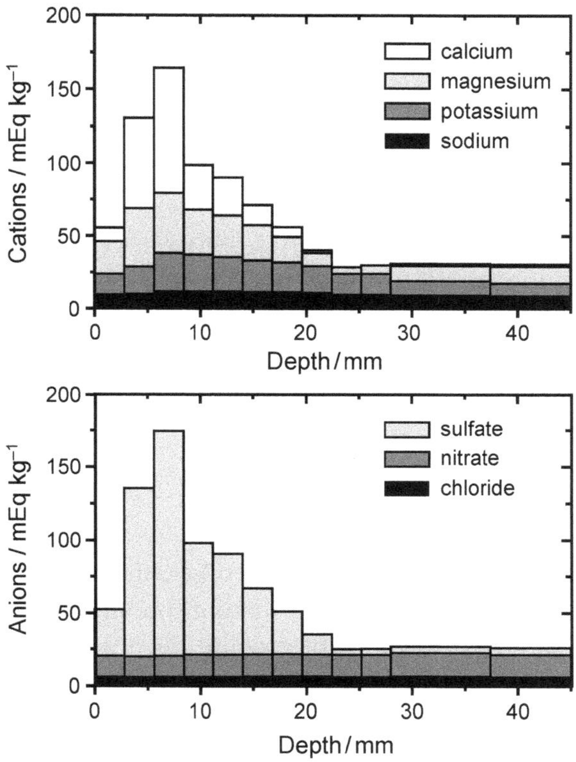

Fig. 3.5 Salt profile in sandstone from Schloss Schillingsfürst (Bavaria).

of local pollution was low during the last two centuries. Therefore, it is expected that there is a significant contribution of acid rain to the total deposition of acidity.

Salt enrichment is observed in a depth of about 5–15 mm with a maximum salt concentration of 1.35% by weight in 7 mm depth. Sulfate is the major anion and a simple mass balance consideration yields a gypsum concentration in that depth of about 0.77% (as $CaSO_4 \cdot 2H_2O$). Because of its low solubility, only a small fraction of the gypsum accumulation may be dissolved even if rainwater penetrates the stone (Steiger *et al.*, 2014). In contrast, the remaining salts, mainly a mixture of sulfate, nitrate, chloride, sodium, potassium and magnesium, are much more soluble and will be completely dissolved if rainwater penetrates into the stone. As water evaporates from the stone, salts re-crystallise again and may cause damage due to crystallisation pressure (Steiger *et al.*, 2014). For the salt mixture given, it can be deduced from available solubility data that, apart from gypsum, magnesium and potassium sulfates, preferably $K_2Mg(SO_4)_2 \cdot 6H_2O$ (schoenite), is the least soluble salt and will precipitate first. This is in accordance with the profile in Fig. 3.5 showing a strong enrichment of magnesium, potassium and sulfate at about 3–10 mm depth. The remaining ions are more evenly distributed and were probably present in the dissolved form at the time of sampling. The profile shown in Fig. 3.5 only reflects an intermediate state. On every wetting–drying cycle, the salts (except gypsum) will be completely dissolved and redistributed.

Crusts on rain-exposed surfaces

The cleansing effect of rainfall is limited because of the relatively low solubility of gypsum. Moreover, the dissolution kinetics might further reduce the maximum amount of gypsum that can be removed with running water. Therefore, a third case that needs to be discussed is the situation where a gypsum crust is formed on a limestone or sandstone surface that is exposed to driving rain or run-off water. This can happen if the rate of gypsum formation exceeds the removal rate with running water or infiltration. In polluted environments, the occurrence of white areas, therefore, is often limited to inclined surfaces, e.g. the top side of cornices. Vertical surfaces, though not sheltered, are less affected by driving rain and the accumulation of airborne particulate matter and gypsum is also frequently observed on exposed vertical surfaces of buildings in polluted urban areas. Considerable enrichment of gypsum or the formation of crusts on façades exposed to moderate rainfall intensities is particularly often found on calcareous stones of high porosity, such as coarse-grained limestones and calcite-cemented sandstones (Sabbioni and Zappia, 1992). Because of the greater saturation times of such materials, rainwater is absorbed and gypsum is transported into the stone. The amount of gypsum in the pore space continuously increases with every wetting and drying cycle and, besides the chemical weathering of calcite, additional damage results from crystal growth of the reaction product.

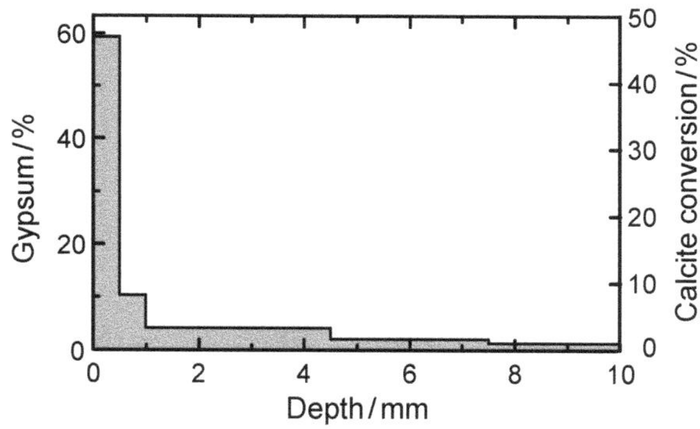

(a) limestone (Halberstadt Cathedral)

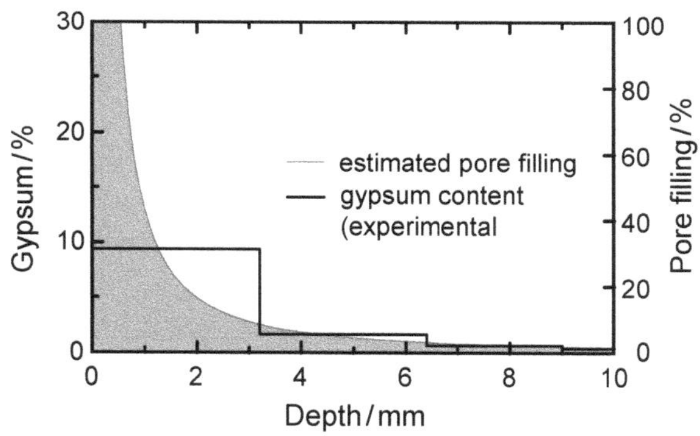

(b) sandstone (Leineschloss, Hanover)

Fig. 3.6 Profiles of gypsum enrichment in limestone (Halberstadt Cathedral) and sandstone (Leineschloss, Hanover) exposed to SO_2 pollution and driving rain.

Typical profiles of gypsum enrichment in limestone and sandstone blocks are depicted in Fig. 3.6. Both samples were taken from unsheltered stone surfaces of buildings located in urban areas, i.e., exposed to both driving rain and dry deposition of SO_2. Obviously, the large rate of gypsum formation exceeded the cleansing effect of driving rain over long periods of time, resulting in significant gypsum enrichment at the stone surfaces. The gypsum content shown in the two profiles was calculated from the sulfate and calcium concentrations determined in drill core samples.

In the case of the limestone profile from Halberstadt Cathedral (Germany), the gypsum content close to the original limestone surface is as high as 60% by weight corresponding to a conversion of calcite into gypsum of nearly 50%. In this case, air pollution damage cannot be described in terms of material loss (surface recession). The total sulfate content as illustrated by the shaded area in Fig. 3.6 is a better measure to quantify damage. The situation is different in the case of the profile in the quartz sandstone (Fig. 3.6b). Gypsum from external sources is enriched in the pore space of the sandstone. From the analysis of thin sections, it is known that the pores close to the surface of sandstones with a gypsum crust are usually completely filled with gypsum (Charola *et al.*, 2007). With the porosity and the density of the sandstone the gypsum content can be expressed as the pore filling ratio (right y-axis in Fig. 3.6b) yielding an average pore filling in the first sample (0–3 mm depth) of 33%. Assuming an exponential decrease with maximum pore filling at the stone surface, the fitted curve is obtained indicating complete pore filling at the stone surface. In this case, the quantification of damage is even more difficult. Though formation and enrichment of gypsum in the pores is the actual cause of deterioration, the damage mechanism *per se* is a mechanical process, i.e. the crystallisation pressure generated during growth of gypsum crystals (Steiger *et al.*, 2014). The evaluation of the state of damage requires more detailed microscopic investigation of the stone fabric. Nonetheless, the total amount of sulfate (or gypsum) enrichment, indicated by the hatched area in Fig. 3.6b, is a useful indicator.

3.4 Surface Recession Rates and Damage Functions

Surface recession on limestone and marble façades exposed to intense rainfall is a direct measure of the integral material loss due to acid deposition and dissolution of calcite and the karst weathering effect. Measurements of surface recession rates on limestone and marble were recently reviewed by Steiger *et al.* (2014). The lowering of the surface can be determined with a high resolution micrometer, using reference points such as lead plugs (Inkpen *et al.*, 2012), inscriptions in tombstones (Attewell and Taylor, 1990), unweathered mineral constituents (Winkler, 1987), or, the thickness of tombstones (Baer and Berman, 1983). In field exposure experiments, the weight loss of exposed specimens is an appropriate quantity from which the surface recession can be calculated. Finally, surface recession rates were calculated from run-off water experiments using both the excess calcium concentrations in the run-off water, i.e. reflecting the influence of dissolution, and total particulate matter suspended in the runoff water, i.e. reflecting the recession via particulate loss (Cooper *et al.*, 1992; Cardell-Fernández *et al.*, 2002).

A selection of experimental surface recession data is listed in Table 3.3. The range of recession rates in the studies mentioned so far shows significant scatter, reflecting the variety of materials, pollutant situations and exposure conditions. In several of the studies, substantial scatter also occurred on a point-to-point basis, again reflecting the complexity of the erosion process due to mineral dissolution involving large variation in local microclimatic conditions, orientation of surfaces, and material inhomogeneities. Considering all these points, the results of the studies appear to be in reasonable accordance and allow an assessment of average recession rates for calcareous materials. One of the major influences appears to be the orientation of the exposed surface, as the incident rainfall intensity is much higher on horizontal than on vertical surfaces.

Though careful interpretation of the data in Table 3.3 is required, it is obvious that recession rates are higher at urban and industrial sites than in rural environments. It also appears that the reduced recession rates determined in the more recent studies reflect the decrease in air pollution levels. Particularly, the measurements carried out at St Paul's Cathedral in central London clearly confirm this trend. This is also in agreement with the results of the UN ECE Program on Effects on Materials (Kucera *et al.*, 2007). In this program, Portland limestone was exposed over a 4-year period at 49 sites, most of them spread all over Europe with additional sites in Russia, Israel, the US, and Canada. The recession rates determined from weight loss data hardly exceeded values of $0.010\,mm\,a^{-1}$ even at the most polluted urban and industrial sites; for rural sites, recession rates were typically $<0.004\,mm\,a^{-1}$.

It is desirable to describe damage to stone as a function of the most important environmental parameters in terms of damage or dose response functions (Livingston, 1997). Several such approaches

Table 3.3 Experimental surface recessions of carbonate stones.

Location/environment	Period	Rate mm a^{-1}	Method[(a)]	Time (a)	Ref.[(d)]
Portland limestone					
UN ECE sites (min)	1997–2001	0.002	wl	4	(1)
UN ECE sites (max)	1997–2001	0.016	wl	4	(1)
UK NMEP sites (avg)	1987–1989	0.014	wl	2	(2)
London (St. Paul)	1718–1980	0.078	mm	262	(3)
	1980–1990	0.049	mm	10	(4)
	1990–2000	0.038	mm	10	(4)
	2000–2010	0.035	mm	10	(4)
London (central)	1955–1965	0.029	wl	10	(5)
	1981–1983	0.016	wl	2	(6)
	ca. 1994	0.014	wl	1	(7)
Watford (rural)	1955–1965	0.010	wl	10	(5)
	ca. 1994	0.012	wl	1	(7)
SE England (rural)	1981–1983	0.010	wl	2	(6)
Dublin (centre)	1985–1986	0.020	ro	1	(8)
Antwerp	1988–1991	0.013	ro	3	(9)
Belgium (rural)	1988–1991	0.010	ro	3	(9)
Marble					
Rural Eastern US	1984–1989	0.018	wl	5	(10)
Chicago	1927–1987	0.026	mm	60	(11)
St. Louis	1974–1977	0.007	wl	2.5	(12)
Durham (UK, min)	1890–1980	0.002	mm	10–100	(13)
Durham (UK, max)	1890–1980	0.010	mm	10–100	(13)
Philadelphia	1873–1979	0.014	th	1–105	(14)
New York		0.018	th		
Long Island (rural)		0.008	th		
Philadelphia (centre)	1790–1985	0.035	th	145–195	(15)
Philadelphia (suburbs)		0.016	th	145–195	(15)
Philadelphia (rural)		<0.005	th	145–195	(15)
Edinburgh	1780–1880	0.09	th	100	(16)
South Australia	1878–1973	0,001	mm	36–95	(17)
Various carbonate stones					
Germany (polluted)	1978–1979	0.047	wl	1	(18)

(Continued)

Table 3.3 (*Continued*)

Location/environment	Period	Rate mm a^{-1}	Method[a]	Time (a)	Ref.[d]
Germany (rural)	1978–1979	0.013	wl	1	(18)
Germany (polluted)	1978–1979	0.030	wl	1	(18)
Germany (rural)	1978–1979	0.012	wl	1	(18)
UK NMEP sites (avg)[b]	1987–1989	0.014	wl	2	(2)
UK NMEP sites (avg)[c]	1987–1989	0.014	wl	2	(2)
Vienna (centre)	1983–1984	0.062	wl	1	(19)
rural Eastern US	1984–1989	0.034	wl	5	(10)

[a] Experimental technique (wl: weight loss, mm: micrometer, ro: run-off, th: thickness of tombstones)

[b] White Mansfield dolomitic sandstone

[c] Monks Park limestone

[d] References: (1) Kucera *et al.* (2007), (2) Butlin *et al.* (1992), (3) Sharp *et al.* (1982), (4) Inkpen *et al.* (2012) (5) Honeyborne and Price (1977), (6) Jaynes and Cooke (1987), (7) Grossi *et al.* (1995), (8) Cooper *et al.* (1992), (9) Cardell-Fernández *et al.* (2002), (10) Baedeker *et al.* (1992), (11) Winkler (1987), (12) Haynie (1983), (13) Attewell and Taylor (1990), (14) Baer and Berman (1983), (15) Feddema and Meierding (1987), (16) Geikie (1880), (17) Cann (1974), (18) Luckat (1981), (19) Weber (1985).

do exist, however, it is not the purpose of this report to discuss all of them. Some of the more common equations were recently discussed by Cardell-Fernández *et al.* (2002), Bonazza *et al.* (2009) and Brimblecombe and Grossi (2009). Some equations are based on theoretical considerations (e.g. Lipfert, 1989a), others are non-linear empirical equations, e.g., damage functions for Portland Limestone (Kucera *et al.*, 2007).

The mathematical form of damage function includes environmental variables (concentrations, rainfall rate, humidity, temperature among others) and a quantitative measure of damage (e.g. mass loss, surface recession). A widely used damage function is the Lipfert function (Lipfert, 1989a). This function expresses the surface recession rate of freely exposed limestone or marble as the sum of karst dissolution, acid rain dissolution and the dry deposition of gaseous SO_2 and HNO_3:

$$-dx/dt = (b_0 + b_1[H^+])R + q_S v_{d,S} c_S + q_N v_{d,N} c_N \qquad (3.14)$$

where dx/dt is the recession rate (in $\mu m\,a^{-1}$), R is the rate of incident rainfall (in $m\,a^{-1}$), $[H^+]$ is the proton concentration in rainwater (in $\mu mol\,L^{-1}$), c_S, c_N, $v_{d,S}$ and $v_{d,N}$ are the concentrations c (in $\mu g\,m^{-3}$) and deposition velocities v_d (in $cm\,s^{-1}$) of SO_2 (S) and HNO_3 (N). The coefficients b_0 and b_1 in Lipfert's model describe the solubility of calcite. b_0 is the surface recession effect of the solubility in rainwater in equilibrium with atmospheric CO_2 and b_1 is used to represent the increased solubility in acidic rainwater. Values suggested by Lipfert (1989a) are $b_0 = 18.8\,\mu m\,a^{-1}$ and $b_1 = 0.016(\mu m\,a^{-1})\,(\mu mol\,L^{-1})$. The coefficients q_S and q_N are stoichiometric coefficients relating the flux of the two acidic gases from the atmosphere and the recession resulting from reaction with and dissolution of calcite. For v_d in $cm\,s^{-1}$ and c in $\mu g\,m^{-3}$ the coefficients are $q_S = q_N = 0.18\,(cm\,s^{-1})\,(\mu g\,m^{-3})\,(m\,a^{-1})$.

Cardell-Fernández *et al.* (2002) compared recession rates for Massangis limestone determined in run-off experiments (see Table 3.3) with values calculated using the Lipfert model and obtained good agreement. Other authors used the model to predict the effects of changes in future climate and air pollution on limestone recession. For such applications, the Lipfert equation offers advantages compared to empirical, non-linear damage functions which may cause problems if extrapolation is required.

Nonetheless, the Lipfert function has to be used carefully. First, the rainfall rate R is a very critical parameter in the model. The rates of driving rain on the surfaces of a building depend on a number of parameters (Blocken *et al.*, 2013) and may show significant small scale variation affected by the local microclimatic conditions and the orientation of surfaces. In particular, the rainfall rates to horizontal surfaces as usually provided from the local weather services cannot be used to calculate the recession of vertical or inclined stone surfaces with different orientations at a building.

Second, the Lipfert function is actually a calcite dissolution model and assumes that a calcite grain is only lost if it is completely dissolved. However, a smaller degree of calcite dissolution may be sufficient to weaken the grain contacts, resulting in a loss of cohesion. In this case, the material loss occurs mainly as sanding. In fact, in run-off water experiments it was noticed by several authors that calculated surface recession based on excess calcium concentrations in run-off solutions underestimate significantly the true surface recession (Baedeker *et al.*, 1992; Vleugels and Van Grieken, 1995). This was the reason why recession data from run-off experiments were only considered for Table 3.3 if calculated from both dissolved calcium and suspended particulate matter (Cooper *et al.*, 1992; Cardell-Fernández *et al.*, 2002). It should be also noted that the values of the stoichiometric coefficients b_0, b_1 and q in Eq. (3.14) provided by Lipfert (1989a) apply for generic calcite and should be corrected if applied to a very porous material with a significantly lower density (Bonazza *et al.*, 2009).

Third, the Lipfert function is based on the assumption that the reaction products of the dry deposition of the acidic gases are completely removed from the stone surface by run-off water. However, as mentioned before, this is only true for stone surfaces exposed to very high rainfall rates. If the enrichment of gypsum (or other reaction products) is the dominant process, surface recession is not appropriate as a quantitative measure of the state of damage. For example, considering the example shown in Fig. 3.6a, the total sulfate content in the limestone may be more appropriate for quantifying the damage caused by conversion of calcite into gypsum. The total sulfate content may be further expressed as an equivalent depth of complete calcite conversion which can then be compared to a surface recession at the rain exposed surfaces. For the profile shown in Fig. 3.6a, this yields an equivalent depth of 0.4 mm. It is hardly possible to estimate annual rates as it is unknown how much sulfate of the layer accumulated at the stone surface was removed by run-off water, wind erosion or by stone cleaning in the past.

Another very critical step in applying a Lipfert type damage function is the selection of appropriate deposition velocities for SO_2 and the nitrogen compounds. This issue is discussed in more detail in the following section.

3.5 Deposition Velocities

A useful measure for quantitative comparisons of the sensitivity of different materials to dry deposition is the deposition velocity, v_d, defined as the ratio of the surface flux, F, i.e. the deposition per unit area and time, to the ambient concentration of the gas, c:

$$v_d = \frac{F}{c} \tag{3.15}$$

It is tempting to assume that v_d is a constant for each combination of a particular pollutant and material under investigation. Then, the flux of the pollutant to that material can be easily calculated in damage functions such as Eq. (3.14) if the ambient concentration of the pollutant is known. For example, constant deposition velocities are often used with the Lipfert function to calculate the dry deposition fluxes of SO_2 and HNO_3. However, the dry deposition of a pollutant involves different processes and several of them may be rate limiting and depend on additional variables. First, the transport of the pollutant to the surface is controlled by aerodynamic factors (Hicks, 1982). Secondly, the pollutant needs to be irreversibly deposited to the surface, either by chemical reaction, or by physical processes such as dissolution or adsorption. Considering that the reciprocal of v_d is the overall resistance r to mass transfer, the influence of the two processes on the deposition velocity can be expressed in terms of an electric analogy as the sum of series resistances (Hicks *et al.*, 1987):

$$\frac{1}{v_d} = r = r_a + r_b + r_s \tag{3.16}$$

Here, r_a is the aerodynamic resistance that is controlled by turbulent exchange in the atmosphere, r_b is the resistance of the quasi laminar boundary layer at the material surface which is largely controlled by molecular diffusivity. Both resistances, r_a and r_b, are strongly affected by wind speed that controls both atmospheric turbulence and the width of the laminar boundary layer. In typical field experiments, the influence of r_a and r_b on the total deposition can be hardly distinguished such that these two resistances are usually combined into a single 'atmospheric' or mass transfer resistance $r_A = r_a + r_b$. Minimum resistances r_A for turbulent conditions and wind speeds of about $10\,m\,s^{-1}$ are 0.25–$0.33\,s\,cm^{-1}$ corresponding to maximum deposition velocities at zero surface resistance of 3–4 $cm\,s^{-1}$ (Lipfert 1989b). It should be noted, however, that pollutant concentrations are usually low under such turbulent conditions. In less well mixed atmospheres (i.e. under stable or neutral conditions), atmospheric resistances of about 0.9–1.8 s cm^{-1} are more realistic. Spiker *et al.* (1992) report values of 0.8–1.4 s cm^{-1} for wind speeds between about 3–5 m s^{-1} as determined in a wind tunnel. Behlen *et al.* (2008) found an average atmospheric resistance of 0.9 ± 0.3 s cm^{-1} corresponding to maximum deposition velocities of 1.1 ± 0.3 cm^{-1} s in field exposure experiments.

3.5.1 *Surface resistance SO_2 deposition*

The surface resistance r_s depends on the specific reaction of the pollutant with a particular material. For example, the deposition of SO_2 involves its dissolution in moisture films present on surfaces, the oxidation of the S(IV) species to S(VI) and the reaction with the mineral constituents of the material. It follows that chemical composition, surface wetness, controlled by the ambient relative humidity, and porosity are important material properties affecting the surface resistance. Apart from the material properties, the surface resistance might be also influenced by the ambient concentrations of other pollutants such as the oxidants ozone and hydrogen peroxide leading to synergistic effects.

Surface resistances have to be determined experimentally and different methods are used for this purpose. The amount of deposited material may be analysed directly by leaching the sulfate (or other sulfur species) that are formed during exposure of a test material in a laboratory chamber under controlled conditions or in a field experiment. The amount of sulfur deposit formed over a period of time is then referred to the average SO_2 concentration in the same period to obtain v_d. Alternatively, the deposition velocity may be determined by mass balance methods, e.g. by comparison of the SO_2 concentrations at the inlet and the outlet of an atmospheric flow chamber containing the test materials. The concentration decrease is then a measure of the amount of SO_2 deposited.

A compilation of literature data of SO_2 deposition velocities was provided by Sabbioni (2003) in the first edition of this book. Results from laboratory chamber experiments carried out under controlled conditions confirm the influence of both wind speed and relative humidity on the SO_2 deposition velocity. Typical values of v_d determined in chamber experiments are in the order of 0.1 cms^{-1} or below. In contrast, the deposition velocities determined in field exposure experiments fall in the range 0.1–1 cms^{-1}. This striking difference might be due to synergistic effects under field conditions where materials are typically exposed to complex mixtures of pollutants including oxidants, other acidic species (e.g. nitrogen oxides, nitric acid) or alkaline substances such as ammonia. Another important parameter, often overlooked, is the influence of the SO_2 concentration on the deposition velocity. For practical reasons, most of the chamber studies are carried out at SO_2 concentrations that are at least one or two orders of magnitude higher than typical ambient concentrations.

In field experiments it was shown that, in fact, there is a strong dependence of the deposition velocity on the SO_2 concentration (Behlen *et al.*, 2008). In this study, experiments were carried out using three different types of stone, a carbonate sandstone, a quartzitic sandstone and a sandstone containing feldspar and clay minerals. These three materials represent the full spectrum of mineral reactivities in natural stone. The stones were exposed rain-protected and deposition velocities were calculated from the average SO_2 concentration and the amount of sulfate formed during the exposure period. Smoothed values of v_d as a function of SO_2 concentration are depicted in Fig. 3.7. The strong influence of SO_2 concentration is evident. At low concentrations the rate limiting step in SO_2 dry deposition is the atmospheric resistance yielding a limiting value of v_d of $1.1 \pm 0.3\ cm^{-1}$ s at very low SO_2 concentrations. With increasing SO_2 concentrations the surface resistance increases and the overall deposition velocity decreases. This effect increases with decreasing content of less reactive minerals and reflects the buffering capabilities of the stone materials.

The strong influence of concentration was clearly detectable in the field experiments although there was substantial variation of other parameters influencing the deposition fluxes during the exposure experiments, for example, wind speed, relative humidity and concentration of other pollutants including acids, bases and oxidants. In a later study, the authors also confirmed this strong influence

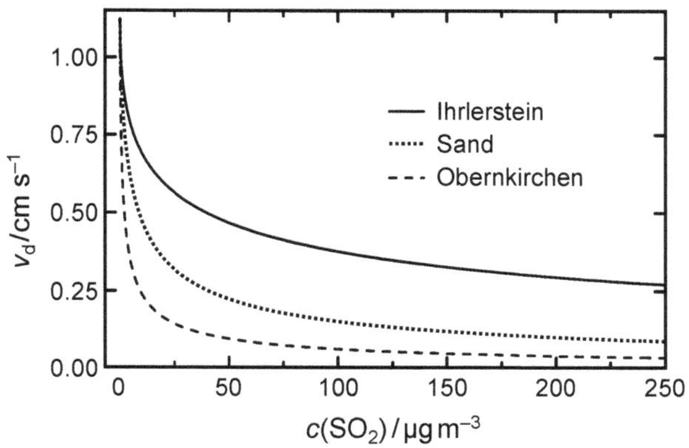

Fig. 3.7 Concentration dependence of SO_2 deposition velocities v_d on Ihrlerstein sandstone (carbonate), Obernkirchen sandstone (quartzitic) and Sand sandstone (feldspar and clay mineral containing).

of the SO_2 concentration in chamber experiments with controlled concentrations of SO_2 in air (without other gases) and with negligible atmospheric resistance (Wiese *et al.*, 2013). The deposition velocities determined in this study at realistically low SO_2 concentrations were in very good agreement with the values determined in the field experiments of Behlen *et al.* (2008).

In order to understand the concentration dependence of v_d, a very simple model was established by Behlen *et al.*, (2008). The absorption of SO_2 in a liquid film on mineral surfaces of porous stones is controlled by the solubility of SO_2 which is strongly pH dependent because of the dissociation equilibria leading to the release of protons. The equilibrium concentrations of S(IV) species in the water film can be easily calculated from the Henry constant and the two equilibrium constants of the dissociation reactions. The second process affecting the pH in the water film is the oxidation of S(IV) to S(VI) as HSO_4^- is a much stronger acid than HSO_3^-. There are a number of possible oxidation reactions and oxidising agents (Section 3.2.1). The only reactions that consume protons are the reactions with the mineral constituents of the stone, i.e. reactions (3.7)–(3.9). Considering that the dissolution and dissociation equilibria are fast reactions and assuming a steady state equilibrium of proton production by SO_2 dissolution, oxidation of S(IV) and dissociation on the one hand, and proton consumption by mineral dissolution reactions on the other hand, the following simple model equation with only one adjustable parameter was derived:

$$r_{\mathrm{s}} = k_{\mathrm{r}}^{-1} p_{\mathrm{S}}^{[1-n/(n+1)]} \qquad (3.17)$$

where p_{S} is the SO_2 partial pressure and k_{r} is an overall rate constant that contains the Henry constant, the equilibrium constants of the dissociation equilibria, an approximate average rate constant of the S(IV) oxidation, an average rate constant for the mineral dissolution reaction and the available surface area for this reaction (Behlen *et al.*, 2008). The exponent n arises from the rate expression of the mineral dissolution reactions and takes a value of $n = 1$ for carbonate stone. This yields the very simple relation $r_{\mathrm{s}} = p_{\mathrm{S}}^{1/2}/k_{\mathrm{r}}$ for carbonate materials, i.e. limestone, marble and carbonate sandstones. For such materials the overall resistance is then:

$$r = r_A + \frac{p_{\mathrm{S}}^{1/2}}{k_{\mathrm{r}}} \qquad (3.18)$$

For sandstones $n = 0.5$ was used, yielding $r_{\mathrm{s}} \sim p_{\mathrm{S}}^{2/3}$. Using this model and treating k_{r} as the only adjustable parameter (with values of $n = 0.5$ for the silicate stones), the concentration dependence of the deposition velocity on different types of stone could be represented to within experimental uncertainty by a single adjustable parameter, i.e. a specific rate constant for each stone (Behlen *et al.*, 2008). The same equation was used successfully to reproduce the deposition velocities to a carbonate sandstone in chamber experiments (Wiese *et al.*, 2013).

Several recent studies confirmed that the assumption of a linear term in the SO_2 concentration with a constant deposition velocity in damage functions may not be a good approximation. For example, Kucera *et al.* (2007) had to use two different dose–response functions for Portland limestone, one describing the behavior at low to moderate SO_2 concentrations, and a second equation valid at high concentrations. Also, Grossi *et al.* (2008) noted larger deviations between different damage functions at very low SO_2 concentrations. Finally, Inkpen *et al.* (2012) pointed out that the decrease in the recession rates of Portland limestone at St Paul's Cathedral in London is less steep than expected from the decrease in SO_2 concentrations. Considering that dry deposition of SO_2 was found to be the major source of calcium dissolution, at least until the end of the eighties of the last century, this is surprising and contradicts the respective term in the Lipfert equation which is linear in the SO_2 concentration. Such inconsistencies may be overcome if the concentration dependence of v_d is included in damage function models (see Brimblecombe and Grossi, 2009).

3.5.2 *Deposition velocities of nitrogen compounds*

Due to their low solubilities, much lower deposition velocities to wet surfaces are expected for the nitrogen oxides compared to sulfur dioxide. Indeed, early attempts to correlate the enrichment of nitrate in stone surfaces with ambient NO_x concentrations were unsuccessful (Luckat and Zallmanzig, 1985). In contrast, nitric acid is known to deposit readily on virtually all material surfaces and was suggested as a major source of nitrate enrichment in calcareous materials (Sikiotis and Kirkitsos, 1992; Fenter *et al.*, 1995). Because of its reactivity, it is assumed that the deposition velocity of HNO_3 is virtually independent of surface chemical processes. Thus, turbulent transport to the material surface and the ambient HNO_3 concentration are the limiting factors controlling the delivery rate of HNO_3 to material surfaces.

Despite the sensitivity of virtually all materials to nitric acid, the degradation potential as compared to sulfur dioxide might be limited due to the rather low atmospheric concentrations (see Table 3.1). In fact, there is experimental evidence from exposure experiments that sulfate accumulates more rapidly in calcareous stones than nitrate does (e.g. Jaynes and Cooke, 1987; Baedeker *et al.*, 1992). However, in view of strongly decreasing SO_2 concentration levels but still increasing NO_x emissions, deposition of oxidised nitrogen species may play a more important role in the deterioration of building materials in the future.

Behlen *et al.* (1996) determined deposition velocities of various nitrogen compounds to a carbonate sandstone at 80% RH in a chamber study. They reported values of v_d of $<0.001 \pm 0.001$ cm s^{-1} for NO, 0.02 ± 0.01 cm s^{-1} for NO_2, 1.8 ± 0.4 cm s^{-1} for HNO_3,

0.12 $\pm$ 0.02 cm s^{-1} for HONO and 0.006 $\pm$ 0.001 cm s^{-1} for PAN. They also carried out field exposure experiments with different stones that were exposed rain-sheltered. During the exposure, the concentrations of NO, NO_2, HNO_3, HONO and PAN were monitored and the values of v_d were used together with the concentrations to determine the deposition of nitrate to the stone specimens. In most cases, the agreement between calculated and experimental nitrate uptake was very good. Considering both the ambient concentrations of the gases and their deposition velocities, Behlen *et al.* found that the total deposition of nitrate is mainly due to HNO_3 and NO_2 with a minor contribution only of HONO. The contributions of PAN and NO were negligible.

It should be noted that the deposition velocities reported by Behlen *et al.* (1996) were measured in a well mixed flow chamber, i.e. under conditions of a very small atmospheric resistance r_A. Their value of $v_d = 1.8$ cm s^{-1} confirms that the surface resistance of HNO_3 deposition must, in fact, be close to zero such that the atmospheric resistance is rate limiting. As discussed before, atmospheric resistances of about 0.9–1.8 s cm^{-1} are more realistic, corresponding to deposition velocities of 0.6–1.1 cm s^{-1}. Also, SO_2 deposition velocities determined in field exposure experiments under conditions of low surface resistance (i.e. low SO_2 concentration) often yield values close to 1 cm s^{-1}.

In conclusion, it appears that the deposition velocity $v_d = 0.32$ for HNO_3 that is used in the Lipfert function might be too small. Moreover, the study of Behlen *et al.* (1996) shows that deposition of NO_2, which is usually not considered in damage functions (e.g. in the Lipfert function), should not be neglected.

3.6 Conclusion

As a general conclusion, it can be seen that building stones exposed to polluted atmospheres are severely damaged by acid deposition. Though the underlying damage mechanisms are reasonably well understood, it is still difficult to predict rates of damage and to relate the observed damage quantitatively to environmental parameters such as pollutant concentrations, rainfall rates etc. Research is still needed to improve quantitative damage functions that can be reliably used to estimate deterioration of buildings in the past and to predict the effects of future changes in air pollution. This requires (1) unambiguous measures to define and quantify damage, (2) appropriate values of the deposition velocities of the most important pollutants and (3) careful assessment of both past pollution and future changes in air pollution.

References

Amoroso, G.G. and Fassina, V. (1983). *Stone Decay and Conservation*, Elsevier, Amsterdam, pp. 1–7.

Attewell, P.B. and Taylor, D. (1990). Time-dependent atmospheric degradation of building stone in a polluting environment, *Environ. Geol. Water Sci.*, 16, pp. 43–55.

Baedeker, P.A., Reddy, M.M., Reimann, K.J. and Sciammarella, C.A. (1992). Effects of acidic deposition on the erosion of carbonate stone — Experimental results from the U.S. National Acid Precipitation Assessment Program (NAPAP), *Atmos. Environ.*, 26B, pp. 147–158.

Baer, N.S. and Berman, S. (1983). 'Marble tombstones in national cemeteries as indicators of stone damage: general methods', in *Preprints 76th Annual Meeting of the APCA*, Air Pollution Control Association, Atlanta, No. 83–5.7.

Behlen, A., Steiger, M. and Dannecker, W. (2008). Deposition of sulfur dioxide to building stones: the influence of the ambient concentration on the deposition velocity, *Environ. Geol.*, 56, pp. 595–603.

Blocken, B., Derome, D. and Carmeliet, J. (2013). Rainwater runoff from building facades: a review, *Build. Environ.*, 60, pp. 339–361.

Bonazza, A., Brimblecombe, P., Grossi, C.M. and Sabbioni, C. (2007). Carbon in black crusts from the tower of London, *Environ. Sci. Technol.*, 41, pp. 4199–4204.

Bonazza, A., Messina, P., Sabbioni, C., Grossi, C.M. and Brimblecombe, P. (2009). Mapping the impact of climate change on surface recession of carbonate buildings in Europe, *Sci. Total Environ.*, 407, pp. 2039–2050.

Brimblecombe, P. and Grossi, C.M. (2009). Millennium-long damage to building materials in London, *Sci. Total Environ.*, 407, pp. 1354–1361.

Camuffo, D., Del Monte, M., Sabbioni, C. and Vittori, O. (1982). Wetting, deterioration and visual features of stone surfaces in an urban area, *Atmos. Environ.*, 16, pp. 2253–2259.

Cann, J.H. (1974). A field investigation into rock weathering and soil forming processes, *J. Geol. Edu.*, 22, pp. 226–230.

Cardell-Fernández, C., Vleugels, G., Torfs, K. and Van Grieken, R. (2002). The process dominating Ca dissolution of limestone when exposed to ambient atmospheric conditions as determined by comparing dissolution models, *Environ. Geol.*, 43, pp. 160–171.

Charola, A.E., Pühringer, J. and Steiger, M. (2007). Gypsum: a review of its role in the deterioration of building materials, *Environ. Geol.*, 52, pp. 339–352.

Cooper, T.P., O'Brien, P.F. and Jeffrey, D.W. (1992). Rates of deterioration of Portland limestone in an urban environment, *Stud. Conserv.*, 37, pp. 228–238.

Davidson, C.I., Tang, W., Finger, S., Tyemezian, V., Striegel, M.F. and Sherwood, S.I. (2000). Soiling patterns on a tall limestone building, *Environ. Sci. Technol.*, 34, pp. 560–565.

Feddema, J.J. and Meierding, T.C. (1987). Marble weathering and air pollution in Philadelphia, *Atmos. Environ.*, 21, pp. 143–157.

Fenter, F.F., Caloz, F. and Rossi, M.J. (1995). Experimental evidence for the efficient "dry deposition" of nitric acid on calcite, *Atmos. Environ.*, 29, pp. 3365–3372.

Geikie, A. (1880). Rock weathering as illustrated in Edinburgh churchyards, *Proc. Roy. Soc. Edinburgh*, 10, pp. 518–532.

Grossi, C.M., Murray, M. and Butlin, R.N. (1995). Response of porous building stones to acid deposition, *Water Air Soil Pollut.*, 85, pp. 2713–2718.

Grossi, C.M., Brimblecombe, P., Esbert, R.M., Alonso, F.J. (2007). Color changes in architectural limestones from pollution and cleaning, *Color Res. Appl.*, 32, pp. 320–331.

Grossi, C.M., Bonazza, A., Brimblecombe, P., Harris, I. and Sabbioni, C. (2008). Predicting twenty-first century recession of architectural limesone in European cities, *Environ. Geol.*, 56, pp. 455–461.

Hall, C. and Kalimeris, A.N. (1982). Water movement in porous building materials. 5. Absorption and shedding of rain by building surfaces, *Build. Environ.*, 17, pp. 257–262.

Halsey, D.P., Dews, S.J., Mitchell, D.J. and Harris, F.C. (1995). Real time measurements of sandstone deterioration: A microcatchment study, *Build. Environ.*, 30, pp. 411–417.

Haynie, F.H. (1983). Deterioration of marble, *Durability Build. Mater.*, 1, pp. 241–254.

Hicks, B.B. (1982). 'Wet and dry surface deposition of air pollutants and their modeling' in *Conservation of Historic Stone Buildings and Monuments*, Committee on Conservation of Historic Stone Buildings and Monuments, National Academic Press, Washington D.C., pp. 183–196.

Hicks, B.B., Baldocchi, D.D., Meyers, T.P., Hosker, R.P. and Matt, D.R. (1987). A preliminary multi resistance routine for deriving dry deposition velocities from measured quantities, *Water Air Soil Pollut.*, 36, pp. 311–330.

Honeyborne, D.B. and Price, C.A. (1977). *Air pollution and the decay of limestones*, Building Research Establishment, BRE Note 117/77, Garston.

Inkpen, R., Viles, H., Moses, C., Baily, B., Collier, P., Trudgill, S.T. and Cooke, R.U. (2012). Thirty years of erosion and declining atmospheric pollution at St Paul's Cathedral, London, *Atmos. Environ.*, 62, pp. 521–529.

Jaynes, S.M. and Cooke, R.U. (1987). Stone weathering in Southeast England, *Atmos. Environ.*, 21, pp. 1601–1622.

Kucera, V., Tidblad, J., Kreislova, K., Knotkova, D., Faller, M., Reiss, D., Snethlage, R., Yates, T., Henriksen, J., Schreiner, M., Melcher, M., Ferm, M., Lefèvre, R.A. and Kobus, J. (2007). UN/ECE ICP Materials dose-response functions for the multi-pollutant situation, *Water Air Soil Pollut.: Focus*, 7, pp. 249–258.

Künzel, H. (1987). Sind wirklich Schadstoffe aus der Luft an der fortschreitenden Zerstörung unserer Denkmäler schuld? *Bautenschutz Bausanierung*, 10, pp. 139–141.

Lee, Y.N. and Schwartz, S.E. (1981). Evaluation of the rate of uptake of nitrogen dioxide by atmospheric and surface liquid water, *J. Geophys. Res.*, 86, pp. 11971–11983.

Lipfert, F.W. (1989a). Atmospheric damage to calcareous stones: comparison and reconciliation of recent experimental findings, *Atmos. Environ.*, 23, pp. 415–429.

Lipfert, F.W. (1989b). Dry deposition velocity as an indicator for SO_2 damage to materials. *J. Air Pollut. Control Assoc.*, 39, pp. 446–452.

Livingston, R.A. (1997). 'Development of air pollution damage functions' in Baer, N.S. and Snethlage, R., *Saving Our Architectural Heritage*, John Wiley & Sons, Chichester, pp. 37–62.

Livingston, R.A. (1992). 'Graphical methods for examining the effects of acid rain and sulfur dioxide on carbonate stones' in Delgado Rodrigues, J., Henriques, F., Jeremias, T., *Proceedings 7th International Congress on Deterioration and Conservation of Stone*, Lisbon, pp. 375–386.

Luckat, S. (1973). Die Wirkung von Luftverunreinigungen beim Steinzerfall, *Staub Reinhalt. Luft*, 33, pp. 283–285.

Luckat, S. (1981). *Quantitative Untersuchungen des Einflusses von Luftverunreinigungen bei der Zesrtörung von Naturstein*, Zollern Institute, Research Report 104 01 018/02, Federal Ministry of the Interior.

Luckat, S. and Zallmanzig, J. (1985). *Research Report 106 08 006*, Zollern-Institut, Bochum, 1985, pp. 58–61.

Mausfeld, S.A. and Grassegger, G. (1992). Abbauprozesse an Feldspäten und Tonmineralen unter den Bedingungen der Bauwerksverwitterung, *Z. Dt. Geol. Ges.*, 143, pp. 23–39.

Mausfeld, S.A. and Grassegger, G. (1994) 'The changing environment of pore solutions in natural building stones during immission accelerated weathering processes' in Zezza, F., Ott, H. and Fassina V (Eds.), *Conservation of Monuments in the Mediterranean Basin. Proceedings of the 3rd International Symposium*, Venice, pp. 129–135.

Neumann, H.-H., Steiger, M., Wassmann, A. and Dannecker, W. (1993). 'Aufbau und Ausbildung schwarzer Gipskrusten und damit zusammenhängender Gefügeschäden von Naturwerksteinen am Beispiel des Leineschlosses (Hannover)', in

Snethlage, R. (Ed.), *Jahresberichte Steinzerfall–Steinkonservierung Band 3–1991*, Verlag Ernst & Sohn, Berlin, pp. 151–167.
Reddy, M.M. (1988). Acid rain damage to carbonate stone: a quantitative assessment based on the aqueous geochemistry of rainfall runoff from stone, *Earth Surf. Proc. Land.*, 13, pp. 335–354.
Riederer, J. (1973). Die Wirkungslosigkeit von Luftverunreinigungen beim Steinzerfall, *Staub Reinhalt. Luft*, 33, pp. 15–19.
Roekens, E. and van Grieken, R. (1989). Rates of air pollution induced surface recession and material loss for a cathedral in Belgium, *Atmos. Environ.*, 23, pp. 271–277.
Sabbioni, C. (2003). 'Mechanisms of air pollution damage to stone', in Brimblecombe, P. (Ed.), *The Effects of Air Pollution on the Built Environment. Air Pollution Reviews — Vol. 2*, Imperial College Press, London, pp. 63–106.
Sabbioni, C. and Zappia, G. (1992). Decay of sandstone in urban areas correlated with atmospheric aerosol, *Water Air Soil Pollut.*, 63, pp. 305–316.
Schäfer, M. and Steiger, M. (2002). A rapid method for the determination of cation exchange capacities of sandstone: preliminary data, *Geol. Soc. Spec. Publ.*, 205, pp. 431–439.
Schiavon, N. (1992). 'Decay mechanism of oolitic limestones in an urban environment: King's College Chapel, Cambridge and St Luke's Church, London', in Webster, R.G.M. (Ed.), *Stone Cleaning and the Nature, Soiling and Decay Mechanisms of Stone: Proceedings of the International Conference Held in Edinburg, U.K.*, Donhead, London, pp. 258–267.
Seinfeld, J.H. and Pandis, S.N. (1998). *Atmospheric Chemistry and Physics*, John Wiley & Sons, New York.
Sharp, A.D., Trudgill, S.T, Cooke, R.U., Price, C.A., Crabtree, R.W., Pickles, A.M. and Smith, D.I. (1982). Weathering of the balustrade on St. Paul's Cathedral, London, *Earth Surf. Process. Land.*, 7, pp. 387–389.
Sikiotis, D. and Kirkitsos, P. (1995). The adverse effects of nitrates on stone monuments, *Sci. Total Environ.*, 171, pp. 173–182.
Skoulikidis, Th. and Charalambous, D. (1981). Mechanism of sulphation by atmospheric SO_2 of the limestones and marbles of the ancient monuments and statues. II. Hypothesis concerning the rate determining step in the process of sulphation, and its experimental confirmation, *Br. Corros. J.*, 16, pp. 70–77.
Smith, B.J., Magee, R.W. and Whalley, W.B. (1994). 'Breakdown patterns of quartz sandstone in a polluted urban environment, Belfast, Northern Ireland' in Robinson, D.A. and Williams, R.B.G. (Eds.), *Rock Weathering and Landform Evolution*, John Wiley & Sons, Chichester, pp. 131–150.
Smith, S.J., van Aardenne, J., Klimont, Z., Andres, R.J., Volke, A. and Delado Arias, S. (2011). Anthropogenic sulfur dioxide emissions: 1850–2005, *Atmos. Chem. Phys.*, 11, pp. 1101–1116.
Spiker, E.C., Hosker, R.P., Comer, V.J., White, J.R., Werbe, R.W., Harmon, F.L., Gandy, G.D. and Sherwood, S.I. (1992). Environmental chamber for study of

the deposition flux of gaseous pollutants to materials surfaces, *Atmos. Environ.*, 26A, pp. 2885–2892.

Steiger, M. and Dannecker, W. (1994). 'Determination of wet and dry deposition af atmospheric pollutants on building stones by field exposure experiments', in Zezza, F., Ott, H., Fassina, V. (Eds.), *Conservation of Monuments in the Mediterranean Basin. Proceedings of the 3rd International Symposium*, Venice, pp. 171–178.

Steiger, M., Wolf, F. and Dannecker, W. (1993). 'Deposition and enrichment of atmospheric pollutants on building stones as determined by field exposure experiments', in Thiel, M.-J. (Ed.), *Conservation of Stone and Other Materials*, E & FN SPON, London, pp. 35–42.

Steiger, M., Charola, A.E. and Sterflinger, K. (2014). 'Weathering and deterioration' in Siegesmund, S. and Snethlage, R. (Eds.), *Stone in Architecture*, 2nd Ed., Springer-Verlag, Berlin, pp. 226–316.

van Breemen, N., Burrough, P.A., Velthorst, E.J., van Dobben, H.F., de Wit, T., Ridder, T.B. and Reijnders, H.F.R. (1982). Soil acidification from atmospheric ammonium sulphate in forest canopy throughfall, *Nature*, 299, pp. 548–550.

Vergès-Belmin, V. (1994). Pseudomorphism of gypsum after calcite, a new textural feature accounting for the marble sulphation mechanism, *Atmos. Environ.*, 28, pp. 295–304.

Vleugels, G. and Van Grieken, R. (1995). Suspended matter in run-off water from limestone exposure setups, *Sci. Total Environ.*, 170, pp. 125–132.

Weber, J. (1985). 'Natural and artificial weathering of Austrian building stones due to air pollution', in Félix, G. (Ed.), *Proceedings of the 5th International Congress on Deterioration and Conservation of Stone*, Presses Polytechniques Romandes, Lausanne, pp. 527–535.

Whalley, B., Smith, B. and Magee, R. (1992). 'Effects of particulate air pollutants on materials: investigation of surface crust formation' in Webster R.G.M. (Ed.), *Stone Cleaning and the Nature, Soiling and Decay Mechanisms of Stone: Proceedings of the International Conference Held in Edinburg, U.K.*, Donhead, London, pp. 227–234.

Wiese, U., Behlen, A. and Steiger, M. (2013). The influence of relative humidity on the SO_2 deposition velocity to building stones: a chamber study at very low SO_2 concentrations, *Environ. Earth Sci.*, 69, pp. 1125–1134.

Winkler, E.M. (1973). *Stone: Properties, Durability in Man's Environment*, Springer-Verlag, New York, pp. 87–88.

Winkler, E.M. (1987). Weathering and weathering rates of natural stone, *Environ. Geol. Water Sci.*, 9, pp. 85–92.

Zehnder, K. (1982). *Verwitterung von Molassesandsteinen an Bauwerken und in Naturaufschlüssen, Beitr. Geol. Schweiz, Geotechn. Ser. 61*, Kümmerly & Frey, Bern.

CHAPTER 4

Composition and Chemistry of Crusts on Stone

Alessandra Bonazza and Cristina Sabbioni

Institute of Atmospheric Sciences and Climate, National Research Council (ISAC-CNR), Italy

4.1 Introduction

General agreement exists that atmospheric pollution is the main agent responsible for the damage layers present on monuments and historic buildings in urban areas (Leysen *et al.*, 1989; Rossval, 1988; Sabbioni, 1995). The earliest and most extensive studies on damage affecting monuments were related to marble and limestone materials, firstly because they have been widely used all over the world in statuary and buildings, and secondly because they represent a more simple system than other stones. This is on account of their chemical homogeneity, being almost entirely composed of calcium carbonate (or, more rarely, Ca and Mg carbonates), and low porosity, which permits the identification of a clearly defined interaction interface between the atmosphere and the undamaged material. In addition to chemical composition, porosity has been recognised as a further driving feature in favoring formation and pattern distribution of crusts on building facades.

In relation to the way rainwater wets monument surfaces, three main damage typologies have been identified for marble and limestone (Sabbioni, 2003):

(1) White areas. These typologies are found on vertical and horizontal external surfaces where rainwater runoff dominates, leading to the dissolution of the carbonate rock and the regression of the original surface, evidencing the colour characteristic of the stone material.
(2) Grey areas. This pattern is typical of sheltered surfaces such as porticoes. Such areas are subject to dry deposition only of atmospheric gas and aerosols, and the interaction between pollutants and materials is very limited.
(3) Black areas. These dark patinas are observed on stone surfaces wetted by rainwater but sheltered from intensive runoff and wash out, and are therefore the areas where atmospheric pollutants accumulate, along with the products of reactions between the stone and atmospheric gases and aerosol. They are typical of areas protected by elements of architectonic relief.

In the case of porous stones, such as sandstone and calcarenite, the weathering pattern is not so directly linked to surface features (geometry). Here, a homogeneous blackening is encountered due to high porosity and superficial roughness, which favour rain wetting and the capture of particulate matter (Sabbioni and Zappia, 1992).

It is well known that soiling and blackening of architectural surfaces occur as result of interaction of carbonate substrate with SO_2 and the deposition and incorporation of carbonaceous particles into the surface layers (Sabbioni, 2003). The reaction of SO_2 with the carbonate material leads to the formation of gypsum (i.e. calcium sulfate dihydrate) through the widely studied sulfation process, while elemental carbon (EC) in the aerosol is recognised as being responsible for soiling and blackening. It adds to the ensuing aesthetic damage on monument surfaces (Grossi and Brimblecombe, 2004).

This chapter will look at the composition of crusts that disfigure our cultural heritage, the particularly those developing on carbonate stones, focusing on their mineralogy, concentration of soluble salts and carbon fractions, with the aim of identifying the major pollution sources causing their formation. The changes in air pollution can be shown when comparison of modern soiling with historic black crusts is done in terms of mineralogy and geochemical characterisation will also be discussed.

4.2 Mineralogy of Crusts

Composition of crusts, developed on monument and building facades, due to gaseous pollutants and aerosol deposition and interaction, has been thoroughly studied by the mineralogical point of view mainly by optical microscopy observations of thin sections in transmitted light and x-ray diffraction and scanning electron microscopy analysis (Camuffo *et al.*, 1982; Del Monte *et al.*, 2001; Amoroso 2002; Fassina *et al.*, 2004; Bonazza *et al.*, 2005; Perez-Monserrat *et al.*, 2012; Ozga *et al.*, 2013).

Dealing with mineralogy of crusts requires first of all to consider two main categories of stones on which damage layers develop: (i) carbonate stone (either methamorphic and sedimentary lithotypes), such as marble and limestone, more prone to undergo pollution impact leading to formation of crusts and damage layers; (ii) silicate stone (both plutonic/volcanic and sedimentary), like granite and sandstone with siliceous cement.

It is well known that crusts form preferably on carbonate stone, being this type of rock more vulnerable towards atmospheric SO_2 dry and wet deposition. Crusts on carbonate stone are mainly composed of gypsum (i.e. calcium sulfate dihydrate), because of the SO_2 reaction with calcite, $CaCO_3$, which is the main mineralogical component of the substrate, following the well-known chemical reaction:

$$CaCO_3 + H_2SO_4 + H_2O \rightarrow CaSO_4 \cdot 2H_2O + CO_2 \qquad (4.1)$$

This deterioration process, basically consisting of transformation of calcite in gypsum, is termed sulfation (Sabbioni, 2003).

4.2.1 *Marble and limestone*

Mineralogical composition of historic black crusts, which have been long disfigured monumental heritage located in city centres and industrial areas, is mainly (80–90%) gypsum with lamellar/needle shape and/or microcrystalline habit. Calcite, linked to the carbonate substrate, is generally always present, as fragments of the underlying

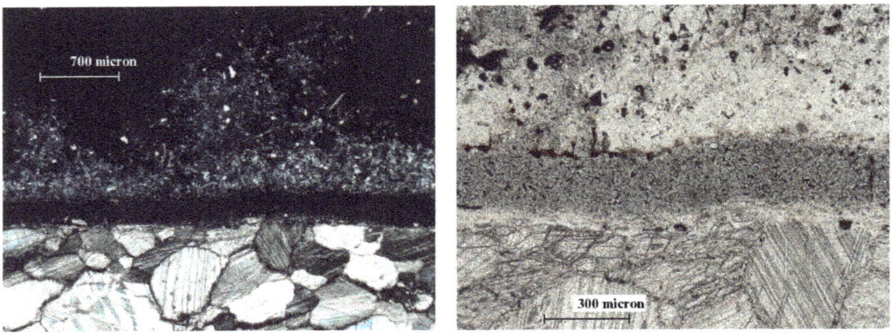

Fig. 4.1 Micrographs (left: cross-polarised light; right: plane polarised light) showing historic thick crust (500 years — estimated age) on Candoglia marble from the Cathedral of Milan (Italy).

material embedded in crusts layer or due to dissolution and secondary crystallisation phenomenon (Bonazza *et al.*, 2005; Ozga *et al.*, 2013). Quartz, feldspars, clay minerals and iron oxides, coming from soil dust deposition, are also present. Generally layers have a dendritic and inhomogeneous morphology and embedded soot, with characteristic carbonaceous particles with sizes reaching 50–60 μm causing blackening and surface darkening (Fig. 4.1). It is well known that combustion of coal and high-S content fuels has in the past been the major cause of blackening to built heritage.

Attention has lately been given to damage layers developed in recent years (from 1950) as results of interaction with modern urban atmosphere. Modern layers on marble and limestone look to be more coherent near the substrate are homogeneously distributed, thinner and take on greyish–brownish tones. From a mineralogical point of view they are composed of a lower percentage (10–30%) of gypsum, generally with a microcrystalline habit. They show a slightly higher content of soil dust, and are characterised by the absence of carbonaceous particles identifiable under optical microscopy, because of their submicron dimensions being mainly emitted by traffic (Fig. 4.2).

4.2.2 *Granite*

In comparison to damage layers on carbonate stone, limited research has been published on damage formation and composition of crusts

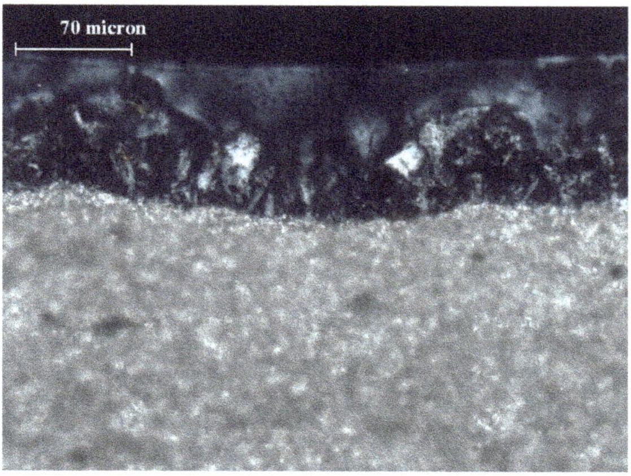

Fig. 4.2 Micrograph (cross-polarised light) showing modern damage layer (50 years — estimated age) on limestone from the north side at S. Maria del Fiore Cathedral in Florence (Italy).

on silicate stones. Although gypsum crystallisation on granitic building stone has also been ascribed to a sulfur dioxide effect, there is little knowledge of the decay mechanisms due to air pollution on stone with a low calcium content. Among silicate stones, granites are widely present in the built heritage in Europe. This is more durable than marble and limestone towards environmental impact due to its chemical composition in addition to low porosity.

Work performed on air pollution impact and damage formation on granite has been published by Schiavon (1994; 1995; 2007), who argued that in humid mesothermic climates, SO_2 from antrophogenic emissions is responsible for accelerated kaolinisation of feldspars in granites. Sulfur dioxide is suggested to play a dual role in the damage of silicate stone being responsible for the sulfation of Ca-bearing minerals (plagioclase for example) leading to formation of gypsum as well as the weathering of feldspars to kaolinite. Both processes cause the weakening of the stone leading to spalling, powdering and exfoliation phenomena. This phenomenon is confirmed by Perez-Monserrat *et al.* (2013), who studied in detail the decay mechanisms of building materials, granite included, at the Royal Palace of Madrid.

4.2.3 *Soot*

Carbonaceous particles deriving from combustion processes embedded in crusts and damage layers are known to be responsible for the darkening of architectural surfaces. Several studies have characterised the atmospheric particles on the basis of their morphology and elemental composition by optical and scanning electron microscopy with energy dispersive x-ray microanalyser with the final aim of identifying the emission sources. Such techniques have made it possible to identify different types of particles, which can be summarised under three general categories:

- Spherical/sub-spherical carbonaceous particles with circular or irregular pores, emitted by oil or distilled oil combustion in domestic heating systems and electric power plants (Sabbioni, 1995), with diameter ranging between 10 and 60 μm (Del Monte *et al.*, 1981), and by vehicle exhaust (Rodriguez Navarro and Sebastian, 1996; Chabas and Lefevre, 2000) (Fig. 4.3);
- Smooth aluminosilicate particles, not porous and spherical in shape produced by coal combustion (Sabbioni, 1995), with diameter ranging between 5 and 30 μm (Del Monte *et al.*, 1981), and by wood burning (Ausset *et al.*, 1998) (Fig. 4.3);
- Rare smooth metal particles, mainly composed of iron or titanium oxides, emitted by coal combustion processes (Sabbioni *et al.*, 1995).

Fig. 4.3 Carbonaceous particle (left) with spongy structure embedded in black crust from S. Maria del Fiore Cathedral in Florence. Aluminosilicate particle (right) in damage layer from Burğ al Klāb tower (also called Gate Tower) in Salé (Morocco).

Whether atmospheric particles simply promote aesthetic damage or play a role in the actual formation of the gypsum crust has been the subject of debate. Simulations of environmental damage processes performed in a flow chamber, with controlled temperature, relative humidity and SO_2 concentration, demonstrated that carbonaceous particles produced by combustion of oil-based fuels contribute significantly to the sulfation of calcareous material (Sabbioni, 1996). Samples of Carrara marble, Travertine and Trani stone were exposed for 150 days in air with 3 ppm of SO_2 concentration at 25°C and 95% relative humidity, and coated with three types of carbonaceous particles collected at the emission points of three oil-fuelled combustion sources: one centralised domestic heating plant and two electricity generating stations. For comparison, "artificial" carbonaceous particles, such as activated carbon and graphite, were also used. Analyses performed (X-ray diffraction, Fourier transform infrared spectroscopy, ion chromatography, scanning electron microscopy) on the samples after the exposure shows that the presence of the three types of atmospheric carbonaceous particles increases the amount of SO_4^{2-} and favours the oxidation of sulfite into sulfate, while the effect of the artificial particles is negligible. The data obtained indicated that the role played by carbonaceous particles in the heterogeneous oxidation of SO_2 into SO_4^{2-} is basically catalytic, on account of their heavy metal content, particularly iron; the role of the specific surface was found to be limited. By contrast, other experiments using fly ash on carbonate stone reached the conclusion that carbonaceous particles play no active role in the sulfation process and actually mask the surface, providing some degree of protection (Hutchinson *et al.*, 1992).

In addition, laboratory experiments have shown how in controlled conditions (high relative humidity, absence of SO_2 and NO_X) carbonaceous particles produced by the combustion of fuel oil in electric power plants nucleate different mineralogical species, such as gypsum, associated with vanadates and amorphous particles of opal (i.e. $SiO_2 \cdot nH_2O$) (Del Monte *et al.*, 1984; Bacci *et al.*, 1983). Furthermore, fly ash, also emitted by coal-fired power plants, has shown crystallite growth of gypsum in laboratory conditions during storage (Fisher *et al.*, 1976; Del Monte and Sabbioni, 1984). The formation of gypsum

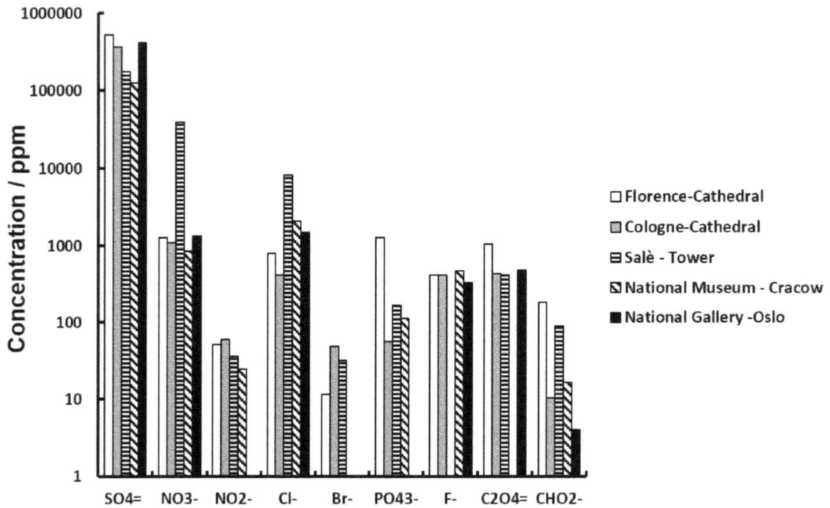

Fig. 4.4 Anion concentration in damage layers collected at S. Maria del Fiore Cathedral in Florence, Cologne Cathedral, Burğ al Klāb tower of ancient ramparts in Salè (Morocco), National Museum of Cracow and National Gallery in Oslo. For details on samples see Ozga *et al.*, (2014).

is attributed to the reaction of calcium and sulfur, which are particularly enriched in the silicate matrix constituting the fly ash. Nevertheless the gypsum nucleated by the particles themselves is negligible compared to the gypsum formed due to atmospheric SO_2 deposition, as shown by simulation experiments (Sabbioni *et al.*, 1996).

4.3 Soluble Salts

Measurements of soluble ions represent a valid tool, taken in combination with carbon fraction and metal content quantification (See Sections 4.4 and 4.5), for the identification of the major stationary and mobile combustion sources of air pollution causing damage layers on built heritage. Figures 4.4 and 4.5 are given as an example of ion soluble composition of damage layers sampled from monuments in different environmental conditions. The results have been achieved as part of the EC 7 FP Project TeACH "Technologies and tools to prioritise assessment and diagnosis of air pollution impact on immovable and movable Cultural Heritage".

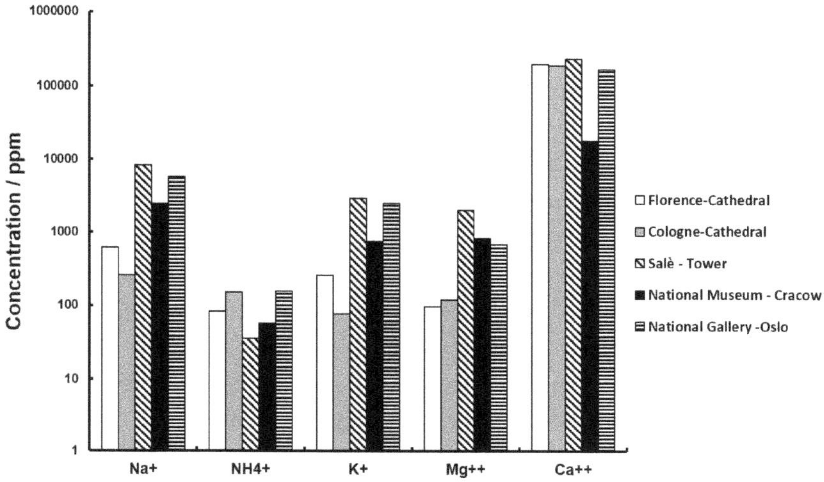

Fig. 4.5 Cation concentration in damage layers collected at S. Maria del Fiore Cathedral in Florence, Cologne Cathedral, Burğ al Klāb tower of ancient ramparts in Salè (Morocco), National Museum of Cracow and National Gallery in Oslo. For details on samples see Ozga *et al.* (2014).

Analyses carried out on crusts, formed on carbonate stones (marble, limestone) at monuments in different European cities, indicate that sulfate is generally the most abundant anion present, reaching a concentration higher than 500000 $\mu g\, g^{-1}$. Lower concentrations are found in recent and modern layers (50000 $\mu g\, g^{-1}$) highlighting that, although sulfation due to SO_2-surface interaction was and remains the most important damage process due to pollution impact, its role as a controlling factor is declining (Bonazza *et al.*, 2005; Ozga *et al.*, 2014). Sulfites are also often detected, demonstrating that sulfur dioxide dry deposition occurs in addition to wet deposition and pointing out the contribution of local pollution sources (mainly vehicular traffic) in damage formation.

Smaller and variable quantities are commonly detected for nitrates and nitrites, with the exception of monuments located in dry environments (see Salè data, Fig. 4.4) (Ozga *et al.*, 2013): nitrogen compounds, deriving mainly from mobile combustion sources (vehicular traffic), form very soluble salts, and their water solubility is responsible for the small quantity found and high variability.

Chloride is commonly linked to marine aerosol deposition and is typically found in higher concentrations in damage layers on sites located in coastal areas (see for example data from Salè, Fig. 4.4), but may be also due to anthropogenic activity, specifically industrial activity or the use of de-icing salts in cold environments (see for example data from Cracow and Oslo, Fig. 4.4).

Small (C1–C2) organic anions, formate (CHO_2^-), acetate ($C_2H_3O_2^-$) and oxalate ($C_2O_4^{2-}$) are normally encountered in considerable amounts. Their role in deterioration of building inorganic materials still requires investigation to be exhaustively understood. A section (4.3.1) is included here, where a database of soluble organic anions is presented. Studies investigating the presence of organic anions on stones, such as limestone and sandstones, indicate three main origins (Sabbioni *et al.*, 2003): (1) deposition of primary and secondary atmospheric pollutants, (2) biological weathering and (3) restoration treatments.

Among the cations, calcium is present with the highest concentration; it may be linked to the carbonate substrate, to soil dust deposition. It can be also attributed to the use of de-icing salts in winter seasons. While NH_4^+ is linked to anthropogenic activity in urban areas, sodium, potassium and magnesium can derive both from marine aerosol and soil dust deposition (Fig. 4.5).

4.3.1 *Soluble organic anions*

The organic fraction of crusts is composed of significant amounts of organic soluble anions (formate, acetate and oxalate), along with other organic compounds in which n-fatty acids, n-alkanes, polycyclic aromatic hydrocarbons (PAH) and benzenecarboxylic acids predominate. These organic constituents are tracers of specific anthropic sources typical of urban areas, such as vehicle exhaust, road dust, industrial combustion and domestic heating systems (Hermosin *et al.*, 2004; Saiz-Jimenez, 2003; Sykorova *et al.*, 2011).

The soluble organic anions, such as oxalate, acetate and formate, are generally measured by ion chromatography (IC), and represent an important fraction of organic carbon (Tavolin, 1998; Macchiarola *et al.*, 1999; Gigli, 2000; Pantani, 2000; Sabbioni *et al.*, 2000; Bonazza *et al.*, 2005). Data of soluble organic anions measured in black crust on carbonate stone from monuments and historic buildings in urban areas are reported in Table 4.1.

Among organic compounds, calcium oxalates have been widely encountered on stone building surfaces and have been identified as whewellite and weddellite, the mono- and dihydrate calcium oxalate, respectively (Alessandrini *et al.*, 1989). Whether their origin is linked to protective treatments, biological weathering or pollutant deposition has been the subject of controversy and at present the data are generally insufficient to exclude any of the origins suggested, except in the case of studies specifically performed on single monuments. Formates and acetates can be due to primary and secondary anthropic aerosol deposition, especially that emitted by combustion process, but can also have also biological origin. In addition, acetates can also derive from the oxidation of the ethanol used as a solvent in protective treatments (Sabbioni *et al.*, 2003; Ghedini *et al.*, 2006).

4.4 Organic and Elemental Carbon

4.4.1 *Carbon in crusts*

In addition to sulfur, carbon has been identified as an important element of anthropogenic origin and its measurement can give clues to the identification of the pollution sources causing crusts formation on building facades.

The analysis of carbon compounds in black crusts carried out so far have evidenced that the total carbon (TC) content in patinas is composed of two main fractions: carbonate (CC) and non-carbonate carbon (NCC):

$$\mathrm{TC} = \mathrm{CC} + \mathrm{NCC} \tag{4.2}$$

The NCC fraction includes two different components, organic carbon (OC), of biogenic and anthropogenic origin, and elemental carbon (EC), which is mainly a product of combustion processes:

$$\mathrm{NCC} = \mathrm{OC} + \mathrm{EC} \tag{4.3}$$

Carbon included in the damage layers on monuments has four different origins (Ghedini *et al.*, 2006 and included references):

(1) calcium carbonate, deriving almost exclusively from the original building materials, both stone and mortars;

Table 4.1 Concentration ($\mu g\,g^{-1}$) of formate, acetate and oxalate measured in damage layers on carbonate stone from monuments and historic buildings in urban sites.

Sample	Site	Monument	Material	CHO_2^-	$C_2H_3O_2^-$	$C_2O_4^{2-}$
VEC6	Venice	Cavagnis Palace	marmorino	412	210	1593
OR2	Siracusa	Column	calcarenite	897	612	512
OR3A	Siracusa	Hist. Building	calcarenite	608	440	1615
OR4	Siracusa	Hist. Building	calcarenite	497	797	24015
OR5A	Siracusa	Hist. Building	calcarenite	378	229	582
OR6A	Siracusa	Hist. Building	calcarenite	804	524	453
BAT1	Pisa	Baptistery	marble	54	123	462
BAT2	Pisa	Baptistery	marble	49	113	534
BAT3	Pisa	Baptistery	marble	62	142	876
BAT5	Pisa	Baptistery	marble	66	152	816
BAT6	Pisa	Baptistery	marble	55	186	830
BAT7	Pisa	Baptistery	marble	68	235	1012
RA1	Ravenna	Rasponi Palace	limestone	96	2532	753
RA6	Ravenna	Rasponi Palace	limestone	6739	1097	10645
RA8	Ravenna	Rasponi Palace	limestone	62	3228	1060
RA9	Ravenna	Adriana Gate	limestone	371	1351	853
RA10	Ravenna	Adriana Gate	limestone	294	1758	1058
RA13	Ravenna	Hist. Building	limestone	296	16616	1363
RA17	Ravenna	Hist. Building	limestone	334	16335	1272
RA18	Ravenna	Hist. Building	marble	456	15277	1525
SV1	Ravenna	S. Vitale Church	limestone	611	13871	9354
SV2	Ravenna	S. Vitale Church	limestone	562	13347	21023
SV3	Ravenna	S. Vitale Church	limestone	651	14192	12750
SV5	Ravenna	S. Vitale Church	limestone	1304	12869	12401
SV6	Ravenna	S. Vitale Church	limestone	700	19399	15259
SV7	Ravenna	S. Vitale Church	limestone	641	13365	1817
SV15	Ravenna	S. Vitale Church	limestone	938	25027	40737
SV17	Ravenna	S. Vitale Church	limestone	946	15912	12140
SV18	Ravenna	S. Vitale Church	limestone	644	13850	16584
SV19	Ravenna	S. Vitale Church	limestone	752	15672	13512
SV21	Ravenna	S. Vitale Church	limestone	815	14808	13873
SV24	Ravenna	S. Vitale Church	limestone	101	13109	4321
COL1	Rome	Coliseum	travertine	171	294	3140
COL2	Rome	Coliseum	travertine	278	375	1191
PTR1	Rome	Tiburtina Gate	travertine	181	2382	209
PTR2	Rome	Tiburtina Gate	travertine	460	2939	773

(*Continued*)

Table 4.1 (*Continued*)

Sample	Site	Monument	Material	CHO_2^-	$C_2H_3O_2^-$	$C_2O_4^{2-}$
PMR30	Rome	Maggiore Gate	travertine	112	76	23
PMR31	Rome	Maggiore Gate	travertine	2102	222	5946
DM1	Milan	Cathedral	marble	649	8890	6065
DM2	Milan	Cathedral	marble	5145	1090	16682
DM3	Milan	Cathedral	marble	16106	8318	12567
DM4	Milan	Cathedral	marble	4420	726	6638
DM5	Milan	Cathedral	marble	463	3543	5219
DF1	Florence	S. Maria del Fiore	marble	4825	935	30179
DF2	Florence	S. Maria del Fiore	marble	1007	1047	25316
DF7	Florence	S. Maria del Fiore	marble	2564	583	28376
CV1	Venice	Corner Palace	limestone	5103	11186	33679
CV2	Venice	Corner Palace	limestone	5314	11487	32092
CV9	Venice	Corner Palace	limestone	3151	5682	19923
CV2	Venice	Corner Palace	limestone	570	12809	29760
VR1	Rome	Vittoriano	limestone	3184	3760	21221
VR3	Rome	Vittoriano	limestone	1260	3891	27272
VR5	Rome	Vittoriano	limestone	4849	9764	13328
EP	Paris	St. Eustache Church	limestone	1536	2995	28889
TP	Paris	St. Eustache Church	limestone	2292	454	9010
CM	Mechelen	St. Rombouts Cathedral	limestone	1215	1334	7965
CS-BG	Seville	Cathedral	calcarenite	1899	1152	28646
CS-PG	Seville	Cathedral	calcarenite	4629	1984	19338
T4a	London	Tower	limestone	6406	476	10344
T5-T6	London	Tower	limestone	2196	482	8264
T11	London	Tower	limestone	1943	66	904
T12g-h	London	Tower	limestone	2640	758	3290
T14	London	Tower	limestone	13560	604	12176

(2) deposition of atmospheric particles containing elemental and organic carbon, as primary and secondary pollutants;

(3) biological weathering due to the action of organisms: among the products of their metabolism, formic, acetic and oxalic acids react with the underlying materials, leading to the formation of calcium formate, acetate and oxalate;

(4) surface treatments, such as oils, waxes, proteins, etc., frequently used in the past to protect monuments and historic buildings.

4.4.2 *Carbonate and non-carbonate carbon in crusts*

The earliest studies on the carbon content of black crusts aimed to discriminate and measure the carbonate and non-carbonate fractions (CC and NCC).

The methodology set up for this purpose follows two different procedures (Bacci *et al.*, 1990; Zappia *et al.*, 1990; 1991; 1993):

(1) On the bulk sample TC is measured by combustion and IR technique (LECO apparatus carbon-sulfur Determinator Mod. CS-44) and CC is quantified using differential and gravimetric thermal analysis (DTA-TGA) simultaneously performed (Netzsch Simultane Thermoanalyse STM Mod. 429); the NCC is then calculated by as the difference between TC and CC.
(2) The NCC fraction is measured by combustion and an IR technique, after treatment of the sample with HCl in the gas phase in order to eliminate the carbonate carbon, while CC is quantified using DTA-TGA as in procedure (1). TC is then calculated as CC + NCC.

Analyses on specially prepared standard samples show that more accurate results are obtained with the second methodology. Procedure (2) has been tested on samples of black crusts collected from marble and limestone monuments in eight sites in Northern and Central Italy: Trento, Milan, Verona, Venice, La Spezia (where sampling was performed on a natural limestone outcrop), Ravenna, Ancona and Rome. The choice of sites was made in order to cover different typologies of urban centres: large (Milan and Rome), medium (Venice), small (Verona and Trento), industrialised (Milan and Ravenna) and maritime (Venice, Ravenna, La Spezia and Ancona). At each site, black crusts were collected from monuments situated in different areas and conditions (town centres, suburbs and streets with varying levels of traffic), so as to ensure that the samples were representative not of a single monument, but of the town taken as a whole. Table 4.2 lists the mean concentrations of TC, CC and NCC

Table 4.2 Mean concentrations (%) of TC, CC and NCC measured in the black crusts sampled on marble and limestone in different Italian cities.

Sites	Material	TC* (%)	CC (%)	NCC (%)
Ancona	marble/limestone	2.78	0.22	2.56
La Spezia	carbonate outcrop	3.90	2.39	1.51
Milan	marble/limestone	1.77	0.17	1.60
Ravenna	marble/limestone	3.63	0.78	2.85
Rome	marble/limestone	4.30	2.35	1.95
Venice	marble/limestone	1.60	traces	1.60
Verona	marble/limestone	3.36	0.99	2.37
Trento	marble/limestone	3.59	2.19	1.40

*TC calculated as CC + NCC

obtained using procedure (2). The results highlight the presence in all the samples analysed of TC concentrations varying from 1.60% (Venice) to 4.30% (Rome). This variability is mainly linked to the CC percentages which in fact presents a minimum in the Venice samples, where only traces were found, and a maximum in Rome (2.35%), due to the amount of original calcareous material removed during sampling. All the specimens also reveal a considerable amount of NCC, varying from 1.40% to 2.85%. The narrower range of variability of NCC indicates a non-substrate origin of the non-carbonate fraction, which turns out to be mainly linked to anthropic sources.

Moreover the NCC stratigraphic concentrations measured from the external layer towards the original rock in samples of black and white areas (Table 4.3) shows that the non-carbonate fraction decreases from the external surface towards the internal layers in the black areas and that the NCC concentration in the underlying rock is approximately two orders of magnitude lower than that measured at the surface. Thus, the results confirm that the stone makes a negligible contribution to the NCC value (Zappia *et al.*, 1993). White areas were also found to contain non-carbonate carbon (mean value 4.0 mg g^{-1}), albeit in modest concentrations, being one order of magnitude lower than in the black crusts (mean value 19.8 mg g^{-1}).

Further studies on different building materials in the historic centre of Bologna were performed to give a better understanding of

Table 4.3 Stratigraphic concentration of NCC (mg g^{-1}) in black and white area samples of layer A (surface deterioration), layer B (underlying deterioration) and layer C (unaltered rock).

Samples	Sample	Layer A (mg g^{-1})	Layer B (mg g^{-1})	Layer C (mg g^{-1})
Black areas	1	25.30	1.68	0.18
"	2	19.77	8.60	0.15
"	3	40.62	30.89	0.28
White areas	1	8.33	5.42	0.22
"	2	8.68	3.00	0.10
"	3	8.92	2.92	0.03

the stone damage mechanism. In particular, black crust specimens sampled on limestone and sandstone surfaces were analysed in order to discriminate and measure the CC and NCC content. Table 4.4 shows the data obtained using procedure (2) (Zappia *et al.*, 1990; Zappia *et al.*, 1991). The behaviour of the two materials towards atmospheric pollution impact is found to be different. In the case of limestone surfaces, the distribution of carbonates and NCC values are similar to those of the damaged limestone investigated in other urban sites (Trento, Milan, Verona, Venice, La Spezia, Ravenna, Ancona and Rome). Conversely, the black crusts on sandstone present more heterogeneous TC and NCC values because of the different physical structure of the stone (high porosity and surface roughness). In particular, a high variability exists in the carbonate data (from trace to 60.9%), and the values of NCC (mean value 0.62 %) are also lower than those encountered in the dark damage layers on marble and limestone.

4.4.3 *Organic and elemental carbon in crusts*

As shown in the literature, since NCC is a consistent and important fraction of the TC present in black crusts, its measurement and discrimination into EC (predominantly a product of combustion processes) and OC (of biogenic and anthropic origin) is essential for a

Table 4.4 Concentration (%) of TC, calcium carbonate and NCC in black crusts on limestone and sandstone monuments sampled in Bologna, Italy.

Sample	Monument or Historic Building	Material	TC* (%)	$CaCO_3^{\circ}$ (%)	NCC# (%)
BO1	S. Petronio Church	limestone		1.4	1.5
BO2	S. Petronio Church	limestone		traces	1.8
BO3	S. Petronio Church	limestone		traces	2.2
BO4	S. Petronio Church	limestone		1.8	3.1
BO5	S. Petronio Church	limestone	2.2	6.8	1.4
BO6	S. Petronio Church	limestone	1.7	1.4	1.5
BO7	S. Petronio Church	limestone	1.7	4.6	1.2
BO8	S. Petronio Church	limestone		1.4	1.7
BO9	S. Petronio Church	limestone	1.9	5.7	1.2
BO10	S. Petronio Church	limestone	2.6	8.6	1.6
BO11	S. Petronio Church	limestone	4.3	14.8	2.5
BO12	S. Petronio Church	limestone		20.0	1.9
Ar1	Building	sandstone		8.1	0.72
Ar2	Building	sandstone		traces	0.49
Ar3	Building	sandstone		10.9	0.78
Ar4	Building	sandstone		35.5	0.95
Ar5	Building	sandstone		20.3	0.78
Ar6	Building	sandstone		26.1	0.58
Ar7	Building	sandstone		59.3	0.28
Ar8	Building	sandstone		59.1	0.34
Ar9	Building	sandstone		60.9	0.63

*calculated; °measured by DTA-TGA; #measured by combustion and IR spectrometry

complete identification and quantification of the main components of the damage layers on monuments and of their origin.

Carbon fractions, including OC and EC, were quantified in black crust samples collected from different Italian and European sites using a chemical-thermal methodology (Ghedini *et al.*, 2006). All of the experimental measurements of carbon in damage layers by this methodology adopt a combustion analysis using a carbon–hydrogen–nitrogen–sulfur–oxygen analyser (CHNSO EA 1108 FISONS Instruments).

For a complete analysis, the analytical methodology requires about 1 g of material and consists of three steps. In the first, total carbon (TC) is determined by burning the bulk sample, while in the second, non-carbonate carbon (NCC) is measured by burning the bulk sample after eliminating its carbonate carbon fraction (CC) by treatment with HCl vapour. CC is then calculated according to the relation:

$$CC = TC - NCC \tag{4.4}$$

The third step allows the determination of the EC content, after the elimination of the organic and inorganic species through a series of alternating basic and acid treatments at about 120°C, and applying CHNSO thermal analysis on the residual part. Finally OC is calculated according to the relation:

$$OC = NCC - EC \tag{4.5}$$

Following results published by Bonazza *et al.* (2005, 2007) and Ozga *et al.* (2013), in general the organic fraction predominates over the elemental one. The higher concentration of OC has been encountered particularly in damage layers collected from monuments sides exposed directly to air pollution due to traffic, which is known to be the current main cause of urban pollution in Western Europe, emitting sub-micronic particles mainly composed of organic compounds (Cachier *et al.*, 2004). Additionally it has also been argued that modern damage layers contain greater OC concentrations than old black crusts in urban areas, where the crust essentially derived from deposits of coal soot (Bonazza *et al.*, 2005; 2007; Bonazza and Brimblecombe, 2015). This suggests that damage layer formation derives mainly from the impact of local pollution sources, with a resulting chemical composition reflecting that of atmospheric combustion sources at each specific site. Table 4.5 reports data obtained after the application of the methodology above described on black crust specimens from monuments and historic buildings in different urban sites directly exposed to road traffic highlighting the constant prevalence of organic carbon over the elemental one.

4.5 Geochemistry of Crusts

Besides measurement of soluble and carbon fraction, geochemical characterisation in crusts can be a useful tool for identifying the major

Table 4.5 TC, CC, NCC, OC and EC concentrations (%) measured in black crusts on carbonate stone from monuments and historic buildings exposed to traffic emissions.

Sample	Site	Monument	Material	TC[a]	CC[b]	NCC[c]	OC[b]	EC[c]
OR2A	Siracusa	Building	calcarenite	2.63	1.29	1.34	0.95	0.40
OR3A	Siracusa	Building	calcarenite	6.58	5.65	0.93	0.77	0.16
OR5A	Siracusa	Building	calcarenite	8.90	7.83	1.08	0.94	0.15
OR6A	Siracusa	Building	calcarenite	1.99	1.22	0.77	0.62	0.15
RA	Ravenna	Building	marble	2.44	0.93	1.51	0.93	0.72
RA1	Ravenna	Building	marble	2.70	0.17	2.51	1.64	0.87
RA13	Ravenna	Building	marble	1.79	0.22	1.56	1.20	0.36
COL1	Rome	Coliseum	travertino	7.24	4.56	2.69	2.33	0.36
COL2	Rome	Coliseum	travertino	3.22	0.73	2.49	2.37	0.12
PTR1	Rome	Tiburtina Gate	travertino	8.07	5.87	2.24	1.94	0.30
PTR2	Rome	Tiburtina Gate	travertino	2.19	0.39	1.80	1.31	0.49
PMR30	Rome	Maggiore Gate	travertino	1.95	0.47	1.48	1.01	0.47
PMR31	Rome	Maggiore Gate	travertino	3.41	1.87	1.54	1.17	0.37
VR1	Rome	Vittoriano	limestone	3.33	0.50	2.83	1.55	1.28
VR3	Rome	Vittoriano	limestone	2.11	0.01	2.10	1.16	0.93
VR5	Rome	Vittoriano	limestone	5.94	0.81	5.13	3.07	2.06
DF7	Florence	S. M. del Fiore	marble	2.22	0.53	1.69	1.06	0.63
CS-BG	Seville	Cathedral	calcarenite	4.54	0.97	3.57	3.11	0.46
CM	Mechelen	S. Rombouts Cathedral	limestone	7.41	5.45	1.95	1.57	0.38
T11	London	Tower	limestone	2.85	2.10	0.75	0.67	0.08
T14k	London	Tower	limestone	3.76	0.71	3.05	2.10	0.95
SW5	Salè	Gate Tower	calcarenite	5.27	3.93	1.34	0.95	0.39
SW7	Salè	Gate Tower	calcarenite	4.41	3.25	1.16	0.80	0.36
SW9	Salè	Gate Tower	calcarenite	5.83	4.65	1.18	0.71	0.47
SW11	Salè	Gate Tower	calcarenite	4.61	3.50	1.11	0.67	0.44
SW12	Salè	Gate Tower	calcarenite	3.99	2.54	1.45	0.80	0.65

[a]measured on bulk samples; [b]calculated; [c]measured on treated samples.

combustion sources responsible for deterioration of built heritage occurring over time at specific sites. Additionally, concentrations of heavy metals can give a clue for establishing the age of deterioration layers with discrimination between old and modern. As example, Pb, Fe and Zn are generally encountered in the inner layers of historic crusts (representing the oldest deposit) and can be linked to the use of leaded gasoline twenty years ago. Del Monte *et al.* (2001) found that the trend of K is mainly correlated to the variations in the quantities of wood burnt over time, through mineralogical and chemical analysis of black crusts on selenite from historic buildings in Bologna.

Recent papers focus on the application of laser ablation inductively coupled plasma mass spectrometry (LA-ICP-MS) for obtaining the profile of trace elements from the inner layers to the external ones in crusts, in comparison also with the unaltered substrate (Belfiore *et al.*, 2013; Barca *et al.*, 2014). On the whole, data obtained suggest that concentrations of heavy metals in the crusts are higher than those in substrates, especially for Fe, Pb, Zn, V, Cu, Cr, Ni and Sb. The results give also evidence that at a specific site, architectural surfaces closer to the ground level (2–5 m) are characterised by highest concentration of heavy metals such as Pd, Zn, Cu and Sn, pointing out a major impact from mobile combustion sources, while areas at greater height (30–40 m) show the effect of background pollution from stationary combustion sources (industrial activity and domestic heating) indicated by higher concentration of Fe, V, Cd and As.

4.6 Air Pollution Changes: Composition of Crusts and Appearance of Buildings

The results suggest that analysis of damage layers from monuments in urban centres can be used to evaluate the effects of emissions from mobile and stationary sources on cultural heritage. It is thus important to gain a knowledge of their composition, through the measurement of the carbonaceous fraction (EC and OC) and soluble salts content (sulfates, nitrates, carbonates, chlorides, formates, acetates and oxalates). Moreover heavy metals content can constitute a marker of specific anthropogenic activity. The discrimination and quantification of these components on monumental surfaces is thus crucial for the identification of the mobile and stationary combustion sources (vehicular traffic/industrial activity/domestic heating) that are major causes of black crust formation. The identification of deterioration cause helps in proposing strategies of preventive conservation aiming at reducing exposure of built heritage to harmful agents by pollutants regulation at regional and local levels.

Sulfur dioxide has decreased from the mid-20th century until today and the decrease is expected to continue. The rate of decrease will likely be more important in the east and south-east of Europe, where concentrations had remained somewhat higher, than in the west of Europe. Thus SO_2, while continuing to be an important damaging agent in the future, is expected to become relatively less significant, implying a change of chemistry and mineralogy of crusts. This change is clearly shown when analyses of recent damage layers composition are carried out. The encountered increasing organic component is argued to be one of the major reasons for a shift from the traditional dark deposits to those with warmer tones in modern soiling. Modifications in air pollution, in addition to climate changes, are consequently responsible for the variation in the appearance of architectural surface, both in colour and distribution pattern (as treated in Chapter 5).

References

Alessandrini, G., Bonecchi, R., Peruzzi, R. and Toniolo, L. (1989). 'Caratteristiche composizionali e morfologiche di pellicole ad ossalato: studio comparato su substrati lapidei di diversa natura' in *Proc. 1st Int. Symposium on The Oxalate Films: Origin and Significance in the Conservation of Works of Art*, Centro Gino Bozza, Milan, pp. 137–150.

Amoroso, G. (2002). *Trattato di Scienza della Conservazione dei Monumenti. Etica della Conservazione, Degrado dei Monumenti, Interventi Conservativi, Consolidanti e Protettivi*, Alinea Editrice s.r.l., Firenze.

Ausset, P., Bannery, F., Del Monte, M. and Lefevre, R. A. (1998). Recording of pre-industrial atmospheric environment by ancient crusts on stone monuments, *Atmos. Environ.*, 32, pp. 2859–2862.

Bacci, P., Sabbioni, C. and Zappia, G. (1990). 'Atmospheric aerosol and damage of monuments and historical buildings', in Masuda, S. and Takahashi, K. (Eds.), *Aerosols Science, Industry, Health and Environment. Proc. of the 3rd Int. aerosol conference*, Japan, pp. 997–1000.

Barca, D., Comite, V., Belfiore, C.M., Bonazza, A., La Russa, M.F., Ruffolo, S.A., Crisci, G.M., Pezzino, A. and Sabbioni, C. (2014). Impact of air pollution in deterioration of carbonate building materials in Italian urban environments, *Appl. Geochem.*, 48, pp. 122–131.

Belfiore, C.M., Barca, D., Bonazza, A., Comite, V., La Russa, M.F., Pezzino, A., Ruffolo, S.A., and Sabbioni, C. (2013). Application of spectrometric analysis to the identification of pollution sources causing cultural heritage damage, *Environ. Sci. and Poll. Research*, 20, pp. 8848–8859.

Bonazza, A., Sabbioni, C. and Ghedini, N. (2005). Quantitative data on carbon fractions in interpretation of black crusts and soiling on European built heritage, *Atmos. Environ.*, 39, pp. 2606–2618.

Bonazza, A., Brimblecombe, P., Grossi, C. M. and Sabbioni, C. (2007). Carbon in black crusts from the Tower of London, *Environ. Sci. and Technol.*, 41, pp. 4199–4204.

Bonazza, A. and Brimblecombe, P. (2015). *Climate and the Changing Appearance of Buildings*, Edipuglia.

Camuffo, D., Del Monte, M., Sabbioni, C. and Vittori, O. (1982). Wetting deterioration and visual features of stone surfaces in an urban area, *Atmos. Environ.*, 16, pp. 2253–2259.

Camuffo, D., Del Monte, M., Sabbioni, C. and Vittori, O. (1982). Wetting, deterioration and visual features of stone surfaces in an urban area, *Atmos. Environ.*, 16, pp. 2253–2259.

Chabas, A. and Lefevre, R.A. (2000). Chemistry and microscopy of atmospheric particulates at Delos (Cyclades-Greece), *Atmos. Environ.*, 34, pp. 225–238.

Cachier, H., Sarda-Esteve, R., Oikonomou, K., Sciare, J., Bonazza, A., Sabbioni, C., Greco, M., Rejes, J., Hermosin, B. and Saiz-Jimenez, C. (2004). 'Aerosol characterization and sources in different European urban atmospheres: Paris, Seville, Florence and Milan', in Saiz-Jimenez, C. (Ed.), *Air Pollution and Cultural Heritage*, A.A. Balkema Publishers, London, pp. 3–12.

Del Monte, M., Sabbioni, C. and Vittori, O. (1981). Airborne carbon particles and marble deterioration, *Atmos. Environ.*, 15, pp. 645–652.

Del Monte, M. and Sabbioni, C. (1984). Morphology and mineralogy of fly ash from coal-fuelled power plant, *Arch. Meteorol. Geophys. Biochem. Sci.*, B35, pp. 93–104.

Del Monte, M., Sabbioni, C., Ventura, A. and Zappia, G. (1984). Crystal growth from carbonaceous particles, *Sci. Total. Environ.*, 36, pp. 247–254.

Del Monte, M., Ausset, P., Forti, P., Lefevre, R.A. and Tolomelli, M. (2001). Air pollution records on selenite in the urban environment, *Atmos. Environ.*, 35, pp. 3885–3896.

Fassina, V., Schilling, M.R., Keeney, J. and Khanijan, H. (2004), 'Weathering of marble in relation to atmospheric pollution and man-made intervention on the church façade of S. Maria del Giglio in Venice', in Saiz-Jimenez (Ed.), *Air Pollution and Cultural Heritage*, A.A. Balkema Publishers, London, pp.117–125.

Fisher, G.L., Chang, D.P.Y. and Brummer, M. (1976). Fly ash collected from electrostatic precipitator: microcrystalline structure and the mystery of spheres, *Science*, 192, pp. 553–555.

Ghedini, N., Sabbioni, C., Bonazza, A. and Gobbi, G. (2006). Chemical-thermal quantitative methodology for carbon speciation in damage layers on building surfaces, *Environ. Sci. Technol.*, 40, pp. 939–944.

Gigli, M. (2000). *Studio Delle Possibili Origini Degli Acidi Organici Responsabili Del Degrado Dei Monumenti,* Tesi di laurea in Scienze Ambientali, Università di Bologna, Italia.

Grossi, C.M. and Brimblecombe, P. (2004). Aesthetics of simulated soiling patterns on architecture, *Environ. Sci. Techno.*, 38, pp. 3971–3976.

Hermosin, B., Gaviño, M. and Saiz-Jimenez, C. (2004). 'Organic compounds in black crusts from different European monuments: a comparitive study' in Saiz-Jimenez, C. (Ed.), *Air Pollution and Cultural Heritage,* A.A. Balkema Publishers, London, pp. 47–55.

Hutchinson, A.J., Johnson, J.B., Thompson, G.E., Wood, G.C., Sage, P.W. and Cooke, M.J. (1992). The role of fly-ash particulate material and other catalysts in stone degradation, *Atmos. Environ.*, 26, pp. 2795–2803.

Leysen, L., Roekensand, E., and Van Grieken, R. (1989). Air pollution induced chemical decay of a sandy-limestone cathedral in Belgium, *Sci. Total Environ.*, 78, pp. 263–287.

Macchiarola, M., Sabbioni, C., Zappia, G. and Riontino, C. (1999). In Brebbia, C.A., and Jager, W. (Eds.), *Structural Studies, Repairs and Maintenance of Historical Buildings VI,* STREMAH VI, WIT Press, London, pp. 191–200.

Ozga, I., Ghedini, N., Giosuè, C., Tittarelli, F., Sabbioni, S. and Bonazza, A. (2014). Assessment of air pollutants sources in the deposit on monuments by multivariate analysis, *Sci. Tot Environ.*, 490, pp. 76–784.

Ozga, I., Bonazza, A., Ait Lyazidi, S., Haddad, M., Ben-Ncerc, A., Ghedini, N. and Sabbioni, C. (2013). Pollution impact on the ancient ramparts of the Moroccan city Salé, *J. Cul. Herit.*, 3, pp. S25–S33.

Pantani, M., Sabbioni, C. and Ghedini, N. (2000). 'Impact of air pollution on architectural monuments: the case of Ravenna', *Procs. of the 5th International Congress on Restoration of Architectural Firenze Heritage 2000,* pp. 1449–1456.

Perez-Monserrat, E.M., Alvarez de Buergo, M., Gomez-Heras, M., Varas Muriel, M.J. and Fort Gonzalez, R. (2013). An urban geomonumental route focusing on the petrological and decay features of traditional building stones used in Madrid, Spain, *Environ. Earth. Sci.*, 69, pp. 1071–1084.

Rodriguez Navarro, C. and Sebastian, E. (1996). Role of particulate matter from vehicle exhaust on porous building stones (limestone) sulfation, *Sci. Total. Environ.*, 187, pp. 79–91.

Rossval, J. (1998). *Air Pollution and Conservation,* Elsevier, Amsterdam.

Sabbioni, C. and Zappia, G. (1992). Decay of sandstone in urban areas correlated with atmospheric aerosol, *Water Air Soil Poll.*, 63, pp. 305–316.

Sabbioni, C. (1995). Contribution of atmospheric deposition to the formation of damage layers, *Sci. Total. Environ.*, 167, pp. 49–56.

Sabbioni, C., Zappia, G. and Gobbi, G. (1996). Carbonaceous particles and stone damage in a laboratory exposure system, *J. Geophys. Res.*, 101, pp. 19621–19627.

Sabbioni, C., Ghedini, N., Gobbi, G., Riontino, C. and Zappia, G. (2000). 'Analytic methodologies for carbon compound identification: Leaning Tower and Baptistery of Pisa', in Fassina, V. (Ed.), *Proc. of Int. Congress on Deterioration and Conservation of Stone*, Elsevier, Venice, pp. 383–390.

Sabbioni, C. (2003). 'Mechanisms of air pollution damage to stone', in Brimblecombe, P. (Ed.), *The Effects of Air Pollution on the Built Environment*, Imperial College Press, Singapore, pp. 63–106.

Sabbioni, C., Ghedini, N. and Bonazza, A. (2003). Organic anions in damage layers on monuments and buildings, *Atmos. Environ.*, 37, pp. 1261–1269.

Saiz-Jimenez, C. (2003). 'Organic pollutants in the built environment and their effect on the microorganisms, in Brimblecombe, P. (Ed.), *The Effects of Air Pollution on the Built Environment*, Imperial College Press, Singapore, pp. 183–225.

Schiavon, N., Chiavari, G., Fabbri, D. and Schiavon, G. (1994) 'Microscopical and chemical analysis of black patinas on granite', in Fassina,V., Ott, H. and Zezza, F. (Eds.), *Proc. 3rd Int. Symp. The conservation of monuments in the Mediterranean basin*, pp. 93–99.

Schiavon, N., Chiavari, G., Schiavon, G. and Fabbri, D. (1995). Nature and decay effects of urban soiling on granitic building stones, *Sci. Tot. Environ.*, 167, pp. 87–101.

Schiavon, N. (2007). Kaolinisation of granite in an urban environment, *Environ. Geol.*, 52, pp. 399–407.

Sykorova, I., Havelcova, M., Zeman, A. and Trejtnarova, H. (2011). Carbon air pollution reflected in deposits on chosen building materials of Prague Castle, *Sci. Total Environ.*, 409, pp. 4606–4611.

Tavolin, M. (1998). *Metodologie Chimico-Fisiche Per La Caratterizzazione Di Materiali Nel Settore Dei Beni Culturali. Tesi Di Laurea In Fisica*, Università di Padova, Italia.

Zappia, G., Sabbioni, C. and Gobbi, G. (1990). 'Dosaggio del carbonio nelle patine di alterazione dei monumenti antichi', *SEP Pollution Città e Ambiente*, pp. 103–110.

Zappia, G., Sabbioni, C. and Gobbi, G. (1991). Damage to stone surfaces in urban environment, *Mater Eng.*, 2, pp. 255–262.

Zappia, G., Sabbioni, C. and Gobbi, G. (1993). Non carbonate content of black and white areas of damaged stone monuments, *Atmos. Environ.*, 27a, pp. 1117–1121.

CHAPTER 5

Soiling and Discolouration

Carlota M. Grossi

School of Health Sciences,
University of East Anglia, United Kingdom

5.1 Introduction

During their history, stone monuments are discoloured by pollution and climate. Pollution was especially intense from the 1700s and in the late 19th and early 20th centuries, the dominant impact of air pollution was the sulfation of surfaces. The parallel deposition of soot caused the blackening of stone surfaces and the formation of black crusts (Brimblecombe and Grossi, 2010). The decrease of sulfur and soot from coal combustion during the the 20th century led to cleaner air in cities, a decrease of pollution-decay rates on building stones and a public desire for cleaner buildings. Changing environments offer a potential for further variation in the colour of buildings, from different biological growth under changing climates or the presence of different pollutants as traditional air pollutants, such as sulfur dioxide and smoke, have been replaced by ozone, nitrogen oxides and particles richer in organic compounds (Brimblecombe and Grossi, 2007; Brimblecombe and Grossi, 2010).

Deposited organic compounds can oxidise in modern urban environments. The resulting products exhibit yellowish or brownish colours rather than black, due to a yellowing process (Grossi and

Brimblecombe, 2008). A change of colouration due to biological staining is also expected. Moreover, sulfur dioxide is phytotoxic, so in urban atmospheres with less sulfur, biological growth is more effective. A greater rate of delivery of nitrate to building surfaces that acts as an *airborne fertiliser* (Adamson *et al.*, 2010) also increases biological colonisation. Depending on climate, different colouration processes (e.g. *greening* or *reddening*) might be expected. Changing biology and colouration in buildings within changing environments is an area of further research. Future brownish/yellowish colours or biological staining may also raise aesthetic issues (Brimblecombe and Grossi, 2010).

5.2 Traditional Soiling of Historical Buildings by Pollution and Blackening

Blackening of building materials is due to the accumulation of carbonaceous fine particles produced by the incomplete combustion of fossil fuels. The main agent responsible for black soiling is particulate elemental carbon (EC), also known as black carbon (BC) due to its high retention on surfaces and high optical absorptivity. In past environments, buildings were blackened by wood burning (Hamilton and Mansfield, 1991) and especially coal soot; but nowadays road traffic emissions, and in particular diesel emissions, are the main source for black carbon (Brimblecombe, 2011). Biological activity, perhaps supported by an ongoing increase in nitrate deposition and organic pollution, may also contribute significantly to stone blackening (Viles and Gorbushina, 2003). Black deposits on surfaces of some Iranian monuments were found to be mainly microbiotic crusts produced by cyanobacterial growth (Bonazza *et al.*, 2007).

The effectiveness or rate of soiling depends on a variety of factors such as atmospheric and surface particle concentration, particle size, colour of the particles relative to the colour of the surface, roughness and orientation of the deposition surface or the efficiency of re-suspension mechanisms by wind and rain (Yocom and Kawecki, 1986; Creighton *et al.*, 1990; Watt *et al.*, 2000; Brimblecombe and Grossi, 2004; Auras *et al.*, 2013).

Blackening can be measured as the reflectance (or lightness) contrast of a darkened surface by comparison with the reflectance of the original surface. Since the pioneering work of Beloin and Haynie (1975), who developed a damage function relating reflectance to particulate concentration, a variety of equations have been proposed to describe the changes in reflectance over time (Bonazza and Brimblecombe, 2013).

The loss of reflectance is a first order process that can be modelled by the expression (Brimblecombe and Grossi, 2004; Watt *el al.*, 2008):

$$R_t = (R_0 - R_c)\exp(-kt) + R_c \tag{5.1}$$

Where R_t is the reflectance or lightness after time t, R_0 is the initial reflectance of the material and k is the rate constant for blackening. According to Mansfield and Hamilton (1989):

$$k = \frac{3(cv_d - Y)}{4r\rho} \tag{5.2}$$

Where c is the concentration of particles in the ambient air, v_d is the deposition velocity, Y is the re-suspension/removal rate, r is the geometric radius of the particles and ρ the density of the particles.

The interpretation of the parameter R_c varies among different researchers. Watt *et al.* (2008, 2009), following Haynie's (1986) model, defined R_c as the reflectance from the particles on the covered surface. They approximated $R_c = 0$ (total blackness) leading to the equation:

$$R_t = (R_0)\exp(-kt) \tag{5.3}$$

This equation was used in several soiling or blackening programs (e.g. Toprak *et al.*, 1997; Kucera, 2005). The soiling constant k was considered to be directly proportional to the amount of deposited particulate matter (see Eq. 5.2) and therefore to atmospheric concentration of PM_{10}. Then, $k = \lambda PM_{10}$; where λ is the dose-response constant. Using this equation, Kucera (2005) gave a value for λ of 0.0088 a^{-1} per $\mu g\, m^{-3}$ of PM_{10}. However, blackened building surfaces still reflect as much as 30%. Brimblecombe and Grossi (2004 and 2009) did not assume complete blackness or zero reflectance. In Brimblecombe and Grossi's models, R_c is the final blackening that would be shown by the material. Brimblecombe and Grossi (2004) considered that

R_c depended on the area covered by soiling or blackening, which is a balance between processes of deposition and removal or re-suspension. Therefore R_c was related to the reflectance of the different soiling particles and the initial reflectance of the surface.

Brimblecombe and Grossi (2009) approached R_c as the reflectance of an old crust that reached a stable level of reflectivity. They assumed that the whole surface would be covered by the crust and the final reflectance of the crust R_c was similar to the reflectance of the deposited particles. This allowed for elemental carbon (EC) and material that was not elemental carbon, such as dust, with lighter colours.

Brimblecombe and Grossi's (2009) model accounted for the flux of elemental carbon to the surface. The flux is defined as $F_{EC} = V_{dEC}EC$; where V_{dEC} is the deposition velocity of EC and EC is the concentration of atmospheric elemental carbon. The model took the form as:

$$R_t = (R_0 - R_c)\exp(-V_{dEC}ECt/\tau) + R_c \tag{5.4}$$

Where τ is a *folding density* (i.e. the surface concentration of elemental carbon required to reduce the reflectivity by a factor e), estimated as 7.3 g m^{-2} from work at the Tower of London (Bonazza *et al.*, 2007). The blackening constant (as λ — or dose-response constant as in Watt *et al.*, 2008) is expressed as V_{dEC}/τ and the initial rate of change in reflectance as:

$$-\frac{dR}{dt} = (R_0 - R_c)\,V_{dEC}EC/\tau \tag{5.5}$$

The first order blackening constant in this case was found to be 0.013 a^{-1} for each μg m^{-3} of EC or 0.0016 a^{-1} for each μg m^{-3} of PM_{10} taking V_{dEC} as 0.003 m.s^{-1} (Nicholson, 1988) and taking 12% of PM_{10} to be as EC. These models consider absolute rather than relative changes in reflectance.

If we consider relative changes in reflectance, the blackening process might be modelled as:

$$R_t/R_0 = (R_0/R_0 - R_c/R_0)\exp(-kt) + R_c/R_0 \tag{5.6}$$

Where the effect of the original surface colour is removed or at least reduced. In this case R_c depends on the area covered by particles as in Brimblecombe and Grossi (2004).

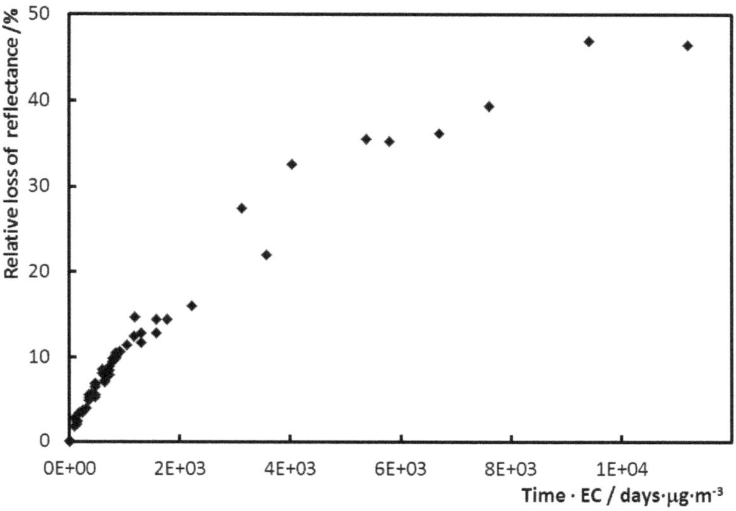

Fig. 5.1 Relative loss of reflectance for light-coloured porous or rough carbonate materials under blackening trials in Rome (Loruso *et al.*, 1996) and Oviedo and Burgos (Grossi *et al.*, 2003).

Figure 5.1 represents relative loss of reflectance versus $EC \cdot t$ in several light porous carbonate stones in different exposure trials. Analysis of the data plotted in Fig. 5.1 yielded to $k = \lambda \cdot EC$ where the blackening constant was found, for this set of data, to be around 0.068 a^{-1} for each microgram.m^{-3} of EC or 0.0081 a^{-1} for each microgram·m^{-3} of PM_{10}.

The asymptotic blackening reflectance R_c can be related to atmospheric concentrations of EC and the surface concentration of EC on buildings (Bonazza *et al.*, 2007). A review of published blackening data for light coloured stones suggests that R_c varied from 75–80% in remote areas (EC $< 1\,\mu g\,m^{-3}$), 60–70% at urban background sites (EC $\approx$ 2–3 $\mu g\,m^{-3}$), 40–60% in trafficked areas (EC $\approx 6 \mu g\,m^{-3}$) to 30–40% in highly trafficked areas and road tunnels (Grossi and Brimblecombe, 2008). Recent research confirmed that current and potential blackening in the future is related to traffic-related particulate matter (Auras *et al.*, 2013). This concentration decreases exponentially with the distance to roads (Fig. 5.2), therefore it is expected that severe blackening will be concentrated in areas adjacent to intense traffic.

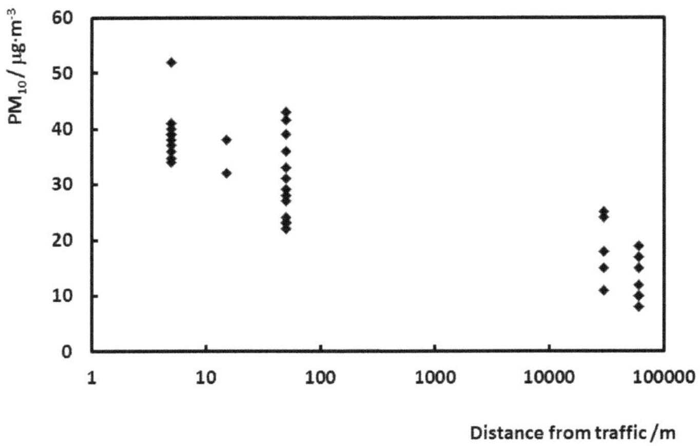

Fig. 5.2 Atmospheric PM_{10} concentration and distance to traffic (Grossi and Brimblecombe, unpublished material).

These blackening models are applicable to light-coloured materials and in the case of stone to rough (or porous) light coloured stones. Recent blackening exposure trials showed a decrease of stone reflectance in agreement with bounded exponential models (Urosevic *et al.*, 2012).

Blackening is difficult to assess in dark stones (Grossi *et al.*, 2003). Similarly, the exponential relationship might not explain the initial loss of reflectance in light coloured hard or polished materials. Here the rate of soiling increases progressively at the beginning and later decreases until saturation (Grossi and Brimblecombe, 2008) potentially following a Hill equation fit as proposed by Ionescu *et al.* (2006) for glass. Low porosity materials, such as smooth stone or glass, are more resistant to the dark particle adhesion so require a slower induction period for blackening.

5.3 New Trends in Climate and Pollution and New Building Colourations

5.3.1 *Yellowing*

Contemporary urban particulate matter, mainly derived by traffic combustion sources, has high organic carbon/elemental carbon ratios (OC/EC) (Sillanpaa *et al.*, 2005; Bonazza and Brimblecombe, 2013). The OC/EC ratio is usually higher at the road-sides, gasoline being the main source for OC. Typical OC/EC ratios are 2.2 and 0.8 for light-duty gasoline and heavy duty diesel vehicles respectively (Brimblecombe, 2011).

The consequent deposits rich in organic materials offer the potential for biological activity and oxidation in modern atmospheres with secondary pollutants such as ozone. This can produce different colours and patterns, with the organic rich deposits becoming brownish-yellow. Bonazza *et al.*, 2005 found OC/EC between 1.5 and 2.2 in deposits of Florence monuments, whereas in older crusts of the Cathedral of Milan, the ratio OC/EC was smaller, varying from 0.1 to 0.7. In a similar manner to soils, organic carbon compounds can be slowly converted to humic-like substances, HULIS (Graber and Rudich, 2006), which are typically yellow or brown rather than black. Some organic compounds in deposits or crusts can also migrate and form a yellow layer on the surface of the stone (Gaviño *et al.*, 2004). Lower sulfur dioxide concentrations of recent decades and increasing deposition of nitrogen and organic compounds have also induced the formation of coloured patinas and encouraged biological activity on buildings (Vázquez-Calvo et al, 2007).

Yellowing on building surfaces is increasingly noticed at sites such as St. John's Cathedral in Norwich, UK or the Tower of London (Grossi and Brimblecombe, 2008) (Fig. 5.3), where more modern crusts have higher organic carbon/sulfur (OC/SO_4) ratios (Bonazza

Fig. 5.3 Yellow and brownish colouration on the Kentish ragstone and Portland stone at the Tower of London.

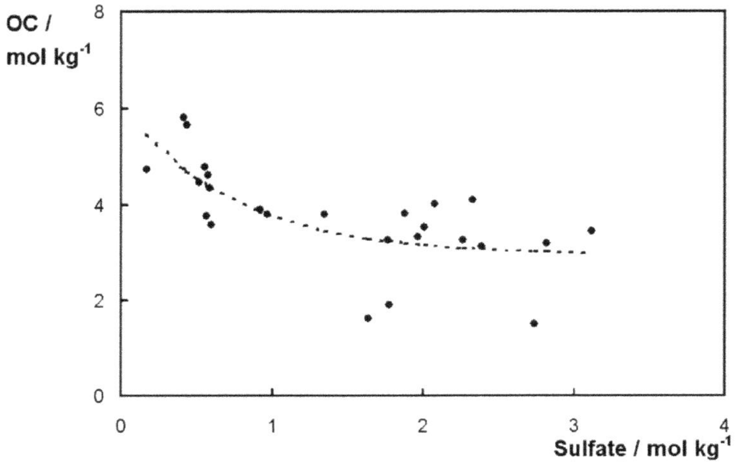

Fig. 5.4 Organic carbon (OC) versus sulfate concentrations from a range of crusts of several European monuments (Grossi and Brimblecombe, 2008; data from Bonazza *et al.*, 2005).

et al., 2007). Figure 5.4 shows measurements from a range of crusts and illustrates that the older crusts with high sulfate concentrations have lower concentrations of organic material. This is the result of the change from high sulfur coal to liquid fuels with lower sulfur content, but with higher organic carbon.

5.3.2 *Biological colouration*

Cleaner atmospheres also mean that climate is a factor of increasing importance in the weathering of monuments (Brimblecombe 2011). Warmer climates promote novel ecosystems on stone surfaces and further changing in the appearance or the colouration of the buildings. Moreover, sulfur dioxide is phytotoxic, so in urban atmospheres with less sulfur, biological growth is more effective. A greater rate of delivery of nitrate to building surfaces that acts as an *airborne fertiliser* (Adamson *et al.*, 2010) also increases biological colonisation. A potential transition from blackening, through yellowing to biological colouration processes during the 20th and 21st centuries is plotted in Fig. 5.5.

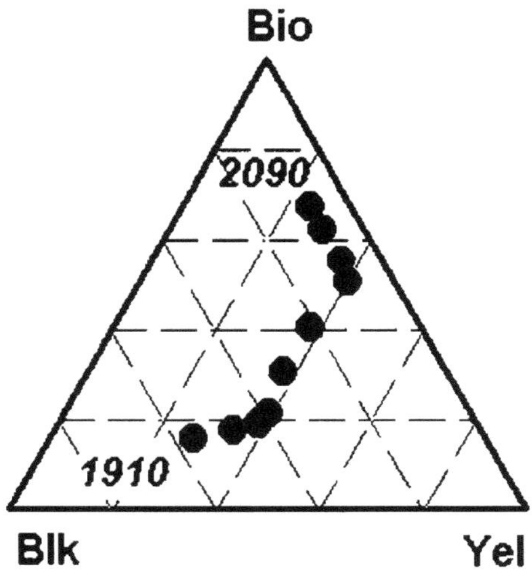

Fig. 5.5 Potential transition from blackening, through yellowing to biological colouration processes during the 20th and 21st centuries (Brimblecombe, 2010).

The Köppen–Geiger climate classification can be a good approximation for some heritage risks (Brimblecombe, 2010 and 2011) and might be used to proxy biological growth. The Köppen–Geiger scheme describes climates in terms of codes. The first letter describes the broad group of climate A (from tropical) to E (polar or alpine), each of them having several types and subtypes. Using the output of climate change models, the climate of Paris (and other European cities in the same climate area) is expected to change from fully humid (Cfb) to a Mediterranean-like type (Csa) (Grossi *et al.*, 2011). Similarly, Hall *et al.* (2011) concluded that moisture regimes and the associated deterioration patterns in the UK are expected to tend towards those found in other Mediterranean areas at the end of the 21st century as a consequence of climate change. This can be translated to changes on biological colonisation of buildings.

Depending on climate type, different colouration processes (e.g. *greening* or *reddening*) might be expected. In Northern Ireland (Adamson *et al.*, 2010) a *greening* effect, due to the growth of green algae, is increasingly observed in locations that experience longer durations of wetness in winter, which, from UKCP09 projections, is

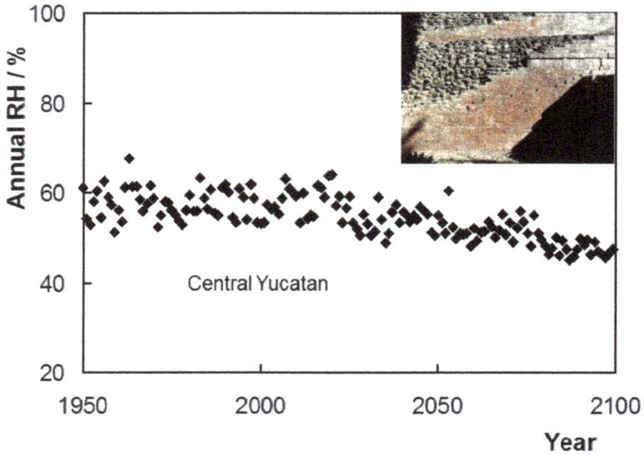

Fig. 5.6 Predicted changes in relative humidity by the Hadley model in the Central Yucatan Peninsula (Mexico) and current reddening of Mayan remains as a result of dryer conditions. (Photo and diagram courtesy of Peter Brimblecombe).

expected to further increase in the future (Smith *et al.*, 2010) . In some locations this greening effect may be favoured by reduction on atmosphere sulfur in cities. Previously high levels might have inhibited colonisations, but this has been replaced by an "airborne fertiliser" in the form of nitrogen oxides.

In other geographic areas experiencing hotter and drier summers, we expect the development of orange-brown or orange-grey weathering layers from the colonisation of algae, bacteria and fungi, characteristics of Mediterranean climates (Urzi and Realini, 1998; García-Vallés *et al.*, 2010). There is also evidence of reddening of Mayan monuments in Calakmul (Mexico Yucatan Peninsula) due to changes to a dryer climate and colonisation by cyanobacteria (Brimblecombe, personal communication, Fig. 5.6).

5.4 Aesthetics Implications

Public reaction to the aspect of the surfaces is relevant to the management of historic buildings. Pressure for cleaning can be related to visitor perceptions of the lightness or reflectance of the facades. Brimblecombe and Grossi (2005) suggested a series of aesthetic thresholds related to soot loading in the atmosphere. At high soot concentrations near busy roads (with EC concentrations around 10 $\mu g\,m^{-3}$) buildings were viewed as unacceptably dark. When concentrations are lower, in the EC range 2–3 $\mu g\,m^{-3}$, the appearance of buildings becomes more acceptable. In the past, it is likely that historic buildings were exposed to much higher concentrations of EC, so the improved air quality in many locations means a decline in blackening (Grossi and Brimblecombe 2007).

The patterns of blackening or the distribution of blackened areas on a facade is also an important factor for public response. While some patterns are considered acceptable, others provoke strong negative reactions. Patterns that obscure the architectural lines of the building are less desirable than those which enhance shadowing effects and increase the contrast of architectural details (Andrew, 1992; Grossi and Brimblecombe, 2004). The appearance of blackening patterns in the future might also vary with changes in pollutant deposition and rainfall patterns.

The mechanism and dynamics of removal of black crusts from buildings over time as pollution decreases is still poorly understood. Blackening patterns have been less frequently investigated than the amount of soiling. However, this is an area of increasing research and includes the development of computer simulation techniques, which can be realistic and verified experimentally. The pioneering work of Davidson *et al.* (2000), Tang and Davidson (2004) and Tang *et al.* (2004) on blackening patterns included experimental and numerical simulation. More recently, there has been an intense research on wind-driven rain on buildings, which involved experimental, semi-empirical methods and numerical methods (Blocken and Carmeliet, 2004; Blocken *et al.*, 2011). Lately, Blocken *et al.* (2010) proposed the use of high resolution whole-building numerical modelling for detailed analysis of the potential consequences of climate change on historic buildings. They used Computational Fluid Dynamics (CFD) to calculate wind-driven rain (WDR) deposition on a monumental tower building in the Netherlands — the St. Hubertus Hunting Lodge.

Until recently, visitors were more sensitive to changes in lightness than in hue or chroma. This is understandable given the large changes in lightness compared to traditionally subtle shifts in chroma.

Nowadays and in the future, with longer experience of cleaner buildings this may change. Changes in colour, from oxidative yellowing or new biological colouration, may be more apparent to the visitor. People are sensitised to colours in the natural word, where seasonal shifts in terms of hue and chroma are familiar in landscapes (Hering, 2007). Similarly, they might identify these on buildings (Grossi and Brimblecombe, 2008).

Complex bio-mineralogical crusts and staining could replace black crusts, especially where vehicular pollution elevates nitrate levels to encourage growth (Smith *et al.*, 2011). If the conditions are wetter, the biological colonisation of historic buildings can be thriving, covering entire facades in a layer of green, black or orange causing unpleasant streaks to run down from roof level to ground (Smith *et al.*, 2011). This can constitute an aesthetic nuisance, with no easy remedy, because of the difficulty to clean with a rapid re-colonisation by microorganisms under appropriate environmental conditions.

References

Adamson, C.S., Mccabe, S., Mcallister, D., Smith, B.J., and Warke, P.A. (2010). 'Mapping the spatial distribution of precipitation, biological soiling and decay of monuments in Northern Ireland: Towards understanding long-term stone response to moisture' in *Proc. 19th Carpathian-Balkan Geological Association Congress, Special volume 99 of the Scientific Annals of the Aristotle University of Thessaloniki Faculty of Sciences*, pp. 183–190.

Andrew, C. (1992). 'Towards an aesthetic theory of building soiling' in Webster, R.G.M. (Ed.), *Stone Cleaning and the Nature*, Donhead, London, pp. 63–81.

Auras, M., Beer, S., Bundschuh, P., Eichorn, J., Mach, M., Scheuvens, D., Schorling, M., von Schummann, J., Snethlage, R. and Weinbruch, S. (2013). Traffic-related immissions and their impact on historic buildings: implications from a pilot study at two German cities. *Environ. Earth Sci.*, 69, pp. 1135–1147.

Beloin, N.J. and Haynie, F.H. (1975). Soiling of building surfaces, *J. Air Pollut. Control Assoc.*, 25, pp. 393–403.

Blocken, B. and Carmeliet, J. (2004). A review of wind-driven rain research in building science, *J. Wind Eng. Ind. Aerodyn.*, 92, pp. 1079–1130.

Blocken, B., Briggen, P.M., Schellen, H.L. and Hensen, J.L.M. (2010). 'Climate change and high-resolution whole-building numerical modelling', in *WTA-Colloquium Effect of Climate Change on Built Heritage*, pp. 195–216.

Blocken, B., Stathopoulos, T., Carmeliet, J., Hensen, J.L.M. (2011). Application of computational fluid dynamics in building performance simulation for the outdoor environment: an overview, *J. Build. Perform. Simu.*, 4, pp. 157–184.

Bonazza, A. and Brimblecombe, P. (2013). Climate and the changing appearance of buildings (unpublished)

Bonazza, A., Brimblecombe, P., Grossi, C.M. and Sabbioni, C. (2007). Carbon in black crusts from the Tower of London, *Environ. Sci. Technol.*, 41, pp. 4199–41204.

Bonazza, A., Sabbioni, C. and Ghedini, N. (2005). Quantitative data on carbon fractions in interpretation of black crusts and soiling on European built heritage, *Atmos. Environ.*, 39, pp. 2607–2618.

Bonazza, A., Sabbioni, C., Ghedini, N., Hermosin, B., Jurado, V., Gonzalez, J.M. and Saiz-Jimenez, C. (2007). Did smoke from the Kuwait oil well fires affect Iranian archaelogical heritage? *Environ. Sci. Technol.*, 41, pp. 2378–2386.

Brimblecombe, P. (2010). 'Heritage Climatology' in Lefevre R.A, and Sabbioni, C. (Eds.), *Climate change and cultural heritage*, Edipuglia, pp. 49–56.

Brimblecombe, P. (2011). 'Environment and architectural stone' in Siegesmund, S. and Snethlage, R. (Eds.), *Stone in Architecture*, Springer, pp. 317–346.

Brimblecombe, P. and Grossi, C.M. (2004). 'The rate of darkening of material surfaces', in Saiz-Jimenez, C. (Ed.), *Air Pollution and Cultural Heritage*, Balkema, Rotterdam, pp. 193–198.

Brimblecombe, P. and Grossi, C.M. (2005). Aesthetic thresholds and blackening of stone buildings. *Sci Total Environ.*, 349, pp. 175–189.

Brimblecombe, P. and Grossi, C.M. (2007). Damage to buildings from future climate and pollution, *Apt Bulletin: Journal Of Preservation Technology*, 38, pp. 2–3.

Brimblecombe, P. and Grossi, C.M. (2009). Millenium-long damage to building materials in London, *Sci. Total Environ.*, 407, pp. 1354–1361.

Brimblecombe, P. and Grossi, C.M. (2010). 'Deposition, transformation and remobilization of soot and diesel particles on building surfaces' in Moreira, D. and Vilhena, M. (Eds.), *Air Pollution and Turbulence*, CRC Press, Boca Raton, pp. 1–13.

Creighton, P.J., Lioly, P.J., Haynie, F.H., Lemmons, T.J., Miller, J.L., Gerhart, J. (1990). Soiling by atmospheric aerosols in an urban industrial area, *J. Air Waste Manag. Assoc.*, 40, pp. 1285–1289.

Davidson, C.I., Tang, W., Finger, S., Etyemezian, V., Striegel, M.F. and Sherwood, S.I. (2000). Soiling patterns on a tall limestone building: changes over 60 years, *Environ. Sci. Technol.*, 34, pp. 560–565.

García-Valles, M., Aulinas, M., López-Melción, J.B. and Moya-Garra, S.I. (2010). Patinas developed in environmental burial conditions: the Neolithic steles of Reguers de Seró (Lleida, Spain), *Environ. Sci. Pollut. R.*, 17, pp. 1287–1299.

Gaviño, M., Hermosin, B., Castillejo, M., Oujja, M., Rebollar, E., Vergès-Belmin, V., Nowik, W. and Saiz-Jimenez, C. (2004). 'Black crust removal: the effect of stone yellowing and cleaning strategies' in Saiz-Jimenez, C. (Ed.), *Air Pollution and Cultural Heritage*, Balkema, London, pp. 239–245.

Graber, E.R. and Rudich, Y. (2006). Atmospheric HULIS: How humic-like are they? A comprehensive and critical review, *Atmos. Chem. Phys.*, 6, pp. 729–753.

Grossi, C.M. and Brimblecombe, P. (2004). Aesthetics of simulated soiling patterns on architecture, *Environ. Sci. Technol.*, 38, pp. 3971–3976.

Grossi, C.M. and Brimblecombe, P. (2007). Effects of long term changes in air pollution and climate on the decay and blackening of European stone buildings, *Geol. Soc. Special Pub.*, 271, pp. 117–130.

Grossi, C.M. and Brimblecombe, P. (2008). Past and future colouring patterns of historic stone buildings, *Materiales de Construccion*, 58, pp. 143–160.

Grossi, C.M., Brimblecombe, P., Menéndez, B., Benavente, D., Harris, I., Déqué, M. (2011). Climatology of salt transitions and implications for stone weathering, *Sci. Total Environ.*, 409, pp. 2577–2585.

Grossi, C.M., Esbert, R.M., Díaz-Pache, F. and Alonso, F.J. (2003). Soiling of building stones in urban environments, *Build. Environ.*, 38, pp. 147–159.

Hall, C., Hamilton, A., Hoff, W.D., Viles, H.A. and Eklund, J.A. (2011). Moisture dynamics in walls: response to micro-environment and climate change, *Proc. R. Soc.*, 467, pp. 194–211.

Hamilton, R. and Mansfield, T.A. (1991). Airborne particulate elemental carbon: its sources, transport and contribution to dark smoke and soiling, *Atmos. Environ.*, 25A, pp. 715–723.

Haynie, F.H. (1986). 'Theoretical model of soiling of surfaces by airborne particles', in Lee, S.D. (Ed.), *Aerosols: Research, Risk Assessment, and Control Strategies*, Lewis Publishers, Boca Raton, pp. 951–959.

Hering, B. (2007). Seasonal changes of dominating hues in landscapes close to nature, *Spoken Version, ICCJ*, Budapest.

Ionescu, A., Lefèvre R.A., Chabas, A., Lombardo, T., Ausset, A., Candau, Y. and Rosseman, L. (2006). Modeling of soiling based on silica-soda-lime glass exposure at six European sites, *Sci. Total Environ.*, 369, pp. 246–255.

Kucera, V. (2005). *Model for multipollutant impact and assessment of threshold levels for cultural heritage*, Swedish Corrosion Institute.

Lorusso, S., Troili, M., Biondi, F. and Marabelli, M. (1996). 'Air quality of historic cities and deterioration of monuments: case-studies in Rome and Tivoli', in *Proceedings of the 8th International Congress on Deterioration and Conservation of Stone*, Berlin, pp. 281–300.

Mansfield, T.A. and Hamilton, R.S. (1989). 'The soiling of materials: models and measurements in a road tunnel' in Brasser, L.J. and Mulder, W.C. (Eds.), *Proceedings of the 8th World Clean Air Congress*, Elsevier Science Publishers B.V., Amsterdam, pp. 353–357.

Nicholson, K.W. (1988). The dry deposition of small particles: a review of experimental measurements, *Atmos. Environ.*, 22, pp. 2653–2666.

Sillanpaa, M., Frey, A., Hillamo, R., Pennanen, A.S. and Salonen, R.O. (2005). Organic, elemental and inorganic carbon in particulate matter of six urban environments in Europe, *Atmos. Chem. Phys.*, 5, pp. 2869–2879.

Smith, B.J., McCabe, S., McAllister, D., Adamson, C., Viles, H.A. and Curran, J.M. (2011). A commentary on climate change, stone decay dynamics and the "greening" of natural stone buildings: new perspectives on "deep wetting", *Environ. Earth Sci.*, 63, pp. 1691–1700.

Tang, W., Davidson, C.I., Finger, S., Vance, K. (2004). Erosion of limestone building surfaces caused by wind-driven rain: 1. Field measurements, *Atmos. Environ.*, 38, pp. 5589–5599.

Tang, W. and Davidson, C.I. (2004). Erosion of limestone building surfaces caused by wind-driven rain: 2. Numerical modeling, *Atmos. Environ.*, 38, pp. 5601–5609.

Toprak, S., Aksu, R., Pesava, P. and Horvath, H. (1997). The soiling of materials under simulated atmospheric conditions in a wind tunnel, *J. Aerosol Sci.*, 28, Suppl. 1, pp. 585–586.

Urosevic, M., Yebra-Rodríguez, A., Sebastián-Pardo, E. and Cardell, C. (2012). Black soiling of an architectural limestone during two-year term exposure to urban air in the city of Granada (S Spain), *Sci. Total Environ.*, 414, pp. 564–575.

Urzi, C. and Realini, R. (1998). Colour changes in Noto's calcareous sandstone as related to its colonisation by microorganisms, *Int. Biodeterioration Biodegrad.*, 42, pp. 45–54.

Vázquez-Calvo, C., Álvarez de Buergo, M. & Fort, R. (2007). Overview of recent knowledge of patinas on stone monuments: the Spanish experience, *Geological Society Special Publication*, 271, pp. 295–307.

Viles, H.A. and Gorbushina, A.A. (2003). Soiling and microbial growth on urban roadside limestone; a three year study in Oxford, England, *Build. Environ.*, 38, pp. 1217–1224.

Watt, J., Hamilton, R., Lefevre, R.A. and Ionescu, A. (2009). 'Soiling' in Watt J., Tidblad, J., Kucera, V., and Hamilton, R. (Eds.), *The Effects of Air Pollution on Cultural Heritage*, Springer, Science + Business Media, LLC, New York, pp. 105–126.

Watt, J., Jarrett, D., Hamilton, R. (2008). Dose-response functions for the soiling of heritage materials due to air pollution exposure, *Sci. Total Environ.*, 400, pp. 415–424.

Watt, J., Massey, S., Kendall, M. (2000). 'Fractal modelling of particulate deposition in the development of black crust on stone' in *9th International Congress on Deterioration and Conservation of Stone*, pp. 637–646.

Yocom, J.E. and Kawecki, J.M. (1986). 'Overview of soiling and materials damage from aerosols', in *Aerosols*, Lewis Publishers, Boca Raton, pp. 913–922.

CHAPTER 6

Air Pollution Damage to Metals

Johan Tidblad

Swerea KIMAB, Sweden

6.1 Introduction

It has been known for several centuries that air pollutants emitted by burning of fossil fuels have a serious impact on metals exposed in the built environment. In addition to the loss of mechanical strength and failure of protective coatings, the release of heavy metals to the environment is of concern. Calculations of the economic impact of sulfur pollutants on technical materials have indicated that the costs for corrosion damage to materials including metals are substantial.

There are many parameters that can influence the atmospheric corrosion of metals — the degradation process is a complex interplay between chemical, physical and biological parameters. The present treatment focuses on air pollutants and their effect on the built environment, i.e. the aspects specific to the urban situation. It is, however, important to recognise that atmospheric corrosion is a process that occurs even in the absence of pollutants and that the interplay between natural and anthropogenic factors determine to which extent urban conditions affects and accelerates the "natural" or background atmospheric corrosion.

The present treatment serves as an overview of the effects of acidifying air pollutants to metallic materials. It is divided in two main sections and three additional sections. Section 6.2 describes the general role of atmospheric pollutants in the corrosion process. Despite the general characteristics of the pollutants, every metal is unique and a selection of metals is described separately in Section 6.3. The quantification of effects, in particular the assessment of life times and calculation of corrosion costs deserves a special treatment given in Section 6.4. Finally, the trends in corrosion including effects of climate change are given in Section 6.5. This chapter is not a review. Instead of a complete list, references have been selected that are suitable for further reading including general references on atmospheric corrosion (Kucera and Mattson, 1987; Kucera and Fitz, 1995; Leygraf and Graedel, 2000; Watt *et al.*, 2009) and are cited once at the most appropriate place.

6.2 Effects of the Environment

The most important pollutants acting as corrosive agents are sulfur and nitrogen compounds including secondary pollutants, and particulate materials. It has clearly been demonstrated that pollutants enhance the natural corrosion process for several metallic materials even if it has so far been difficult to quantify the direct contribution of individual pollutants other than sulfur dioxide (SO_2) in the weathering process.

Depending on the way pollutants are transported from the atmosphere to the corroding surface, two types of deposition processes are recognised in atmospheric corrosion — dry deposition and wet deposition. Wet deposition refers to precipitation whereas dry deposition refers to the remaining processes, including gas-phase deposition and particle deposition.

6.2.1 *The multi-pollutant situation*

In many European countries, the concentration of SO_2 has decreased significantly during the 20th century while the concentrations of nitrogen pollutants, ozone (O_3) and particulates remain at high levels. Nitrogen dioxide (NO_2) has also decreased significantly but not as much as SO_2. Traditionally and with right, SO_2 has been regarded as the most important corrosion stimulator. With the reduced levels, SO_2

is no longer the only main pollutant. In addition to the added effects of many individual pollutants, synergistic effects of e.g. SO_2 in combination with NO_2 or O_3 need to be considered. This complex pollution situation with effects of a number of important single-pollutants in combination with several synergistic effects, described in the following, is named the "multi-pollutant situation". The general acceptance of the new multi-pollutant situation was manifested in the multi-effect, multi-pollutant Protocol to Abate Acidification, Eutrofication and Ground-level Ozone adapted in 2000 in Gothenburg within the Convention of Long-range Transboundary Air Pollution of the United Nations Economic Commission for Europe.

6.2.2 *Climate*

Climate has both indirect and direct effects on atmospheric corrosion of metals. Indirect effects include for example the less mixing and higher concentration levels of pollutants present when there is a temperature inversion. The present treatment only describes direct effects. The dominating part of the research activities has been performed in Europe and North America and therefore the effect of acidifying pollutants has mainly been investigated in regions situated in temperate climate. The recent focus on possible effects of climate change is described in Section 6.5.

Atmospheric corrosion is an electrochemical process and proceeds only when the surface is sufficiently wet, the degradation process may be considered as discontinuous. This is illustrated in Fig. 6.1 where it is seen that the instantaneous corrosion rate varies strongly with time, several orders of magnitude. The time of wetness (TOW) is a commonly used concept for metals that refers to the time when corrosion occurs, i.e., when a moisture layer is present.

This concept is useful when describing the degradation process and when classifying different climatic regions from a corrosion point of view but is difficult to calculate from readily available meteorological data. Therefore, recently developed dose-response functions include annual averages of temperature and relative humidity, which are more readily available.

Dry deposition of pollutants is greatly influenced by temperature (T) and relative humidity (RH). T and RH are important since they

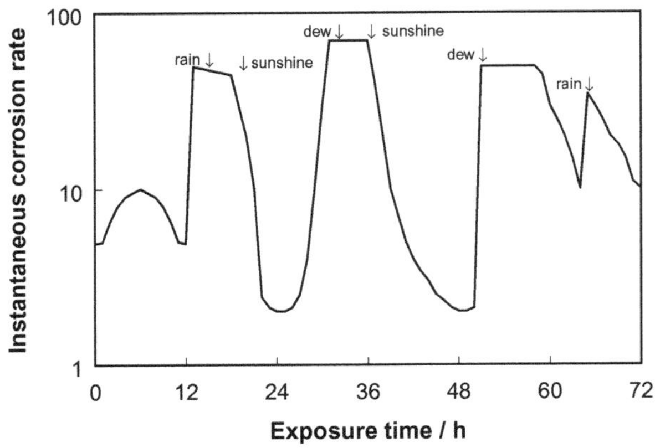

Fig. 6.1 Instantaneous corrosion rate during a few days of exposure (adapted from Kucera and Mattson, 1987).

are the main factors that determine the thickness of the moisture layer in the absence of rainfall. The corrosion rate increases with RH, starting from a critical humidity value, where the adsorbed water layer begins to act as an electrolyte. At low temperatures, the corrosion rate increases with increasing annual temperature since the fraction of time above the water freezing point also increases. At higher temperatures the corrosion in non-marine areas decrease with increasing temperature, which has been attributed to periods with a surface temperature above the ambient temperature, partly related to solar radiation. An elevated surface temperature leads to a faster evaporation of moisture after rain or condensation periods and to a decrease in the thickness of the adsorbed water layer and, consequently, to a decrease of the time when the metal surface is wet. Figure 6.2 shows an example of the temperature effect for zinc based on a dose-response function presented in Section 6.3 (Eq. 6.2).

6.2.3 *Gaseous pollutants*

Trace gases of importance for atmospheric corrosion include O_3, H_2O_2, SO_2, H_2S, COS, NO_2, HNO_3, NH_3, HCl, Cl_2, HCHO and HCOOH. The present treatment focuses on SO_2, NO_2, O_3 and HNO_3 — pollutants that are especially important in urban outdoor environments. Carboxylic acids have also earned some attention due

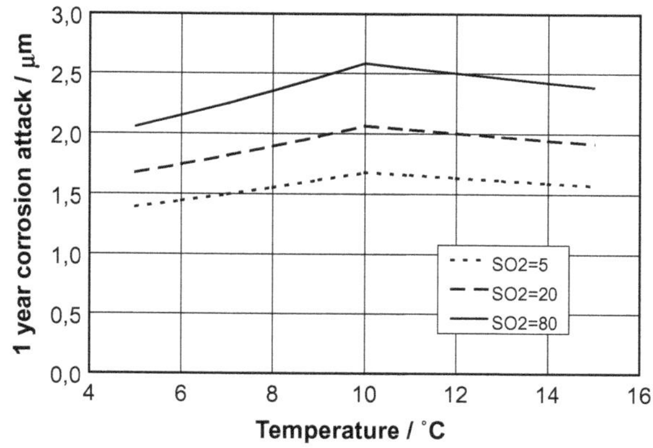

Fig. 6.2 Corrosion attack of unsheltered zinc vs. temperature calculated from Eq. 6.2 with Rh=75% and a constant wet deposition term, 0.029 Rain$[H^+]$, corresponding to 0.6 μm. Curves are shown for the SO_2 concentrations 5, 20 and 80 $\mu g/m^3$.

to the increased use of methanol and ethanol as alternative fuels. Especially formic acid is a strong acid, which makes the potential effect high. However, data is scarce on estimated future pollutant levels and the limited amount that is available suggests that outdoor concentrations are relatively low compared to indoor concentrations that can be obtained in, for example, museum environments (not included in the present treatment).

6.2.3.1 *SO_2*

In the field of atmospheric corrosion, SO_2 is the single most investigated gaseous pollutant. Its detrimental effect on materials in general has since long been an indisputable fact. Sulfur compounds emitted to the atmosphere were suspected to be the cause of acidification already in the end of the 19th century. Systematic laboratory exposures in the 1930's demonstrated the corrosive effect of SO_2 on metals. This was later also proven by field exposures and SO_2 was for a long time considered to be the main corrosive pollutant. Even with decreased levels it is important to remember that SO_2 is still one of the main actors in the new multi-pollutant situation, although not as dominant as in the past.

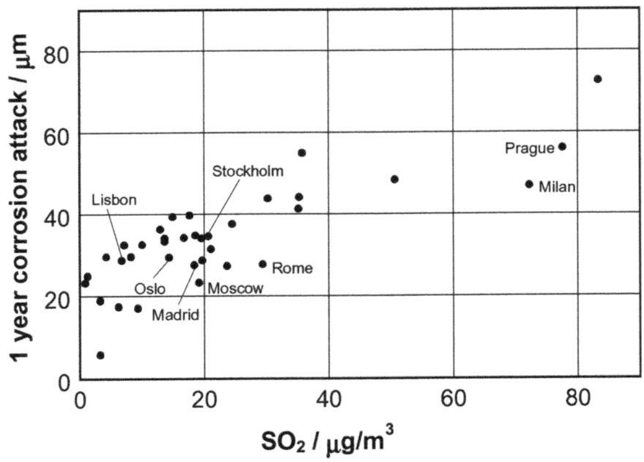

Fig. 6.3 Corrosion attack of unsheltered carbon steel exposed in the UN ECE exposure programme during the period September 1987 to August 1988 vs. SO_2 concentration.

SO_2 is dissolved in the moisture layer forming sulfite and, after oxidation, sulfate:

$$SO_2(g) + H_2O \rightarrow H^+ + HSO_3{}^- \rightarrow \ldots \rightarrow 2H^+ + SO_4^{2-} \qquad (6.1)$$

This process results in an acidification of the moisture layer, which enhances the corrosion process. Sulfate is also frequently present in corrosion products. The SO_2 deposition rate depends mostly on the material, it is often higher for sensitive materials, and varies between 0.01 and 2 cm/s.

Figure 6.3 shows a plot of 1-year corrosion values of carbon steel in the 1980s, with selected cities in Europe highlighted. It is worth noting that the SO_2 pollution levels and corrosion rates today in most European cities are significantly lower (see Section 6.5). Quantitative relationships, i.e. dose-response functions, expressing the corrosion attack as a function of SO_2 and other parameters exists for several metals (see Section 6.3).

6.2.3.2 *Synergistic effects of NO_2 and O_3 in combination with SO_2*

In contrast to SO_2, the effects of NO_2, and especially O_3, are not well documented (Arroyave and Morcillo, 1995; Tidblad and

Kucera, 1996). Field exposure programmes have failed to detect effects attributable to NO_2 or O_3 with the exception of O_3 and copper (see Section 6.3.4). Laboratory exposures of several metals including zinc, copper and nickel have led to the following picture regarding synergistic effects.

The term *synergistic effect* means that the corrosion attack for a material exposed to a mixture of gases is greater than the sum of effects from individual single-gas exposures. Significant synergistic effects have been found in the laboratory for both O_3 and NO_2; however, the relative magnitude of the effects depends on the metal. For both O_3 and NO_2 the synergistic effect is attributed to the promotion of reaction 6.1, i.e. the oxidation of S(IV) to S(VI). The reaction between SO_2 and O_3 is most likely stoichiometric while the reaction between SO_2 and NO_2 needs further investigation. With few exceptions, there seem to be synergistic effects. The synergistic effects increase with relative humidity for most metals. However, at high relative humidity, aluminium and iron show no SO_2+NO_2 synergism. When comparison is possible, the synergistic effect of SO_2+O_3 is stronger than SO_2+NO_2, except for nickel, which is illustrated in Fig. 6.4.

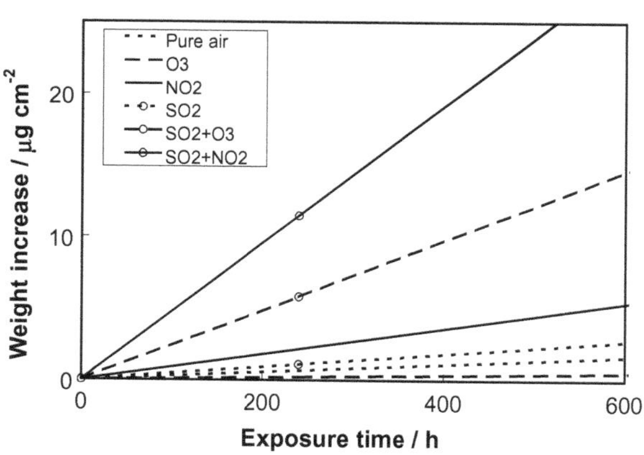

Fig. 6.4 Weight increase of nickel exposed in air at 70 % Rh based on results after 240 h and assuming a linear time dependence. The influence of additions of O_3 (100 ppb), NO_2 (100 ppb) and SO_2 (90 ppb) including combinations are shown.

6.2.3.3 *HNO_3*

HNO_3 is a secondary pollutant formed by the oxidation of NO_2. It can reach appreciable concentration levels in urban areas — reports are usually in the interval 1–7 $\mu g/m^3$ but in some cases the concentration can be above 10 $\mu g/m3$. HNO_3 is a strong acid with a high deposition velocity that is almost independent on the humidity, which makes it relatively more harmful for dry and warm climates. The magnitude of its effect relative to SO_2, at equal concentrations, is suggested to be very high, with a factor between 2 and 20 depending on material and with the highest value for zinc (Samie *et al.* 2007).

6.2.4 *Particles*

Particles normally increase but may in some cases decrease the deterioration rates of metals. The effect of particles on corrosion can also be either direct or indirect. Anthropogenic particles can be divided into primary and secondary. The primary particles are directly emitted from combustion, have a relatively short life span and deposit near the source. Secondary particles are smaller, less than two micrometers, long-lived, and are the result of chemical reactions amongst other pollutants e.g. SO_2, NO_2, volatile organic compounds and ammonia. Compared to the many effects, described in the following, that particles can have on the atmospheric corrosion, there is a lack of quantitative assessments especially the comparison to effects of gaseous pollutants.

Droplets of sulfuric acid resulting from the oxidation of SO_2 can be neutralised by ammonia forming particles of ammonium sulfate. Similarly, particulate ammonium nitrate can form through the neutralisation of HNO_3. The particles thus enhance the corrosion by providing corrosion stimulators, especially sulfate is frequently found in corrosion products. Also, when water is adsorbed and then evaporates, gaseous ammonia is released to the atmosphere resulting in an acidification of the aqueous layer that may increase the corrosion rate.

Particles may in some cases also decrease the corrosion rate if they are basic, by neutralising the surface water film formed on the degraded material.

Salt containing particles play an important role in atmospheric corrosion and this is related to their ability to increase the time of wetness. They are hygroscopic and start to absorb water when the relative humidity exceeds a critical level, determined ideally by the equilibrium properties of a saturated solution of the salt. In practice, however, the particles deposited on the surface are a mixture of different compounds making it a difficulty to define a critical relative humidity. The ionic content of the particles prolongs the time of wetness also by reducing the freezing point of the adsorbed water below 0 °C. Particulate matter, together with HNO_3, was the focus of the multi-pollutant exposure program within ICP Materials, which resulted in dose-response functions for the multi-pollutant situation (Section 6.3).

6.2.5 *Wet deposition*

SO_2 and NO_2 both contribute to the increased acidity of wet deposition, in addition to their direct effects. The effect of wet deposition is relatively well quantified for many metals. Dose-response functions based on field exposures that includes the effect of acid rain exists for several metals. On the other hand, as with particles, the effect of wet deposition can be either detrimental or beneficial, depending on the conditions. In principle, the wet deposition can have two effects on the corrosion process. First, it transports chemically active compounds present in the rain to the surface, thereby increasing the corrosivity of the moisture layer. Second, it washes away compounds previously deposited on the surface, with the opposite effect. Thus, for a specific material and environment, the choosing of a sheltered exposure condition rather than unsheltered may, or may not, increase the corrosion rate as is illustrated in Table 6.1, where it is shown that aluminium has a higher corrosion rate in sheltered position in contrast to the other metals.

The effect of wet deposition has traditionally been regarded as minor compared to the effect of dry deposition. However, as is illustrated in Fig. 6.5, the effect depends strongly on pH and amount of precipitation. At about pH 4.5 and lower, the effect cannot be neglected.

Table 6.1 Corrosion attack in μm after 8 years of exposure in the UN ECE exposure programme in unsheltered and sheltered positions. Intervals are shown corresponding to lower and upper quartiles (25- and 75-percentiles).

Metal	Unsheltered	Sheltered
Weathering steel	37–59	27–54
Zinc	9.2–18	2.2–3.4
Copper	3.4–5.3	1.8–3.5
Bronze	2.9–5.4	1.6–3.0
Aluminium	1.0–1.7	1.1-3.1

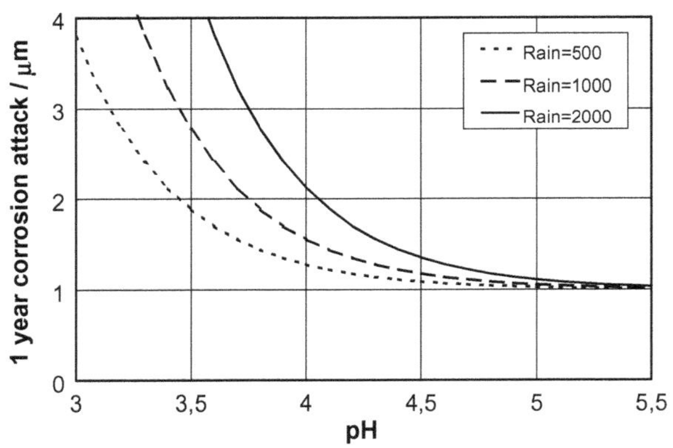

Fig. 6.5 Corrosion attack of unsheltered copper vs. pH of precipitation calculated from Eq. 6.3 and a constant dry deposition term, $0.0027[SO_2]^{0.32}[O_3]^{0.79}Rh\cdot e^{f(T)}$, corresponding to 1.0 μm. Curves are shown for the precipitation (rain) values of 500, 1000 and 2000 mm.

In addition to compounds deposited on the surface, corroded metal ions are removed by the rain-wash. This release of metals to the environment is an environmental concern due to the potential effect that the heavy metals can have on ecosystems. However, when assessing the impact it is important to recognise first that the metal release is significantly lower than the total corrosion rate, which also includes metal ions left on the corroded surface in the corrosion products, and second that only part of the released metal is biologically active. Also, the metal run-off depends on the amount of precipitation reaching the surface, which means that the amount of precipitation as well as the inclination of the surface is of utmost importance when assessing run-off values.

6.3 Effects on Metallic Materials

In its unprotected form, examples of use for carbon steel include mounting and fittings and forged objects. Cast iron has been used in fences and gates. The main use is, however, in its protected form, which makes the economic importance of unprotected steel relative low. Zinc is commonly used as a pigment in zinc-rich coatings and in hot dip galvanising. An alternative use is rolled zinc sheet for roofing and rainwater drains. In buildings, aluminium is mainly used as a roof and facade material. Already in 1897, the San Gioacchino church in Rome was covered with aluminium. The main uses of copper include roofs and window ledges. Historic and cultural monuments made of cast bronze are common especially in statues.

6.3.1 *Effects on metals in general*

A typical metal exposed to dry air spontaneously forms a metal oxide of thickness 1 to 5 nm. Thus, it is the nature of the thin metal oxide, whether it is passive or active, stable or unstable, protective or not protective, that governs the atmospheric corrosion behaviour of that metal, rather than the properties of the base metal. The relative humidity when adsorbed water have properties equal to those of bulk water varies from 50% to 90% depending on the metal and this interval coincides with that of the *critical relative humidity*, above which corrosion rates starts to be significant (see Section 6.2.2). Atmospheric corrosion of metals is an electrochemical process in which the overall reaction can be divided into an anodic and a cathodic reaction. Under most atmospheric conditions, the anode reaction, i.e. metal dissolution, is the rate-limiting step. With time, a film of corrosion products form that usually hinders the transport of ions. This lowers the anodic reaction rate and, hence, the atmospheric corrosion rate. The initial corrosion products are oxides, hydroxides, and even oxyhydroxides. During prolonged exposure, however, a variety of corrosion products may form, including compounds originating from the atmosphere, i.e., sulfate, nitrate and chloride. The atmospheric corrosion of metals is a complex process and the extent of deterioration as well as the mechanism varies considerably depending on the metal.

For several metals that are frequently used in buildings, including objects of cultural heritage, dose-response functions on damage have been obtained. The functions are of importance for the development of systems for classification of corrosivity of environments, for mapping of areas with increased risk of corrosion, and for calculation of cost of damage caused by deterioration of materials. A dose-response function links the dose of pollution, measured in ambient concentration and/or deposition, to the rate of material corrosion. Three major sets of dose-response functions based on international exposure programs have been obtained. The first set, called "dose-response functions for the SO_2 dominating situation" was developed based on data (1987–1994) from the ICP Materials program (Tidblad *et al.*, 2001). The second "dose-response functions for the multi-pollutant situation" was also developed based on data (1997–2001) from the ICP Materials program (Kucera *et al.*, 2007). The third "dose-response functions for ISO classification" was developed based on data (1984–1995) from the ISOCORRAG and MICAT exposure programs (Mikhailov *et al.*, 2004). The functions for the SO_2 dominate over a wide range of different metals while the multi-pollutant functions (carbon steel, zinc and bronze) and ISO functions (carbon steel, zinc, copper and aluminium) are for a more limited set of metals. Therefore, in the following treatment for individual materials, dose-response function examples are given for the SO_2-dominant situation.

Metals dealt with specifically in this chapter include ferrous metals (Graedel and Frankenthal, 1990), zinc and galvanised steel (Graedel 1989a; Odnevall, 1994; He, 2000), copper (He, 2000; Graedel *et al.*, 1987; Strandberg, 1997), bronze (Strandberg, 1997) and aluminium (Graedel, 1989b).

6.3.2 *Ferrous metals*

The corrosion rate of carbon steel is generally high, about 20–60 μm after one year of exposure in non-marine areas, the value to a

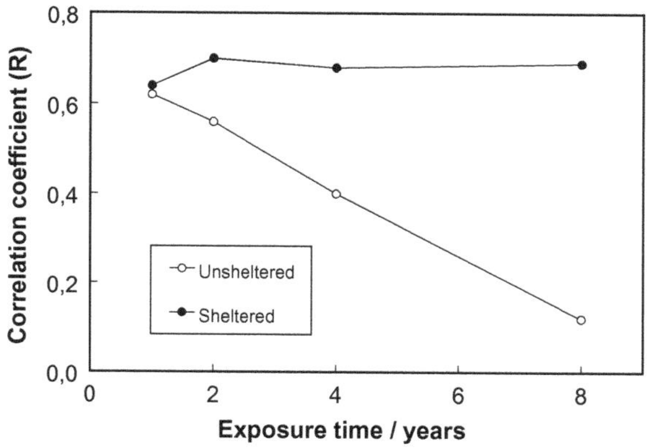

Fig. 6.6 Correlation coefficient, R, vs. exposure time. R is calculated by comparing SO_2 concentration with corrosion attack of weathering steel exposed in unsheltered or sheltered position in the UN ECE exposure programme.

great extent depending on the SO_2 concentration (see Fig. 6.3). Weathering steel behaves in a different way compared with ordinary carbon steel under prolonged exposure and the corrosion rate is stabilised at fairly low levels since a protective rust layer forms during the first 2–4 years of exposure. The decreasing influence of SO_2 with time for unsheltered weathering steel is known for this material, which is illustrated in Fig. 6.6. The practical implication is that weathering steels are a very useful material for unsheltered exposure in SO_2 polluted atmospheres. They have, however, similar corrosion rates to carbon steel in sheltered positions as the creation of a protective rust layer needs regular washing and drying at rain events.

6.3.3 *Zinc and galvanised steel*

As with many other metals, SO_2 is the most important corrosion stimulator for zinc and galvanised steel. In addition, wet deposition may result in dissolution of protective corrosion products, which leads to an increase in the deterioration rate. In marine areas, the deposition of chlorides is also a determining factor. The corrosion attack after one year of exposure is about 1–2 μm in non-marine areas.

Zinc is the only one of the common engineering metals for which a carbonate plays an important role in atmospheric corrosion. After an initial formation of zinc hydroxycarbonate, a zinc hydroxy-sulfate is formed in the urban environment. Deposition of chlorides may eventually also lead to the formation of a zinc chlorohydroxy-sulfate.

Zinc samples were exposed in the UN/ECE materials exposure programme and a dose-response function

$$\mathrm{ML} = 1.4[\mathrm{SO_2}]^{0.22}\mathrm{e}^{0.018\mathrm{Rh}}\mathrm{e}^{\mathrm{f(T)}}\mathrm{t}^{0.85} + 0.029\mathrm{Rain}[\mathrm{H^+}]\mathrm{t} \qquad (6.2)$$

was obtained where ML is the mass loss in g/m^2, $[SO_2]$ is the SO_2 concentration in $\mu g/m^3$, Rh is the relative humidity in %, f(T) is a function of temperature in °C, equal to 0.062(T-10) when T is lower than 10 °C and −0.021(T-10) when T is higher than 10 °C, t is the time in years, Rain is the amount of precipitation in mm/year and $[H^+]$ is the hydrogen ion concentration in precipitation in mg/l.

6.3.4 *Copper and bronze*

The corrosion chemistry of copper is complex. The corrosion products consist of a mixture of corrosion products where cuprite (Cu_2O) and basic sulfates including anthlerite and brochantite that constitute the dominant product on surfaces exposed long-term in the urban environment. The formation of cuprite is especially difficult to quantify in terms of environmental parameters and the formation rate can initially be very high especially in rural areas, which have high O_3 concentrations. Copper is the only metal where field exposures have proven that O_3 as well as SO_2 have an important role. After longer exposure times the relative importance of cuprite decreases in favour of sulfate and the degradation characteristics become similar to those of bronze. In Europe, the corrosion attack on copper and bronze after one year of exposure is about the same, 1–2 μm, or slightly lower compared with zinc. Results from exposures outside Europe in Asia and in Africa within the exposure program on regional air pollution in developing countries (Tidblad *et al.*, 2007) show, however, that after one year of exposure, copper corrosion in these areas are generally higher than the zinc corrosion.

Copper is one of the metals that were exposed in the UN ECE materials programme and a dose-response function

$$\mathrm{ML} = 0.0027[\mathrm{SO_2}]^{0.32}[\mathrm{O_3}]^{0.79}\mathrm{Rh}\cdot\mathrm{e}^{\mathrm{f(T)}}\mathrm{t}^{0.78} + 0.050\mathrm{Rain}[\mathrm{H^+}]\mathrm{t}^{0.89} \qquad (6.3)$$

was obtained where ML is the mass loss in g/m^2, $[SO_2]$ is the SO_2 concentration in $\mu g/m^3$, $[O_3]$ is the O_3 concentration in $\mu g/m^3$, Rh is the relative humidity in %, f(T) is a function of temperature in °C, equal to 0.083(T-10) when T is lower than 10 °C and −0.032(T-10) when T is higher than 10 °C, t is the time in years, Rain is the amount of precipitation in mm/year and $[H^+]$ is the hydrogen ion concentration in precipitation in mg/l.

Bronze used for sculptures usually contain copper, 1–10% tin, 1–10% zinc and 0–6% lead. Bronze is also one of the metals that were quantified in the UN/ECE materials programme as:

$$\begin{aligned} \mathrm{ML} = {} & 0.026[\mathrm{SO_2}]^{0.44}\mathrm{Rh} \cdot e^{f(T)} t^{0.86} \\ & + (0.029\mathrm{Rain}[\mathrm{H^+}] + 0.00043\mathrm{Rain}[\mathrm{Cl^-}])\, t^{0.76} \end{aligned} \tag{6.4}$$

where ML is the mass loss in g/m^2, $[SO_2]$ is the SO_2 concentration in $\mu g/m^3$, Rh is the relative humidity in %, f(T) is a function of temperature in °C, equal to 0.060(T-11) when T is lower than 11 °C and −0.067(T-11) when T is higher than 11 °C, t is the time in years, Rain is the amount of precipitation in mm/year, $[H^+]$ is the hydrogen ion concentration in precipitation in mg/l and $[Cl^-]$ is the chloride ion concentration in precipitation in mg/l. Figure 6.7 shows observed vs. predicted values based on the dose-response function.

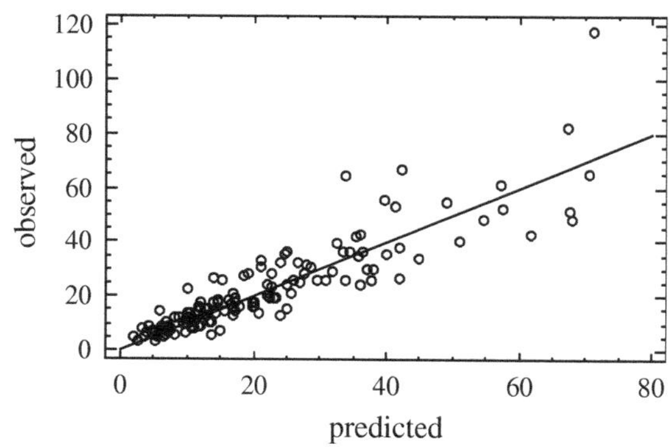

Fig. 6.7 Observed vs. predicted values for mass loss of unsheltered bronze using Eq. 6.4.

6.3.5 *Aluminium*

The corrosion attack on aluminium is generally lower compared to the other metals, about 0.1 to 0.2 μm after one year of exposure in non-marine areas. The low corrosion rate is due to a protective aluminium oxide film that forms initially, γ-Al_2O_3. The film may grow and also be transformed into other compounds, such as hydrated aluminium oxides, hydroxides and sulfates, during the atmospheric corrosion process. As with many other metals, the corrosion rate of aluminium is affected by SO_2 to a great extent but aluminium is particularly sensitive to chlorides, which may result in increased pitting. Sulfate species dominate in corrosion products formed in urban areas.

Looking at Table 6.1, aluminium differs from the other metals in having higher corrosion attack values in sheltered position compared to unsheltered. A possible time development in sheltered position is illustrated in Fig. 6.8 where particles may accelerate corrosion by adsorbing moisture and acidifying compounds from the atmosphere thus producing an acid medium on the surface. Due to the lack of cleansing by rain, these conditions prevail for long periods and the corrosion rate may increase with time initially.

6.4 Lifetime Assessments and Cost Calculations

A damage function is needed to calculate costs. A physical damage function links the rate of material corrosion due to the pollution

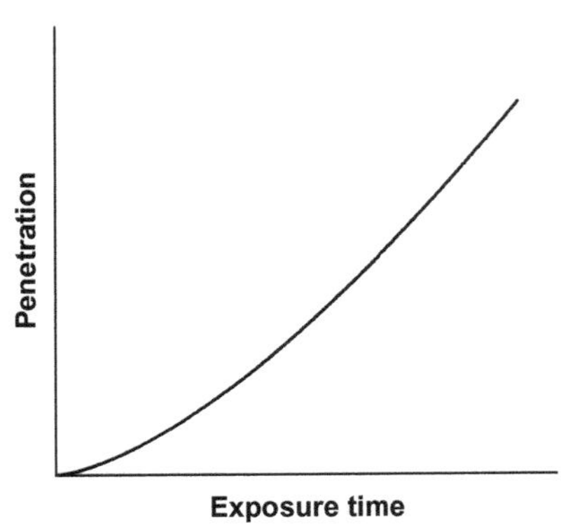

Fig. 6.8 Sketch of corrosion-time curve for rain-sheltered aluminium in a heavy industrial atmosphere (adapted from Kucera and Mattson, 1987).

exposure given by the dose-response function to the time of replacement of maintenance of the material. Performance requirements determine the point at which replacement or maintenance is considered to become necessary. If the performance requirements can be described in terms of a critical degradation level it is possible to transform a dose-response function into a damage function (Kucera *et al.*, 1996; Tidblad *et al.*, 2001). Effects of air pollution on cultural heritage including significant sections on economic damage to cultural heritage was recently reviewed by Watt *et al.*, 2009.

Direct costs of corrosion damage caused by air pollutants can be assessed using a model developed and used first in three cities in Europe: Stockholm, Sarpsborg and Prague, and subsequently for a rough estimate for the whole Europe (Kucera *et al.*, 1993; UNECE Workshop, 1997). The estimated total cost savings were \$9.5 billion per year for SO_2 induced corrosion in Europe resulting from the implementation of the second Sulfur Protocol within UN ECE. The model is shown in Fig. 6.9. The estimated economic damage can be calculated according to the equation

$$K_a = K \cdot S \cdot (L_p^{-1} - L_c^{-1}) = K \cdot S \cdot f \qquad (6.5)$$

where K_a is the additional cost for maintenance/replacement, K is the cost per surface area of material, S is the surface area of material, L_p is the maintenance interval (lifetime) in polluted areas and L_c is the maintenance interval in clean areas. The maintenance frequency, f, is the fraction of material that is subject to actions each year as a result of increased pollution. Equation 6.5 has to be applied for each material individually. Furthermore, each component in Eq. 6.5 can, in principle, have a spatial dependence, i.e., the cost, exposed areas and lifetimes may each depend on the geographical location. In order to make a cost estimation, each block in Fig. 6.9 needs to be addressed. Depending on the available data, the treatment of individual boxes will be more or less comprehensive.

6.5 Corrosion Trends and Effect of Climate Change

The assessment of trend effects have been an integral part of the UN ECE exposure programme on materials (ICP Materials) and the results from this program after 25 years of research were recently

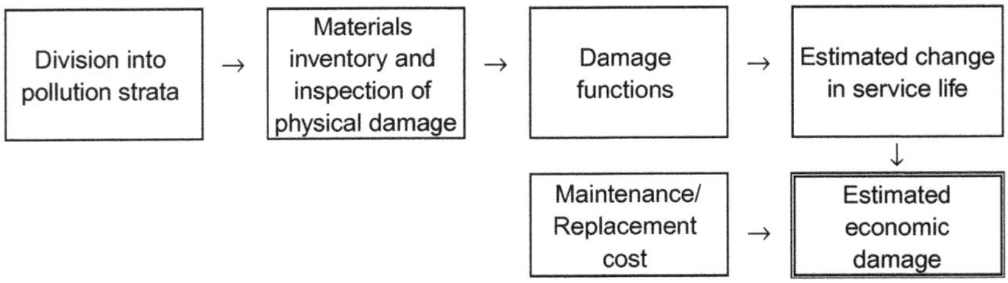

Fig. 6.9 General approach for assessing cost of corrosion damage (Kucera *et al.*, 1993).

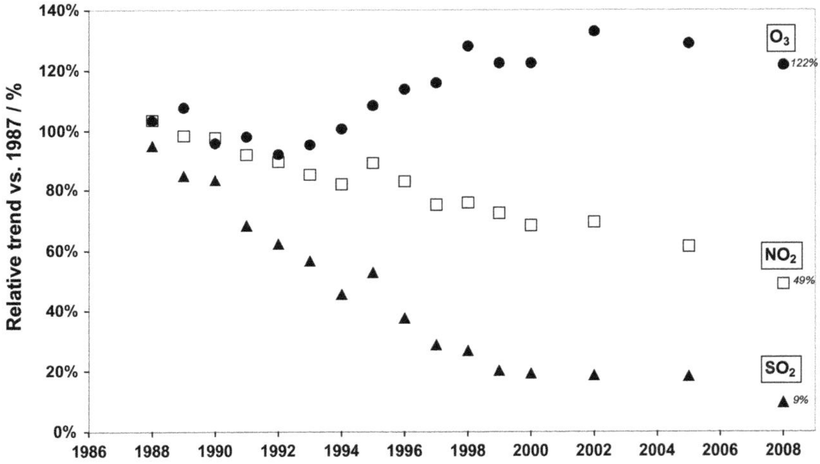

Fig. 6.10 Trends of SO_2, NO_2 and O_3. All values are expressed relative to the initial (1987/88) value (with permission from Tidblad *et al.*, 2012).

summarised (Tidblad *et al.*, 2012) and the following is a brief summary of the findings.

The aim of the trend exposures of ICP Materials is to elucidate the environmental effects of pollutant reductions achieved under the Convention on Long-Range Transboundary Air Pollution and identify extraordinary environmental changes that result in unpredicted materials damage. The trend exposure consisted of repeated 1-year exposures of steel and zinc and the last trend exposure was performed in 2011/2012 and a new trend exposure is expected to start in 2014.

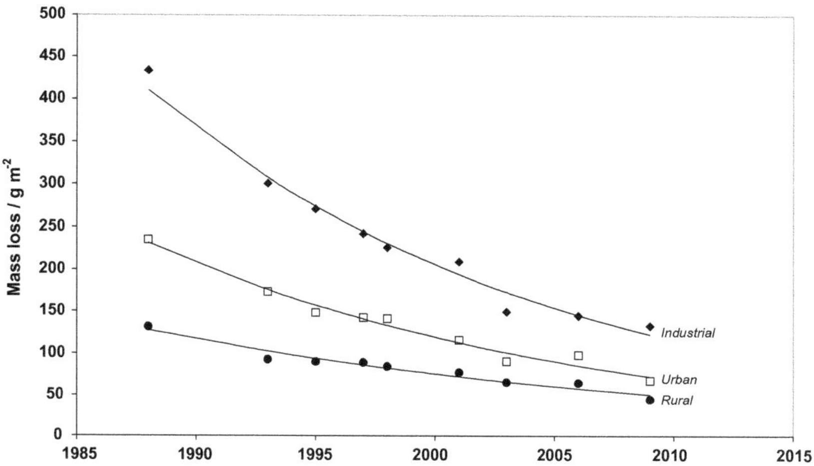

Fig. 6.11 Average mass loss of unalloyed carbon steel for different categories of test sites (industrial, urban and rural) (with permission from Tidblad *et al.*, 2012).

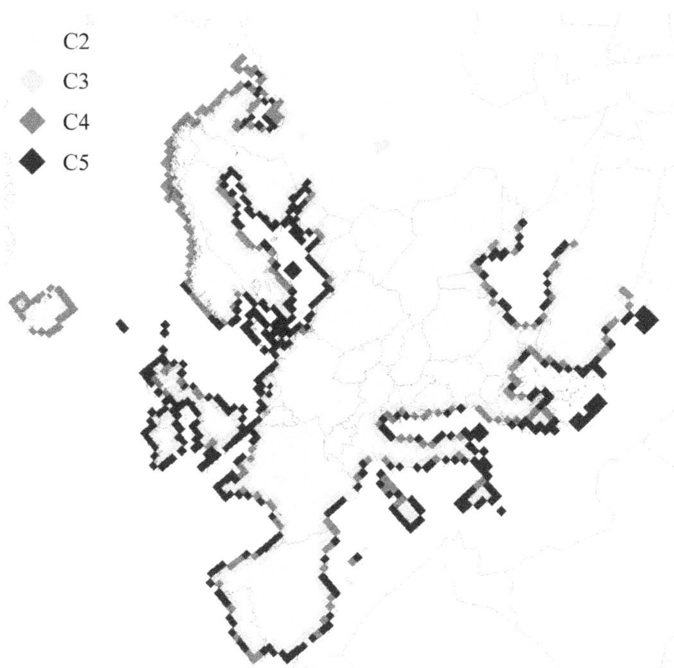

Fig. 6.12 Maximum corrosivity category C2-C5 (2070–2099) estimated from the ISO functions for carbon steel and zinc and HadRM3 temperature, relative humidity, SO_2 and chloride deposition data. (with permission from Tidblad, 2012).

Figure 6.10 shows an overview of pollutant trends during the period 1987 to 2008. In order to present the results on a common scale, all values are compared to the initial value, corresponding to the 1987/88 period (100%). SO_2 has the strongest trend and NO_2 follows. For O_3 no decreasing trends were observed. The decreasing trend in the concentration of acidifying air pollutants has resulted in decreasing corrosion rates of the exposed materials but after the turn of the century, the decrease is limited. Both carbon steel and zinc show decrease in corrosion rate in unsheltered as well as in sheltered positions. The decrease is evident for areas that were initially polluted as well as for areas with initially relatively low pollution (rural areas), as is illustrated in Fig. 6.11.

In addition, the effect of climate change deserves to be mentioned since it has a potentially large impact on possible future trends in corrosion. Metals were one of many materials under investigation in the NOAH's ARK project on effects on climate change on cultural heritage. Figure 6.12 shows a map for the far future (2070–2099) where the possible effect of climate change have been taken into account. The map is based on the recently revised ISO 9223:2012 categories C1-C5 of zinc and carbon steel. Chloride deposition in costal and near-coastal areas dominates the picture and in coastal areas of southern Europe, corrosion is expected to be higher than today (Tidblad, 2012).

References

Arroyave, C. and Morcillo, M. (1995). The effect of nitrogen oxides in atmospheric corrosion of metals, *Corros. Sci.*, 37, pp. 293–305.

Graedel, T.E., Nassau, K. and Franey, P. (1987). Copper patinas formed in the atmosphere I. Introduction, *Corros. Sci.*, 27, pp. 639–657.

Graedel, T.E. (1989a). Corrosion mechanisms for zinc exposed to the atmosphere, *J. Electrochem. Soc.*, 136, pp. 193C–203C.

Graedel, T.E. (1989b). Corrosion mechanisms for aluminum exposed to the atmosphere, *J. Electrochem. Soc.*, 136, pp. 204C–212C.

Graedel, T.E. and Frankenthal, R.P. (1990). Corrosion mechanisms for iron and low alloy steels exposed to the atmosphere, *J. Electrochem. Soc.*, 137, pp. 2385–2394.

He, W. (2000). *Corrosion Rates and Runoff Rates of Copper and Zinc as Roofing Materials*, Licentiate Thesis, Royal Institute of Technology, Stockholm, Sweden.

Kucera, V. and Mattson, E. (1987). 'Atmospheric corrosion', in F. Mansfeld (Ed.), *Corrosion Mechanisms*, Marcel Dekker, New York, pp. 211–284.

Kucera, V., Henriksen, J., Knotkova, D. and Sjöström, C. (1993) Model for calculations of corrosion cost caused by air pollution and its application in three cities, *10th Eur. Corros. Congr.*, Barcelona, Paper No. 084, p. 12.

Kucera, V. and Fitz, S. (1995). Direct and indirect air pollution effects on materials including cultural monuments, *Water Air Soil Pollut.*, 85, pp. 153–165.

Kucera, V., Tidblad, J., Leygraf, C., Henriksen, J., Kreislova, K., Ashall, G. and Stöckle, B. (1996). Dose-response relations from the UN ECE project as a tool for air pollution abatement strategies, *Proc. 13th Int. Corros. Congr.*, Melbourne, Australia.

Kucera, V. *et al.* (+13 coauthors) (2007). UN ECE ICP materials dose-response functions for the multi-pollutant situation, *Water Air Soil Pollut.*: *Focus*, 7, 249–258.

Leygraf, C. and Graedel, T.E. (2000). *Atmospheric Corrosion*, John Wiley & Sons, Inc., New York.

Mikhailov, A.A., Tidblad, J. and Kucera, V. (2004). The classification system of ISO 9223 standard and the dose-response functions assessing the corrosivity of outdoor atmospheres, *Protection of Metals*, 40, pp. 541–550.

Odnevall, I. (1994). *Atmospheric Corrosion of Field Exposed Zinc*, Thesis, Royal Institute of Technology, Stockholm, Sweden.

Samie, F., Tidblad, J., Kucera, V. and Leygraf, C. (2007). Atmospheric corrosion effects of HNO_3. A comparison of laboratory and field exposed copper, zinc and carbon steel, *J. Electrochem. Soc.*, 154, pp. C249–C254.

Strandberg, H. (1997). *Perspectives on Bronze Sculpture Conservation. Modeling Copper and Bronze Corrosion*, Thesis, Göteborg University, Sweden.

Tidblad, J. and Kucera, V. (1996). *The Role of NO_x and O_3 in the Corrosion and Degradation of Materials*, Swedish Corrosion Institute, Stockholm, Sweden.

Tidblad, J., Kucera, V., Mikhailov, A.A., Henriksen, J., Kreislova, K., Yates, T., Stöckle, B. and Schreiner, M. (2001). UN ECE ICP Materials. Dose-response functions on dry and wet acid deposition effects after 8 years of exposure, *Water Air Soil Pollut.*, 130, pp. 1457–1462.

Tidblad, J. *et al.* (+11 coauthors) (2007). Exposure programme on atmospheric corrosion effects of acidifying pollutants in tropical and subtropical climates, *Water Air Soil Pollut.*: Focus, 7, pp. 241–247.

Tidblad, J. *et al.* (+22 coauthors) (2012). Effects of air pollution on materials and cultural heritage: ICP Materials celebrates 25 Years of research, *International Journal of Corrosion*, 2012, Article ID 496321, 16 pages.

Tidblad, J. (2012). Atmospheric corrosion of metals in 2010–2039 and 2070–2099, *Atmos. Environ.*, 55, pp. 1–6.

UN ECE Workshop (1997). 'Economic evaluation of air pollution damage to materials' in Kucera, V., Pearce, D. and Brodin, Y.W. (Eds.), *Proceedings of the UN ECE*

Workshop on Economic Evaluation of Air Pollution Abatement and Damage to Buildings Including Cultural Heritage, Report 4761, Swedish Environmental Protection Agency, Stockholm, Sweden.

Watt, J., Tidblad, J., Kucera, V., Hamilton, R. (Eds.) (2009). *The Effects of Air Pollution on Cultural Heritage,* Springer.

CHAPTER 7

Air Pollution Damage to Glass

Monica de Bardi and Manfred Schreiner

*Institute of Science and Technology in Art,
Academy of Fine Arts, Vienna, Austria*

7.1 Introduction

When we think about glass our thought goes most of the time to a stable material with high transparency, brittle and widely used in our daily life for windows, hollow glasses and many other objects. Glass is a material used for disparate applications in different fields because of its good formability and high performances. It is a material whose properties are changed by composition and production technology and it is therefore subject of systematic evaluation since some hundred years ago.

A wide variety of glass is nowadays produced both, organic and inorganic; in this chapter the particular case of silicate glass is discussed, being the most common type of glass nowadays but also since 3500–4000 years ago (Tite *et al.*, 2002). The first man-made glass artefacts were core-formed, melting a mixture of sand (SiO_2), natron (Na_2CO_3) and lime ($CaCO_3$) (Frank, 1982). The objects were of poor quality and small size, but from the 2nd to 1st century BC, the technique of glassblowing was introduced. This technological change made the production of glass easier and it became a more common and affordable material used also for domestic purposes. Glass production technology evolved until the 5th century AD when, after the decline of the West Roman Empire, the know-how was lost especially in the area south of the Alps (Macfarlane and Martin, 2002). Only afterwards between the 10th and 15th century the production of glass resulted in masterpieces such as the medieval stained glass of the cathedrals in Chartres or St. Denis in France and Canterbury in England.

While modern glass is considered a stable material (e.g. used to contain acids or to stabilise nuclear wastes), historical glass and especially medieval glass suffers from atmospheric pollution, predominantly in the last centuries, due to the increase of the acidifying gases and airborne particulate matter in the ambient atmosphere. It is therefore important to understand the role of different parameters on the chemical and physical properties of glass and it could suggest the best conditions for conservation and eventually a proper restoration treatment (Morey and Merwin, 1932).

7.1.1 *What is glass? A chemical point of view*

Common silicate glass is formed from silica and other alkali and earth-alkali metal oxides, heated to the melting temperature region to form a liquid moldable into the required shape. Glass can be defined as an amorphous (and therefore non-crystalline but solid) material with no long range order. Zachariasen (1932) suggested a model for pure silicate glass based on a network composed by the repetition of units of tetrahedra formed by a silicon atom at the center and four oxygen atoms located in each corner (SiO_4, Fig. 7.1); the tetrahedra are connected to each other forming a three-dimensional structure sharing oxygen atoms which are therefore bridging two tetrahedra. This theory was confirmed several years later with the possibility of measuring interatomic distances and to show the presence of rings in the structure (Poggemann *et al.*, 2003; Frischat *et al.*, 2004; Raberg *et al.*, 2005).

Pure silica is characterised by a high melting temperature of 1720°C circa, which was quite difficult to reach in ancient times. To lower this temperature and to obtain a better workable liquid, some network modifiers (cations) are added to act as fluxes and

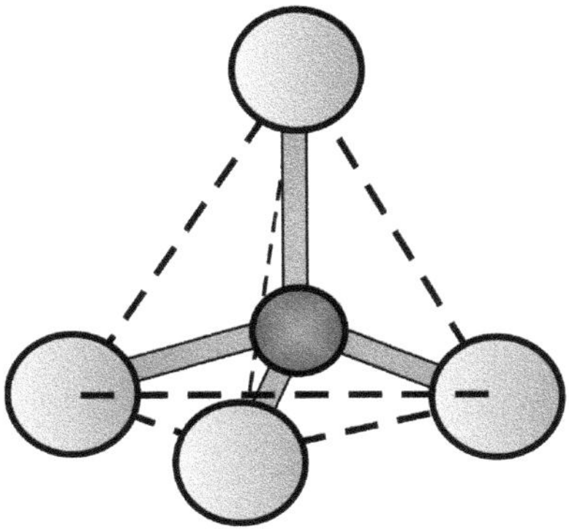

Fig. 7.1 Silica tetrahedron.

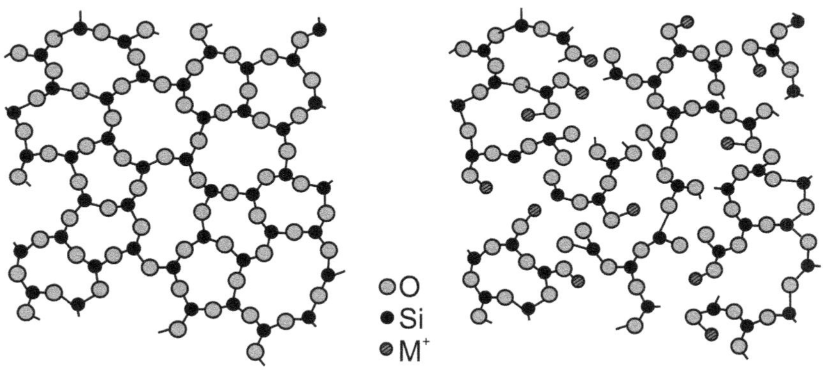

Fig. 7.2 Schematic structure of pure silica glass (left) and after the addition of network modifiers (right).

stabilisers. As a result, the softening and melting temperature region can be reached with less complex heating technologies and the melt becomes less viscous; therefore, the production of glass became easier. The addition of network modifiers leads to the formation of non-bridging oxygen atoms with negative charges compensated by the ionic bond to network modifier cations. The silica network is broken up and the properties change compared to pure silica glass (Fig. 7.2).

The network modifiers can be monovalent cations such as sodium (Na^{+}) or potassium (K^{+}); these alkali-metal ions bear only a single charge and therefore form a terminal point in the structure and consequently reduce the viscosity of the batch. The addition of such ions is responsible for the low chemical durability of glass as explained in the following section.

The network modifiers can also be divalent cations such as calcium (Ca^{2+}) and magnesium (Mg^{2+}); these alkaline-earth ions bear a double charge and can therefore bridge two non-bridging oxygen atoms and are much more tightly bound compared to the monovalent cations. They prevent the crystallisation during the cooling of the batch and at the same time increase the chemical stability.

Other ions can be added to the glass (less than 1%) in order to modify the colours, the transparency, optical and physical properties such as resistance to high temperature, hardness, and brittleness.

7.2 Mechanisms of Deterioration and Corrosion

Nowadays a considerable amount of ancient glass is part of museum collections. This might be finds from excavations and have therefore a past of burial environments, but might also be part of historical collections or part of monument decoration (window panels etc.) and have never been buried. In this chapter, attention will be focused on glass, which was subject of weathering and leaching due to atmospheric phenomena.

Different kinds of damage are usually visible on glass objects. The most common alterations are dulling (loss of the original clarity and transparency) and iridescence (colouration due to light interference resulting in vivid colours). Moreover, the formation of corrosion products can interfere, resulting in cracking, flacking and pits formation (Frenzel, 1985).

The reasons leading to damage of glass objects are distinguishable as mechanical and chemical. Due to a poor manufacturing or to impacts with harder materials the glass breaks or is scratched; thermal shocks and vibration are also responsible for damage. Furthermore, a particle deposition of various materials of the ambient atmosphere (dust, candle soot etc.) occurs on the outside as well as the inside surfaces of the stained glass panels, which can — in combination with condensation of water — initialise chemical reactions affecting not only the glass surface but also the bulk glass.

7.2.1 *Leaching of glass*

As explained in Chapter 7.1.1, the network modifier cations present in the glass are only weakly bound to oxygen atoms. When water is present on the surface and the pH is acidic (pH <7), they are extracted from the glass to the solution and — in order to maintain the electrical neutrality in the glass — the cations are replaced by H^+ or hydrogen bearing species such as H_3O^+ (Eq. 7.1).

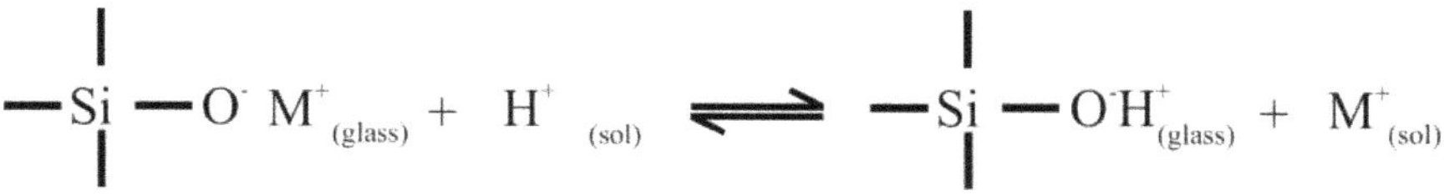

(7.1)

As a result the surface is leached of network modifier cations and a hydrated silica rich layer is formed (Ernsberger, 1977; Clark *et al.*, 1979; Clark and Yen-Bower, 1980; Hench *et al.*, 1980; Newton, 1985).

Several studies (Sanders and Hench, 1973; Hench, 1975; Clark *et al.*, 1979) have shown that the rate of extraction is proportional to the square root of time and after a certain thickness, it becomes slower and linearly dependent on time, as shown by the following equation (Eq. 7.2)

$$Q = a \cdot \sqrt{t} + b \cdot t \tag{7.2}$$

where Q is the amount of leached ions, t the leaching time and a and b are model parameters.

In contrast to that leaching process, when glass is in contact with an alkaline aqueous environment (pH >7 or mainly above 9), the glass network is attacked at the Si-O bonds. The silicate network breaks down and the glass can therefore dissolve (Eq. 7.3).

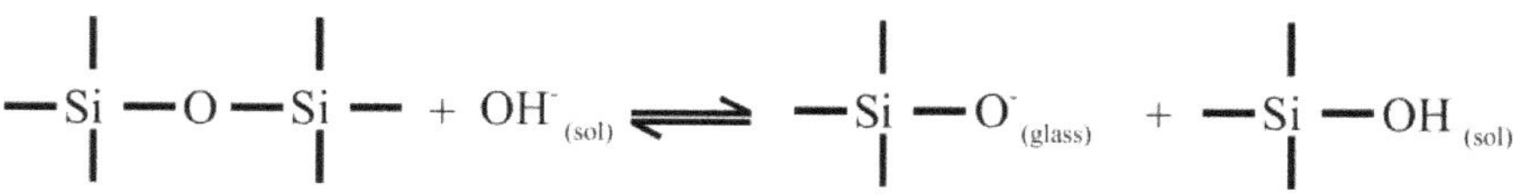

(7.3)

7.2.2 *Weathering of glass*

The weathering of glass is due to its interaction with the ambient atmosphere including relative humidity, acidifying gases and airborne particulate matter. It involves several simultaneous steps and it is therefore complex to study. Even when the glass is mounted in vertical position a thin layer of water is present on its surface because of the air moisture. This allows the ion exchange as explained previously. Additionally, acidifying gases (e.g. CO_2, NO_x, SO_2 and O_3) may dissolve in the thin water film, increasing the acidity of the solution and accelerating the ion diffusion of the leaching processes.

Two types of weathering are distinguished (Clark *et al.*, 1979): One includes a sufficient quantity of moisture to form visible droplets which might lead to a runoff. The other type involves a thin condensation layer with no visible droplets formation. The two types can occur simultaneously on the same object and their alteration in dry conditions can further increase the corrosion rate. A thick layer (up to some mm) of weathering products may be formed on the surface because of the precipitation of crystalline compounds, yielding a loss of transparency. The crusts may either have a soft and powdery consistency or be very hard and adhere to the glass. The formation of pits may also occur and they are probably related to the size and distribution of the water droplets on the surface. Micro-defects or fissures are also thought to be preferential sites for the formation of pits (Schreiner, 2004).

7.2.3 *Parameters involved in the corrosion processes*

The reason for the formation of cracks on the surface is that the atoms replaced during the leaching process have a volume bigger than the H^+ that substitute them. Moreover, the H^+ or H_3O^+ ions are more strongly bound to oxygen and the new leached layer is subject to shrinking and cracking when the water evaporates (Ernsberger, 1980 and 1986).

The atmospheric environment plays a very important role in the corrosion of glass. Two main cases of studies can be distinguished: indoor and outdoor.

Concerning the outdoor exposure of glass, several studies have focused on the effect of SO_2 and CO_2, two acidifying gases present as pollutants in the ambient atmosphere. It was found that these compounds do not attack the glass directly, but dissolve in the superficial water layer or in the rain water producing hydroxides and changing the pH. In particular, the hydroxides lead to the production of carbonates or sulfates. Even where carbonates are formed, they are converted into sulfates (Newton, 1979 and 1987). Therefore, mainly gypsum ($CaSO_4{\cdot}2H_2O$) or syngenite ($K_2SO_4{\cdot}CaSO_4{\cdot}H_2O$) or arcanite ($K_2SO_4$) can be identified as crystalline weathering products on the surface of e.g. medieval stained glass.

Because of the leaching process and the substitution of network modifier cations by hydrogen-bearing species from the aqueous solution, hydroxyl ions (OH^-) are formed. These ions migrate to the solution and, if it is not renewed, an accumulation of hydroxyl and alkali ions will increase the pH. An important parameter to consider is therefore the ratio between the surface area of the glass and the volume of the aqueous solution involved in the process (SA/V) (Ethridge *et al.*, 1979). When this value increases, the conditions for an increase of pH takes place and the breakdown of the silicate network and dissolving of silica increases. The two mechanisms (leaching and dissolving) are in competition.

In addition to the pH of the thin water film on the glass surface, many parameters are involved in the weathering processes such as the glass composition, temperature, surface treatment, orientation, and affinity of the glass surface for water. A very critical parameter is the presence of condensation/drying cycles, which might lead to more degradation than a constantly high level of humidity (Simpson, 1953). Moreover, the water film may evaporate, leading to the precipitation of weathering products on the glass surface.

7.2.4 *Other mechanisms of corrosion*

It has to be kept in mind that in addition to the corrosion/weathering mechanisms described, there are other various processes that can induce changes of the glass. One of them is the bio-attack which should also be considered especially in situations with high temperature and humidity conditions. It is due to the presence of lichens, fungi, bacteria, yeasts, and algae which require only a support substratum to grow (Tennent, 1981; Staudigel *et al.*, 1995; Gorbushina and Palinska, 1999; Gallien *et al.*, 2001); the attack is not due to the need to extract nutrients from the glass (these are provided by photosynthesis): Glass itself cannot be used as an organic nutrient but it can be a source of minerals and trace elements. These organisms trap moisture and water which contain CO_2 due to their respiration, introducing therefore a micro climate where the corrosion may start as explained in the previous sections.

Another mechanism is the so-called solarisation, which consists of the change in the colour to purple after a long exposure to sunlight due to an interaction between iron and manganese oxides present in some glass of the medieval and also of the nineteenth century (Brill, 1963).

7.3 Analytical Approach

There is a wide variety of modern surface analytical techniques that are useful in gaining information from glass artefacts. The non-invasive techniques are preferred in case of studies on cultural heritage, but micro invasive techniques are also well accepted for the capability to get precious information from a very small sample. In the last decades many studies were performed in order to determine the chemical composition of the glass and interest was given mainly by the need of high performance of modern glass under extreme temperature and pressure (Bridgman and Simon, 1953; Arndt and Stöffler, 1969). Nevertheless, several studies were carried out on glass of historical interest in order to understand the corrosion/weathering processes occurring on glass and the factors involved and to be able to predict the chemical durability of an object or suggest proper conservation treatments.

The availability of synthetic glass with a chemical composition similar to historical glasses gives the possibility to perform accelerated artificial ageing using extreme conditions to speed up the corrosion mechanisms. However, it has to be mentioned that between these results and natural degradation processes a gap may exist, as mainly just two or three component glasses (soda-silica or soda-lime-silica) were used in experimental studies. These may be far away from the real chemical composition of historical glasses made by an empirical knowledge, with an abundance of impurities due to the raw materials. Moreover, the experiments are performed in controlled and laboratory conditions which cannot properly simulate or exactly correlate with the complex system of reactions occurring on glass in natural environments and for long periods. Nevertheless, these experiments are useful to explain the natural processes which might be expected, when glass is exposed to burial, weathering and display conditions.

7.3.1 *Information achievable*

Different techniques are used in fields such as archaeology or art history. It is possible to get information about approximate date of manufacture, provenience and therefore evidence of trading, manufacturing techniques and authenticity. In some cases it is even possible to get information about the use and eventually about the chemical composition of an object such as vases. For conservators and restorers, chemical analyses are useful to understand the reasons for the degradation and to suggest conservation and restoration treatments. In the following sections, some analytical techniques are briefly explained.

7.3.2 *Analytical techniques*

A preliminary visual examination of the glass object is usual, using visible light that is almost parallel to the glass surface to enhance the surface irregularities. Moreover, transmitted light is also applied to determine the translucent or opaque quality and the presence of inhomogeneity or bubbles.

7.3.2.1 *XRF (X-Ray Fluorescence analysis)*

As a first step, the chemical composition, especially the silica content and the amount of network modifiers such as Na, K or Ca, is important for the corrosion/weathering stability, and also for the temporal or regional authenticity (Hall *et al.*, 1964). XRF has been applied widely, as it is non-destructive and in most cases also non-invasive and can be carried out in air. The availability of portable instruments makes it perfect for in-situ investigations in museums or archaeological excavation sites (Hall, 1960; Hall *et al.*, 1964). As a drawback, XRF has a low sensitivity for elements with low atomic numbers such as Na, Al and Mg, especially in air, all of interest in studying glass. Nevertheless, it is frequently used in art and archaeology e.g. applied in the study of coins, pigments and paint layers (Mantler and Schreiner, 2000).

XRF is based on the ionisation of an atom as a consequence of a primary x-ray radiation: It causes the ionisation of an inner shell, creating a vacancy which has to be overfilled to return to a stable state. Due to a relaxation process, the ion emits secondary x-ray photons or Auger electrons (according to certain quantum mechanical rules) which are characteristic for certain elements. X-rays are detected using WD (wavelength dispersive) or ED (energy dispersive) detectors. The latter is more commonly applied because of the shorter measurement time, but with lower resolution and detection limits compare to WD systems. XRF gives qualitative information due to the fact that the x-ray emission is characteristic for the target elements. Quantitative estimations are possible using standard materials and the application of numerical procedures.

7.3.2.2 *Infrared and Raman spectroscopy*

Infrared and Raman spectroscopy give information about structure and bonding of functional groups (Griffiths and De Haseth, 2007). These techniques are also non-invasive and can be performed in-situ with portable devices in a rapid and inexpensive way.

Infrared spectroscopy was often applied in the past for the analysis of glass (Sweet and White, 1969; Doremus, 1980). Beside the first experiments performed on crushed glass and in transmission mode (Ferris and Pederson, 1988), FTIR (Fourier Transform Infrared) spectroscopy is mostly used in the reflection mode without the need for sampling. The potentials of this technique were well described and widely applied to studies on glass corrosion (Sanders *et al.*, 1974; Hench *et al.*, 1979; Doremus, 1994). Infrared spectroscopy is very useful to monitor the Si-O-Si (1100–1020 cm^{-1}) and Si-O^-X^+ (950–920 cm^{-1}) vibration bands of glass and therefore the amount of bridging and non-bridging bonds, whereas Raman spectroscopy delivers complementary results based on the inelastic scattering of light at molecules. This gives information about the connectivity of the glass network excited by an impinging laser (Colomban, 2003; Colomban *et al.*, 2006; Robinet *et al.*, 2006).

In both techniques, changes in intensity are related to changes in composition and it is therefore possible to monitor the leaching processes with them (Smith and Clark, 2004; De Bardi, Wiesinger *et al.*, 2013).

7.3.2.3 *EPXMA (Electron Probe X-ray Micro Analysis)*

EPXMA has become an indispensable tool for the characterisation of the surface morphology and the chemical changes of weathered glass as well as the identification of weathering products (Ruska, 1986). In general, a finely focused electron beam scans the sample, causing emissions via both, elastic and inelastic processes. Secondary electrons (SE) can be detected and topographical and morphological images of the surface obtained with high lateral resolution (up to approx. 3 nm). Backscattered electrons (BE) give information about the relative difference in atomic weight and therefore show an image contrast between areas with different compositions. X-ray photons can also be produced and detected (similar to XRF) so that the elemental composition of the selected domain (or the scanned area) can be determined. Due to the high absorption of electrons in air, the measurements have to be performed in vacuum (approximately 10^{-5} mbar), limiting the application to samples with a size fitting into the vacuum chamber and for non-vacuum-sensitive materials (with regard to outgassing, water content etc). The latter problem was nevertheless solved in the 1980s with the introduction of ESEM (Environmental Scanning Electron Microscope), allowing the measurement of samples in a H_2O-vapor atmosphere in the vicinity of the sample yet keeping the microscope column under high vacuum conditions.

The technique combines the possibility of getting simultaneous imaging and x-ray analysis of micro-domains or scanned areas. Its application in materials science and also on the study of cultural heritage objects was therefore clear since the beginnings (Hornblower, 1962 and 1963).

7.3.2.4 *AFM (Atomic Force Microscopy)*

The atomic force microscope enables imaging of the glass surface down to the atomic scale. A sharp tip mounted on a soft lever is scanned across the sample and its deviation due to the sample topography is measured by means of laser beam deflection (Frommer, 1992; Arribart and Abriou, 2000). AFM can operate in different modes depending on the sample, and insulators can also be imaged without any treatment. It does not require vacuum conditions, enabling analyses under controlled atmospheric conditions (different amounts of relative humidity or acidifying gases) and also under liquids; it is therefore especially useful for the studies of corrosion from the early stages (Schmitz *et al.*, 1995).

7.3.2.5 *ToF-SIMS (Time of Flight Secondary Ion Mass Spectrometry)*

ToF-SIMS is an analytical technique that enables the acquisition of mass spectra, detecting simultaneously all the atomic masses including light elements such as H, which are hard to detect with other techniques. SIMS is based on the detection of secondary ions coming from the effect of a highly energetic primary ion beam impinging on the sample surface. Ions are accelerated into a mass spectrometer (time of flight). The high sensitivity allows isotope studies and the detection of trace elements. Moreover, it can operate with a high lateral resolution (up to 100 nm) and, because of the sputtering possibility, depth profiles of every element (e.g. H, K, Na etc.) can be acquired (Dowsett and Adriaens, 2004; Lenzen *et al.*, 2005; De Bardi, Hutter *et al.*, 2013). It is therefore possible to gain information about the in-depth distribution of elements layer by layer and to monitor the ion exchanges through the glass due to the corrosion mechanisms previously described.

7.3.2.6 *Ion beam and other techniques*

Beside SIMS, NRA (Nuclear Reaction Analysis) has been applied to detect H in glass and for quantitative profiles (Lanford *et al.*, 1976). An incident proton penetrates the nucleus of the sample atoms, which decay rapidly, emitting specific gamma photons or decay products which are characteristic of the atoms involved. In the case of glass, the reaction considered is $^{1}H(^{15}N,\alpha\gamma)^{12}C$. The advantages of NRA are the possibility of quantification and the absence of matrix effect due to the re-deposition of hydrogen bearing molecules from the residual gas in the chamber (as in SIMS measurements). Measurements show that H is present up to 40 at% in the leached layer (Schreiner *et al.*, 1988).

LIBS (Laser Induced Breakdown Spectroscopy) is another technique for the analysis of glass, based on the detection of light emission induced by a laser ablation (ng-µg). It is used for the capability to determine the composition of glass (quantitative analysis) (Carmona *et al.*, 2007) and of the corrosion products (Carmona *et al.*, 2005).

Other techniques such as PIXE (Particle Induced X-ray Emission) are similar to XRF, but the ionisation of an inner shell is due to an incident particle (protons of 2–3 MeV); the x-ray photon emitted is detected similar to XRF. Complementary information (especially concerning light elements) are given by PIGE (Particle Induced Gamma-ray Emission), a technique based on ion-gamma reactions, which are not attenuated by the sample, and the signal originates from a much higher depth (Jembrih *et al.*, 2001; Calligaro, 2008).

RBS (Rutherford Backscattering Spectroscopy) is based on the elastic scattering of incident ions (usually He^{2+}) which are backscattered. It allows the identification of atomic species. The same principle is used for ERDA (Elastic Recoil Detection Analysis) but the incident ions are heavier than the target so that the lighter ions recoil in the forward direction and are registered (Janssens and Van Grieken, 2004; Janssens, 2013). Other techniques such as XANES (Quartieri *et al.*, 2005; Figueiredo *et al.*, 2006) and methods for the measurement of the natural radioactivity of glass have been also applied in a few cases (Sartowska and Kunicki-Goldfinger, 2001).

Careful evaluation of the results is important especially when a large amount of data is available, to understand how the data are related. For these reasons, suitable multivariate statistical procedures have been applied, such as PCA (Principal Component Analysis), FA (Factor Analysis), HCA (Hierarchical Cluster Analysis) or MLR (Multiple Linear Regression) (Melcher and Schreiner, 2004 and 2005).

7.4 Potash-lime-silica Glass, a Special Case of Study

Most of the studies performed on glass weathering focus on a particular kind of composition typical for the medieval times. Great examples are several cathedrals in Europe embellished with magnificent stained glass between the 10th to the 15th century, such as Chartres or St. Denis in France or Canterbury in England. The coloured glass panels were kept together by lead structures to form luminous pictures with figures depicting biblical scenes. During this period, due to the large number of cathedrals and churches under construction, the demand for glass increased considerably. This request needed to be supported by an adequate supply of fuel and raw materials, which was not always possible due to the change and the interruption of trade, especially between Egypt, where the most important raw material natron came from (Wadi el-Natrun), and south Europe and Europe north of the Alps. This probably also caused the transfer of glassmaking sites from the coast to regions with abundance of wood. The glass that was produced north of the Alps thus shows a chemical composition different from ancient or modern glasses, because of the use of beechwood ash instead of natron as a flux, commonly used for the stable Mediterranean soda-lime-silica glasses. This substitution causes a very low chemical stability to environmental conditions, leading to a fast degradation of the glass and the formation of corrosion products on the surfaces of medieval stained glass windows.

Potash-lime-silica glass has a lower durability compared to soda-lime-silica glass (with the same Si content) (Rana and Douglas, 1961) probably because of the different diameters of K and Na. When potassium atoms are leached into the solution, they leave a more open structure in the network of glass and hence more water molecules can interact with the glass compared to soda glass.

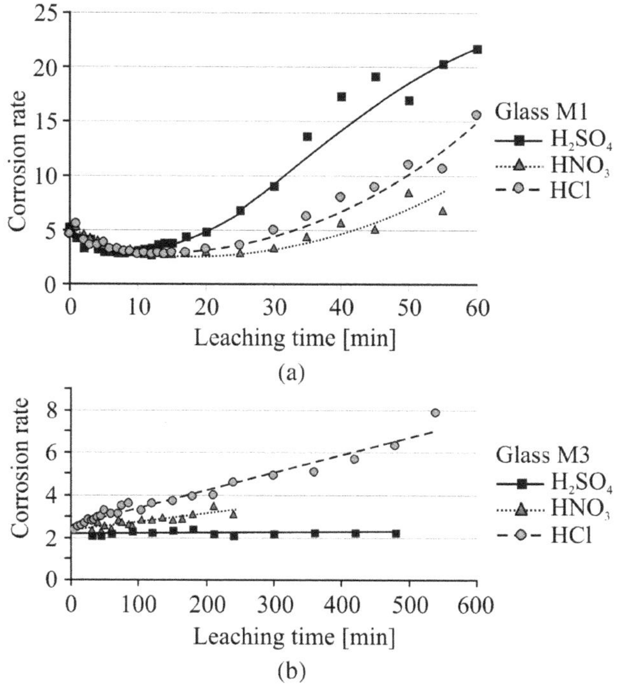

Fig. 7.3 Corrosion rates for two types of glasses in different acidic solutions; taken from De Bardi *et al.*, 2013.

Another typical characteristic of medieval glass is the low silica content. It was demonstrated that below 60 mol% of silica, the glass is extremely unstable because of the fact that there is always a network modifier in between every Si-O group, leading to a high mobility in the glass network. Figure 7.3 presents the results of leaching experiments on potash-lime-silica glasses with chemical compositions similar to medieval stained glass (**M1**: SiO_2 48.0%, K_2O 25.5%, CaO 15.0%, MgO 3.05%, Na_2O 3.0%, Al_2O_3 1.5%, P_2O_5 4.0%; **M3**: SiO_2 60.0%, K_2O 15.0%, CaO 25.0% wt) with different types of acids, but keeping the pH constant in the solutions. It is obvious that the corrosion rate (related to the leached layer thickness) for every acid is different for glasses with different compositions. Sulfuric acid is more corrosive than nitric and hydrochloric acids for glass M1, but for glass M3 the most corrosive is hydrochloric acid (De Bardi, Wiesinger *et al.*, 2013).

Fig. 7.4 SE image of glass M1 leached for 60 min in H_2SO_4, pH = 3; taken from De Bardi, Hutter *et al.*, 2013.

It is therefore important to know the chemical composition of the glass exposed and the types of acids present in the atmosphere (e.g. that may come from acid rain) to identify for example the part of stained panels which are more sensitive to weathering. A glass with composition similar to M1 can develop a leached layer of thickness up to 0.5 μm after 30 min of exposure to a solution with pH=3 (De Bardi, Hutter *et al.*, 2013). The leached layer may be subject of a decrease of volume due to condensation processes and cracks may form. The SE image (Fig. 7.4) shows the formation of a brittle and flaky layer. The EPXMA revealed the typical composition of a leached layer in the micro-domain A (enriched in Si) and the composition of the bulk in area B, where the layer is removed exposing a fresh glass surface to a corrosive environment (i.e. the leached layer is not protective).

In Fig. 7.5 it is possible to notice that the interface between the leached layer and the non-corroded bulk glass is quite sharp and, considering the superficial roughness of the sample and using delicate sputtering conditions, it is possible to achieve information every 4–5 nm of the sample. The figure depicts the depth profile of a potash-lime-silica glass (M1) leached for 1h in HNO_3 (10^{-1}M). The leaching (depletion) of K and Ca in the surface domain of the glass is clearly visible. It is also possible to notice a preferential leaching of

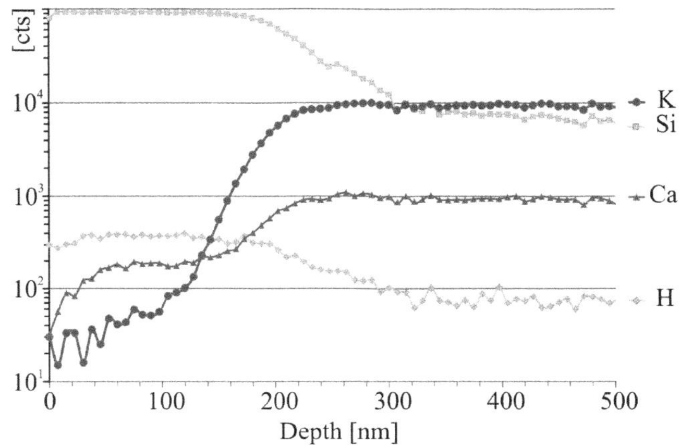

Fig. 7.5 ToF-SIMS depth profile of the potash-lime-silica glass M1 leached for 1h in HNO_3 (10^{-1}M).

the monovalent ions (K and Na) compared to the divalent ions (Ca and Mg) probably due to their double charges. Moreover, the enrichment of H in the leached layer proved also with NRA analyses is evident (Schreiner *et al.*, 1988).

Concerning the weathering, it was proven that the presence of water in the ambient atmosphere is essential for the development of the corrosion processes. Results obtained by in-situ AFM studies on fresh cleaved potash-lime-silica glass samples revealed that under dry conditions no changes can be observed (Fig. 7.6a), whereas in nitrogen with specific amount of humidity the formation of round features growing with time occurred (Fig. 7.6b). Already after 10 min of weathering in N_2 with 70% RH the features had a diameter between 100 and 200 nm (Schmitz *et al.*, 1997).

EPXMA on samples with medieval composition weathered under natural conditions — in the frame of the UN/ECE project, ICP-Materials (Woisetschläger *et al.*, 2000; Melcher and Schreiner, 2004, 2005 and 2006) — showed the formation of corrosion products consisting of sulfates, mainly syngenite ($K_2SO_4 \cdot CaSO_4 \cdot H_2O$), gypsum ($CaSO_4 \cdot 2H_2O$) and arcanite ($K_2SO_4$) (Fig. 7.7). In general, the formation of thick crusts occurred especially on glass with a very low content of silica, whereas in glass with a higher amount of SiO_2, the phenomenon of pitting (50–150 μm in diameter and depth) is more

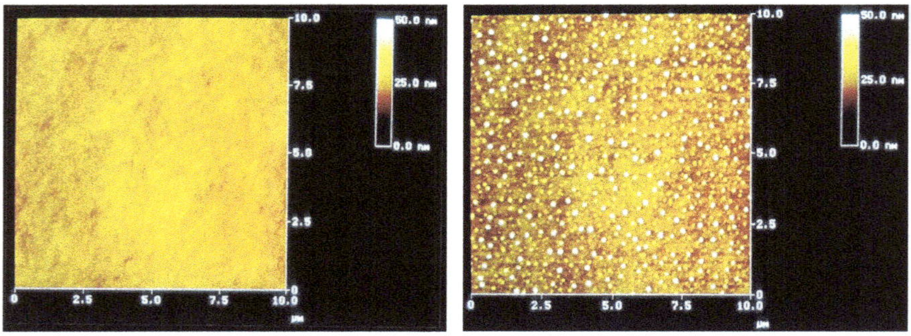

(a) (b)

Fig. 7.6 AFM images of a potash-lime-silica glass under dry nitrogen (a) and after 10 min of exposure to humid N_2 (70% RH) (b); taken from Schreiner, 2004.

frequent. Moreover, cracks can also occur and weathering products are usually located in the cavities (Perez y Jorba *et al.*, 1980; Cox and Gillies, 1986; Schreiner, 1988).

Within the UN/ECE project a correlation of the degree of weathering and the environmental parameters measured at the more than 50 exposure sites was carried out by using statistical methods (MLR — multiple linear regression) (Melcher and Schreiner, 2005 and 2006). The most significant factors influencing the weathering of potash-lime-silica glasses with medieval composition are relative humidity, temperature, SO_2 and NO_x of the ambient atmosphere. In particular, NO_x and temperature have shown a much higher influence on the weathering of glass compared to the other factors (Henriksen *et al.*, 2004).

Airborne particulate matter (APM) also has an important influence on the degradation of glass (Pio *et al.*, 1998). It has been found that the degree of glass deterioration is much higher on samples treated with dust that leads to the formation of sulfates even if the samples were not exposed to SO_2. A much larger effect was found with relative humidity above 70%.

7.5 Reducing Damage

To prevent a further degradation of medieval stained glass objects and stabilise their present conditions, many treatments have been

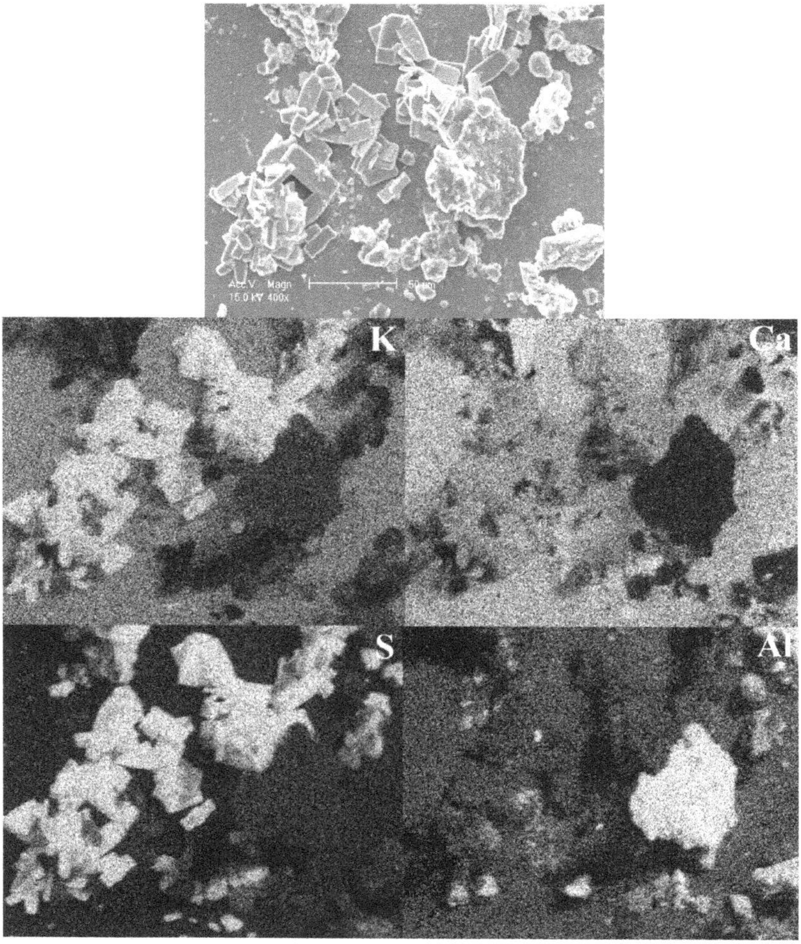

Fig. 7.7 SE image of a weathered surface of glass M3 (top) and X-ray mappings of K, Ca, S, and Al; taken from Schreiner, 2004.

proposed and applied, especially to outdoor objects exposed to the ambient atmosphere. The control of environmental conditions is most important (e.g. relative humidity, temperature, and pollutants). Furthermore, conservation treatments of the glass surfaces have been proposed, which should increase the chemical and mechanical resistance of the glass, maintaining the glass transparency and not altering the appearance of the object. At the same time such treatments have to be reversible.

7.5.1 *Passive conservation*

Protective glazing can control the environmental conditions of an outdoor object and conserve stained glass in situ and to protect it from mechanical and chemical damages. It is a non-invasive method, which provides protection against external pollution and vandalism without interfering with the medieval glass. It could be shown that the glazing system is responsible for the reduction of the SO_2 and O_3 content in the interspace (space between the new protective glazing and the original glass) compared to the outdoor conditions (Godoi *et al.*, 2006) and that it reduces the possibility of condensation on the internal side of the historical glass as well as the total time of the high relative humidity values (Bernardi *et al.*, 2006). Nevertheless, the interspace has to be monitored during time especially concerning the ventilation. If not under control, the temperature and the relative humidity could reach high levels creating a microclimate which can increase the deterioration rate (Vogel and Achilles, 1993; Römich, 2004).

7.5.2 *Active conservation*

It involves different levels of interference with the objects and consists of the application of protective layers directly on the glass. The earliest treatments used shellac and waxes but they were opaque or colored and therefore not always applicable to the objects (Newton and Davison, 1989). A better result was obtained with epoxy or polyester resins, acrylic or vinyl polymers (Davison, 2003; Abd-Allah, 2007). Organic protective coatings are widely used to modify the chemical composition making them highly adaptable to different requirements but, on the other hand, they are sensitive to UV light and to ageing processes: polymeric chains can break, causing changes of some important properties such as the density and elasticity, the colour and the transparency. As a consequence, the organic protective layer may appear yellowed, no longer transparent and it can easily enclose dust particles; moreover, due to the chemical changes occurring to the polymer, its solubility changes may compromise the reversibility of the treatment. Swelling phenomena can lead to stress between the glass and the coating, increasing the tension at the interface between the two materials. Nevertheless, the easy synthesis, the flexibility and the possibility to choose the more convenient precursors can be used to differentiate the physical and chemical properties of the final coating with the aim of improving some properties: good adhesion, low permeability and hydrophobicity, morphology and optical properties (Dave *et al.*, 2004).

On the other hand, new developments involve the inorganic coating especially using the silica sol-gel technique (Carmona *et al.*, 2004; Dal Bianco and Bertoncello, 2008; Dal Bianco, Bertoncello *et al.*, 2008). Even if their elasticity is not comparable to the organic ones, the high similarity between the coating and the substrate can guarantee a good adhesion and compatibility. In Fig. 7.8, the ToF-SIMS depth profile of a glass coated with a sol-gel silica layer is presented: the coating does not inhibit completely the ion migration

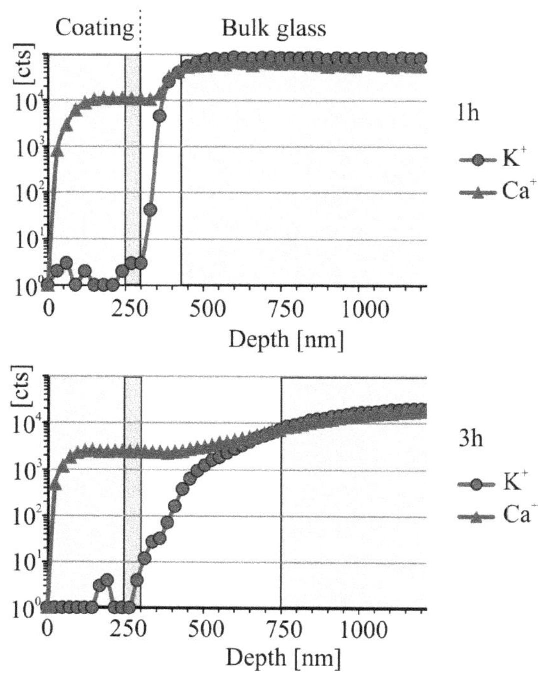

Fig. 7.8 ToF-SIMS depth profile of glass M3 coated with a sol-gel silica layer after 1 and 3 h of leaching in HCl, pH = 1; taken from De Bardi, Hutter *et al.*, 2014.

from the bulk under long leaching conditions, but the kinetic of the process is slowed down and no cracks or damages are visible on the coated surface (De Bardi *et al.*, 2014). The chemical and mechanical resistance is much higher and it cannot undergo degradation processes.

Recently, hybrid inorganic-organic sol-gel coatings have been developed. They are composed by an inorganic structure cross-linked or substituted by organic groups to blend the properties of the two in the same coating (Sanchez *et al.*, 2005; De Ferri *et al.*, 2013). Nanotechnologies are also discussed in the literature (Kasemann and Schmidt, 1994; Ershad-Langroudi and Rahimi, 2009): The aim is the modification of the surface in order to limit the contact between glass and water making the glass surface hydrophobic. Moreover, the thickness of the coating — in the nanometer scale — must not interact with light in the range of approximately 400–700 nm in order to be not visible.

7.6 Conclusion

Summarising the experience of the last 50–70 years of research on the degradation/weathering of medieval stained glass objects, it has to be mentioned that the air pollutants and acidifying gases, which are part of the today's ambient atmosphere, influence the stability and hence the appearance of these artifacts. The natural processes of decay are accelerated by the aggressive environment, risking the loss of this artistic and cultural heritage, especially for glass with a composition low in silica and rich in potassium.

An interdisciplinary collaboration between art historians, curators and conservators on one side and glass scientists on the other side can give the chance to protect such historical monuments from damage. Much information is obtained with modern surface analytical techniques shortly described in this chapter and should help to identify the most dangerous parameters and to suggest proper exposure conditions, in order to reduce the risk for medieval but also ancient glass exposed to the outdoor environment.

References

Abd-Allah, R. (2007). In-situ glass conservation: a case study from the archaeological site of Barsinia, Jordan, *Jordan Adumatu Journal*, 16, pp. 25–36.

Abd-Allah, R. (2007). Stabilization and treatment of corroded glass objects displayed in the museum of Jordanian heritage, *Mediterranean Archaeology and Archaeometry*, 7, pp. 19–28.

Arndt, J. and Stöffler, D. (1969). Anomalous changes in some properties of silica glass densified at very high pressures, *Phy. Chem. Glasses*, 10, pp. 117–124.

Arribart, H. and Abriou, D. (2000). Ten years of atomic force microscopy in glass research, *Ceram-Silik.*, 44, pp. 121–128.

Bernardi, A., Becherini, F., Bassato, G. and Bellio, M. (2006). Condensation on ancient stained glass windows and efficiency of protective glazing systems: two French case studies, Sainte-Chapelle (Paris) and Saint-Urbain Basilica (Troyes), *J. Cult. Herit.*, 7, pp. 71–78.

Bridgman, P.W. and Simon, I. (1953). Effect of very high pressures on glass, *J. Appl. Phys.*, 24, pp. 405–413.

Brill, R.H. (1963). Ancient Glass, *Sci. Am.*, 209, pp. 120–130.

Calligaro, T. (2008). PIXE in the study of archaeological and historical glass, *X-Ray Spectrom.*, 37, pp. 169–177.

Carmona, N., Villegas, M.A. and Navarro, J.M.F. (2004). Protective silica thin coatings for historical glasses, *Thin Solid Films*, 458, pp. 121–128.

Carmona, N., Oujja, M., Rebollar, E., Römich, H. and Castellejo, M. (2005). Analysis of corroded glasses by laser induced breakdown spectroscopy, *Spectrochim. Acta B*, 60, pp. 1155–1162.

Carmona, N., Oujja, M., Gaspard, S., Garcia-Hers, M., Villegas, M.A. and Castellejo, M. (2007). Lead determination in glasses by laser-induced breakdown spectroscopy, *Spectrochim. Acta B*, 62, pp. 94–100.

Clark, D.E., Pantano, C.G. and Hench, L.L. (1979). *Corrosion of Glass*, Books for Industry, The Glass Industry, New York.

Clark, D.E. and Yen-Bower, E.L. (1980). Corrosion of glass surface, *Surf. Sci.*, 100, pp. 53–70.

Colomban, P. (2003). Polymerization degree and Raman identification of ancient glasses used for jewelry, ceramic enamels and mosaics, *J. Non-Cryst. Solids*, 323, pp. 180–187.

Colomban, P., Etcheverry, M.-P., Asquier, M., Bounichou, M. and Tournié, A. (2006). Raman identification of ancient stained glasses and their degree of deterioration, *J. Raman Spectrosc.*, 37, pp. 614–626.

Cox, G.A. and Gillies, J.S. (1986). The X-ray fluorescence analysis of medieval durable blue soda glass from York Minster, *Archaeometry*, 28, pp. 57–68.

Dal Bianco, B. and Bertoncello, R. (2008). Sol-gel silica coatings for the protection of cultural heritage glass, *Nucl. Instr. and. Meth. in Phys. Res. B*, 266, pp. 2358–2362.

Dal Bianco, B., Bertoncello, R., Bouquillon, A., Dran, J.-C., Milanese, L., Roehrs, S., Sada, C., Salomon, J. and Voltolina, S. (2008). Investigation on sol-gel silica coatings for protection of ancient glass: interaction with glass surface and protection efficiency, *J. Non-Cryst. Solids*, 354, pp. 2983–2992.

Dave, B.C., Hu, X., Devaraj, Y. and Dhali, S.K. (2004). Sol-gel derived corrosion-protection coatings, *J. Sol-Gel Sci. Technol.*, 32, pp. 143–147.

Davison, S. (2003). *Conservation and Restoration of Glass*, Butterworth and Heinemann, New York.

De Bardi, M., Wiesinger, R. and Schreiner, M. (2013). Leaching studies of potash-lime-silica glass with medieval composition by IRRAS, *J. Non-Cryst. Solids*, 360, pp. 57–63.

De Bardi, M., Hutter, H. and Schreiner, M. (2013). ToF-SIMS analysis for leaching studies of potash–lime–silica glass, *J. Appl. Surf. Sci.*, 282, pp. 195–201.

De Bardi, M., Hutter, H., Schreiner, M. and Bertoncello, R. (2014). Sol-gel silica coating for potash–lime–silica stained glass: applicability and protective effect, *J. Non-Cryst. Solids*, 390, pp. 45–50.

De Ferri, L., Lottici, P.P., Lorenzi, A., Montenero, A. and Vezzalini, G. (2013). Hybrid sol-gel based coatings for the protection of historical windows glass, *J. Sol-Gel Sci. Technol.*, 66, pp. 253–263.

Doremus, R.H. (1980). Infrared spectroscopy of surfaces of glass containing alkali ions, *J. Non-Cryst. Solids*, 41, pp. 145–149.

Doremus, R.H. (1994). 'Chemical durability: reaction of water with glass', in *Glass Science*, John Wiley & Sons, New York, pp. 215–240.

Dowsett, M. and Adriaens, A. (2004). The role of SIMS in cultural heritage studies, *Nucl. Instr. and. Meth. in Phys. Res. B*, 226, pp. 38–52.

Ernsberger, F.M. (1977). Molecular water in glass, *Journal of The American Ceramic Society*, 60, pp. 91–92.

Ernsberger, F.M. (1980). The role of molecular water in the diffusive transport of protons in glass, *Phys. Chem. Glasses*, 21, pp. 146–149.

Ernsberger, F.M. (1986). Proton transport in solids, *J. Non-Cryst. Solids*, 87, pp. 408–414.

Ershad-Langroudi, A. and Rahimi, A. (2009). Synthesis and characterisation of nano silica-based coatings for protection of antique articles, *Int. J. Nanotechnol.*, 6, pp. 915–925.

Ethridge, E.C., Clark, D.E. and Hench, L.L. (1979). Effects of glass surface area to solution volume ratio on glass corrosion, *Phys. Chem. Glasses*, 20, pp. 35–40.

Ferris, K.F. and Pederson, L.R. (1988). In situ characterisation by Fourier transform infrared spectroscopy of the reactions between glass and water, *Phy. Chem. Glasses*, 29, pp. 9–12.

Figueiredo, M.O., Silva, T.P. and Veiga, J.P. (2006). A XANES study of the structural role of lead in glazes from decorated tiles, XVI to XVIII century manufacture, *Appl. Phys. A*, 83, pp. 209–211.

Frank, S. (1982). *Glass and Archaeology*, Academic Press, London.

Frenzel, G. (1985). The restoration of medieval stained glass, *Sci. Am.*, 252, pp. 100–106.

Frischat, G.H., Poggemann, J.-F. and Heide, G. (2004). Nanostructure and atomic structure of glass seen by atomic force microscopy, *J. Non-Cryst. Solids*, 345–346, pp. 197–202.

Frommer, J. (1992). Scanning tunneling microscopy and atomic force microscopy in organic chemistry, *Angew. Chem.*, 31, pp. 1298–1328.

Gallien, J.-P., Gouget, B., Carrot, F., Orial, G. and Brunet, A. (2001). Alteration of glasses by micro-organisms, *Nucl. Instr. and Meth. in Phys. Res. B*, 181, pp. 610–615.

Godoi, R.H.M., Kontozova, V. and Grieken, R.V. (2006). The shielding effect of the protective glazing of historical stained glass windows from an atmospheric chemistry prospective: case study Saint Chapelle, Paris, *Atmos. Environ.*, 40, pp. 1255–1265.

Gorbushina, A.A. and Palinska, K.A. (1999). Biodeteriorative processes on glass: Experimental proof of the role of fungi and cyanobacteria, *Aerobiol.*, 15, pp. 183–191.

Griffiths, P.R. and De Haseth, J.A. (2007). *Fourier Transform Infrared Spectrometry*. (John Wiley & Sons, New Jersey, USA.

Hall, E.T. (1960). X-ray fluorescent analysis applied to archaeology, *Archaeometry*, 3, pp. 29–35.

Hall, E.T., Banks, S. and Stern, J.M. (1964). Uses of X-ray fluorescence analysis in archaeology, *Archaeometry*, 7, pp. 84–89.

Hench, L.L. (1975). Characterisation of glass corrosion and durability, *J. Non-Cryst. Solids*, 19, pp. 27–39.

Hench, L.L., Newton, R.G. and Bernstein, S. (1979). Use of infrared reflection spectroscopy in analysis of durability of medieval glasses, with some comments on the conservation procedure, *Glass Technol.*, 20, pp. 144–148.

Hench, L.L., Clark, D.E. and Yen-Bower, E. (1980). Corrosion of glasses and glass-ceramics, *Nucl. Chem. Waste Man.*, 1, pp. 59–75.

Henriksen, J.F., Arnese, K. and Ferm, M. (2004). *ICP of Effects on Material Including Historic and Cultural Monuments*, Norwegian Institute for Air Research, Report No. 21, 40 and 50.

Hornblower, A.P. (1962). Archaeological applications of electron probe microanalyser, *Archaeometry*, 5, pp. 108–112.

Hornblower, A.P. (1963). Some examples of work with the electron probe micronalyser, *Archaeometry*, 6, pp. 37–42.

Janssens, K. and Van Grieken, R. (2004). *Non-Destructive Microanalysis of Cultural Heritage Materials*, Elsevier, Amsterdam.

Janssens, K. (2013). *Modern Methods for Analysing Archaeological and Historical Glass*, John Wiley & Sons, West Sussex.

Jembrih, D., Neelmeijer, C., Schreiner, M., Mäder, M., Ebel, M. and Peev, M. (2001). Iridescent Art Nouveau glass — IBA and XPS for the characterisation of thin iridescent layers, *Nucl. Instr. and Meth. in Phys. Res. B*, 181, pp. 698–702.

Kasemann, R. and Schmidt, H. (1994). Coatings for mechanical and chemical protection based on organic–inorganic sol-gel nanocomposites, *New J. Chem.*, 18, pp. 1117–1123.

Lanford, W.A., Trautvetter, H.P., Ziegler, J.F. and Keller, J. (1976). New precision technique for measuring the concentration versus depth of hydrogen in solids, *Appl. Phys. Lett.*, 28, pp. 566–568.

Lenzen, M., Collins, R.E. and Prince, K. (2005). Method of measuring depth profile of hydrogen in soda-lime glass, *J. Non-Cryst. Sol.*, 351, pp. 317–322.

Macfarlane, A. and Martin, G. (2002). *The Glass Bathyscaphe: How Glass Changed the World*, Profile Books, London.

Mantler, M. and Schreiner, M. (2000). X-Ray fluorescence spectrometry in art and archaeology, *X-Ray Spectrom.*, 29, pp. 3–17.

Melcher, M. and Schreiner, M. (2004). Statistical evaluation of potash–lime–silica glass weathering, *Anal. Bioanal. Chem.*, 379, pp. 628–639.

Melcher, M. and Schreiner, M. (2005). Evaluation procedure for leaching studies on naturally weathered potash–lime–silica glasses with medieval composition by scanning electron microscopy, *J. Non-Cryst. Solids*, 351, pp. 1210–1225.

Melcher, M. and Schreiner, M. (2006). Leaching studies on naturally weathered potash–lime–silica glasses, *J. Non-Cryst. Solids*, 352, pp. 368–379.

Morey, G.W. and Merwin, H.E. (1932). The relation between the composition and the density and optial properties of glass, *J. Opt. Soc. of Am.*, 22, pp. 632–662.

Newton, R.G. (1979). 'Sulphur dioxide and medieval stained glass' in *Sulphur Emission and the Environment of London*, Society of Chemical Industry, Water and Environmental Group, London, pp. 311–313.

Newton, R.G. (1985). The durability of glass: a review, *Glass Techn.*, 26, pp. 21–38.

Newton, R.G. (1987). Air pollution damage, *Profess. Stained Glass*, 7, pp. 14–21.

Newton, R. and Davison, S. (1989). *Conservation of Glass*, Butterworth, London.

Perez y Jorba, M., Dallas, J.P., Bauer, C., Bahezre, C. and Martin, J.C. (1980). Deterioration of stained glass by atmospheric corrosion and micro-organisms, *J. Mater. Sci.*, 15, pp. 1640–1647.

Pio, C.A., Ramos, M.M. and Duarte, A.C. (1998). Atmospheric aerosol and soiling of external surfaces in an urban environment, *Atmos. Environ.*, 32, pp. 1979–1989.

Poggemann, J.-F., Heide, G. and Frischat, G.H. (2003). Direct view of the structure of different glass fracture surfaces by atomic force microscopy, *J. Non-Cryst. Solids*, 326–327, pp. 15–20.

Quartieri, S., Riccardi, M.P., Messiga, B. and Boscherini, F. (2005). The ancient glass production of medieval Val Gargassa glasshouse: Fe and Mn XANES study, *J. Non-Cryst. Sol.*, 351, pp. 3013–3022.

Raberg, W., Ostadrahimi, A.H., Kayser, T. and Wandelt, K. (2005). Atomic scale imaging of amorphous silicate glass surfaces by scanning force microscopy, *J. Non-Cryst. Solids*, 351, pp. 1089–1096.

Rana, M.A. and Douglas, R.W. (1961). The reaction between glass and water. Part 1, Experimental methods and observations, *Phy. Chem. Glasses*, 2, pp. 179–195.

Robinet, L., Coupry, C., Eremin, K. and Hall, C. (2006). The use of Raman spectroscopy to predict the stability of historical glasses, *J. Raman Spec.*, 37, pp. 789–797.

Römich, H. (2004). Evaluation of protective glazing systems, *e-Preserv.Sci.*, 1, pp. 1–8.

Ruska, E. (1986). The development of the electron microscope and of electron microscopy, *Bioscience Rep.*, 7, pp. 607–629.

Sanchez, C., Julian, B., Belleville, P. and Popall, M. (2005). Applications of hybrid organic–inorganic nanocomposites, *J. Mater. Chem.*, 15, pp. 3559–3592.

Sanders, D.M. and Hench, L.L. (1973). Mechanism of glass corrosion, *J. Am. Ceram. Soc.*, 56, pp. 373–377.

Sanders, D.M. and Person, W.B. (1974). Quantitative analysis of glass structure with the use of infrared reflection spectra, *Appl. Spectr.*, 28, pp. 247–255.

Sartowska, B. and Kunicki-Goldfinger, J. (2001). Determination of chemical homogeneity of historical glassware using autoradiography of potassium natural radioactivity, *Rad. Measurem.*, 34, pp. 581–583.

Schmitz, I., Prohaska, I., Friedbacher, G., Schreiner, M. and Grasserbauer, M. (1995). Investigation of corrosion processes on cleavage edges of potash–lime–silica glasses by atomic force microscopy, *Fresen. J. Anal. Chem.*, 353, pp. 666–669.

Schmitz, I., Schreiner, M., Freidbacher, G. and Grasserbauer, M. (1997). Phase imaging as an extension to tapping mode AFM for the identification of material properties on humidity-sensitive surfaces, *Appl. Surf. Sci.*, 115, pp. 190–198.

Schmitz, I., Schreiner, M., Friedbacher, G. and Grasserbauer, M. (1997). Tapping-mode AFM in comparison to contact-mode AFM as a tool for in-situ investigations of surface reactions with reference to glass corrosion, *Anal. Chem.*, 69, pp. 1012–1018.

Schreiner, M. (1988). Deterioration of stained medieval glass by atmospheric attack, *Glastechn. Ber.*, 61, pp. 223–230.

Schreiner, M., Grasserbauer, M. and March, P. (1988). Quantitative NRA and SIMS depth profiling of hydrogen in naturally weathered medieval glass, *Fresen. J. Anal. Chem.*, 331, pp. 428–432.

Schreiner, M. (2004). 'Corrosion of historic glass and enamels' in Janssens, K. and Van Grieken, R. (Eds.), *Non-destructive Micro Analysis of Cultural Heritage Materials*, Elsevier, pp. 713–754.

Simpson, H.E. (1953). Some factors affecting the testing of surface durability of glass, *J. Am. Ceram. Soc.*, 36, pp. 143–146.

Smith, G.D. and Clark, R.J.H. (2004). Raman microscopy in archaeological science, *J. Archaeol. Sci.*, 31, pp. 1137–1160.

Staudigel, H., Chastain, R.A., Yayanos, A. and Bourcier, W. (1995). Biologically mediated dissolution of glass, *Chem. Geol.*, 126, pp. 147–154.

Sweet, J.R. and White, W.B. (1969). Study of sodium glasses and liquids by infrared reflectance spectroscopy, *Phy. Chem. Glasses*, 10, pp. 246–251.

Tennent, N.H. (1981). Fungal growth on medieval glass, *J. Br. Soc. Master Glass Painters*, 17, pp. 64–68.

Tite, M., Shortland, A. and Paynter, S. (2002). The beginnings of vitreous materials in the near east and Egypt, *Acc. Chem. Res.*, 35, pp. 585–593.

Vogel, N.A. and Achilles, R. (1993). *The Preservation and Repair of Historic Stained and Leaded Glass*, US Department of the Interior, National Park Service, Cultural Resources, Preservation Assistance, U.S. Government Printing.

Woisetschläger, G., Dutz, M., Paul, S. and Schreiner, M. (2000). Weathering phenomena on naturally weathered potash–lime–silica glass with medieval composition studied by secondary electron microscopy and energy dispersive microanalysis, *Mikrochim. Acta*, 135, pp. 121–130.

Zachariasen, W.H. (1932). The atomic arrangement in glass, *J. Amer. Chem. Soc.*, 54, pp. 38–41.

CHAPTER 8

Air Pollution and Changes to the Biology of Urban Fabric

Beatriz Prieto

Departamento de Edafología y Química Agrícola, Facultad de Farmacia, Universidad de Santiago de Compostela, Spain

Patricia Sanmartín

Departamento de Edafología y Química Agrícola, Facultad de Farmacia, Universidad de Santiago de Compostela, Spain; Laboratory of Applied Microbiology, School of Engineering and Applied Sciences, Harvard University, United States

8.1 Introduction

Cities are artificial environments in which man-made surfaces are susceptible to microbiological colonisation. Most urban habitats are affected by elevated concentrations of pollutants, which interact with organisms. Current studies regarding air pollution effects on the biology of urban fabric mainly focus on sulfur oxides (SO_x), nitrogen oxides (NO_x), heavy metals, particle matter (PM), carbon oxides (CO_x), NH_3 and persistent organic pollutants (POPs), following this order of importance. In 2013, the Intergovernmental Panel on Climate Change (IPCC, 2013) reported increased levels of greenhouse gases, such as carbon dioxide (CO_2), methane (CH_4), and nitrous oxide (N_2O), since 1750 due to human activity. Urban activities like industry, vehicle traffic, power stations, waste treatment, domestic heating and electronic waste (such as rechargeable lithium batteries or cathode-ray tubes from televisions and computer monitors) generate pollutants to which many organisms are sensitive. In 2011, the concentrations of the aforementioned greenhouse gases were 391 ppm, 1803 ppb, and 324 ppb, and exceeded the pre-industrial levels by about 40%, 150%, and 20%, respectively. Concentrations of CO_2, CH_4, and N_2O now substantially exceed the highest concentrations recorded in ice cores during the past 800,000 years. The mean rates of increase in atmospheric concentrations over the past century are, with very high confidence, unprecedented in the last 22,000 years (IPCC, 2013).

The interaction between air pollutants and organisms is highly variable and complex. A specific air pollutant, at a given dose, may be stimulatory, neutral or detrimental to the growth and development of a particular organism. However, most of the changes in the atmospheric chemistry that are derived from air pollutants have a negative impact on the biology of the urban fabric. Therefore, and among the many strategies available, air quality biomonitoring became a standard method for air pollution assessment in the past decades, because of its low-cost and high accuracy.

In this chapter, studies relating three multidisciplinary topics (air pollution, biological organisms and civil engineering structures) are reviewed to examine the impact of air pollution on the biodiversity of urban pollution-sensitive organisms and on the biodeterioration of buildings and structures. Our aim is to synthesise the current knowledge on this topic and highlight the state of the art procedures and its tendencies.

8.2 Airborne Pollutants and Bioindicators

Pollutants or environmental stress inducers can be defined as chemical compounds released into the environment as a result of human activities, causing damage to living organisms (Moriarty, 1999). Sensitivity to pollutants of certain organisms makes them suitable as bioindicators, so they may be used for the identification and qualitative determination of human-generated environmental factors. They respond to environmental pollution with changes in their vital functions or by pollutant accumulation in their tissue. In some cases, bioindicators can be used as biomonitors, organisms that can give quantitative information of contaminants, and indicate long-term trends including recovery (Conti and Cecchetti, 2001). Bioindicators may be of great importance for the heritage field, since the damage caused by air pollutants to material heritage is a problem that can be reduced by establishing a preventive conservation strategy that includes the use of bioindicators. Indeed, because of their sensitivity to pollutants, they can help guide urban management practices like temporary restriction of traffic and pedestrianisation in historical urban centers, which subsequently reduces the ocurrence of pollutants on monument surfaces, thus avoiding degradation processes on surfaces (e.g. soiling, concretions, films and black crusts, and deposits). That kind of actions will result in a cost-benefit preventive conservation strategy, aimed at avoiding and minimising future deterioration or loss due to pollutants.

Some biological organisms are more sensitive than others to air pollution. Lichens, for example, are considered excellent biomarkers of the effects of air pollution and were the first to be used for that purpose as they absorb and accumulate atmospheric pollutants far better than vascular plants (Loppi *et al.*, 1997a). Despite the first report on the use of epiphytic lichens as bioindicators by Nylander (*apud* Conti and Cecchetti, 2001) in 1866, and the fundamental studies on bioindicators and/or biomonitors dated from 1960s, the most noticeable advances were made in the past two decades. These studies reported the impact of pollutants on the oxidative damage of tissues and defense systems of organisms. Some of those studies are listed in Table 8.1. Furthermore, an estimate of the proportional use of

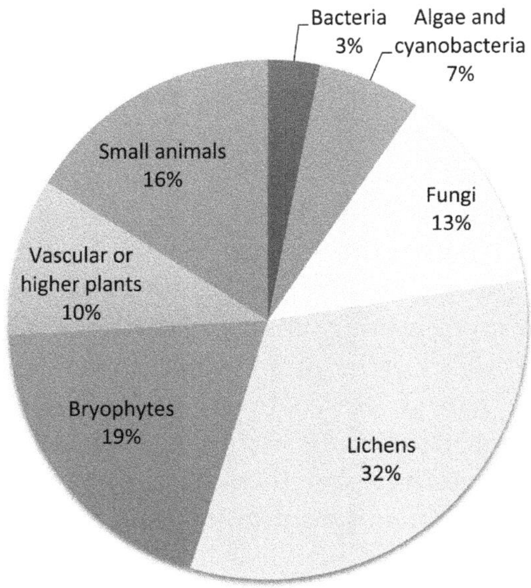

Fig. 8.1 Relative distribution of species used as bioindicators and/or biomonitors of air pollution from the studies outlined in this chapter.

the living organisms as indicators and/or monitors of air pollution at present is shown in Fig. 8.1.

8.2.1 *Bacteria*

Literature regarding the use of bacteria as bioindicators and/or biomonitors of air pollution of urban fabric is scarce. The few available studies mostly focus on the differences in airborne bacteria concentration in urban areas versus non-urban areas, as observed for the first time by Bovallius *et al.* (1978).

Studies that present some changes produced in bacteria as a result of urban pollution, for example by Kessler *et al.* (2012), used genetically modified *Escherichia coli* to demonstrate the value of bacterial reporter strains to assess the nature of biological stress caused by atmospheric PM (particle matter). Kessler also pointed out that most bacterial genotoxicity assays successfully performed in the past have used *Salmonella typhimurium* (also known as *Salmonella typhi* or *Salmonella enterica*).

Table 8.1 Site, period of study, pollutants and changes recorded on biomarkers of pollution.

Reference	Place under study	Period of study	Pollutant	Biomarkers of pollution	Changes registered
Mitchell and Gu (2000)	Urban area in Cambridge, Massachusetts (USA)	Not covered	Sulfur and hydrocarbon air pollutants	Fungi, and heterotrophic and chemolithotrophic bacteria	Presence of atmospheric pollutants inhibited both the size and diversity of the biofilm microbial community
Davies *et al.* (2007)	London (UK)	2002–2003	NO_x and a lesser extent SO_2 and PM	3000 epiphytes from 334 *Fraxinus excelsior* L. (European ash) representing 74 lichen, 14 moss, 7 fungal and 3 algal species	Diversity of epiphytic species increased from 1970s to 2000s Orange colour of *Xanthoria parietina* was inversely related to NO_x
Isocrono *et al.* (2007)	Urban area of Turin (Italy)	2004	SO_2, Pb, total suspended particles, NO_x and feasibly NH_3	35 epiphytic lichens. The most common: *Phaeophyscia orbicularis* and *Physcia adscendens*	Number of species increased significantly over the past 25 years

Cicek *et al.* (2009)	7 historical buildings in Erzurum (Turkey), viz. 'Yakutiye Madrasah', 'Double Minarets', 'Ahmediye Madrasah', 'Silver Tomb', 'Aziziye and Mecidiye Bastions' and 'Citadel Masjid'	Not covered	SO_2, NO_x and PM	10 lichens: *Acarospora cervina; Acarospora fuscata; Aspicilia calcarea; Caloplaca decipiens; Candelariella vitellina; Dimelaena oreina; Lecidea fuscoatra; Lecanora dispersa; Lecanora muralis* and *Xanthoria elegans*	High pollution levels limited lichen growth to few resistant species
Radwan *et al.* (2010)	Industrial, residential and traffic areas in Alexandria city (Egypt)	May 2007	Heavy metals (Zn, Cu, Pb and Cd)	Land snails *Theba pisana* (Müller)	Oxidative stress and heavy metal accumulation were reported on *T. pisana*
Lisowska (2011)	Skawina town (Poland)	2006–2007	SO_2, NO_x and PM	34 epiphytic lichens	Lichens have recolonised the former 'lichen desert' in the town centre

(*Continued*)

Table 8.1 (*Continued*)

Reference	Place under study	Period of study	Pollutant	Biomarkers of pollution	Changes registered
					Nitrogen- and dust-tolerant species have expanded and dominate, while some acidophytes are in decline
Kessler *et al.* (2012) (Laboratory study)	Roof of the Earth Sciences Institute in the Hebrew University Edmond Safra Campus in Jerusalem (Israel)	Not covered	Atmospheric PM samples	2 genetically engineered bacterial (*Escherichia coli*) strains: RFM443 and DE112	Oxidative stress, respiratory inhibition, excess metal concentrations and iron deficiency
Abdel-Halim *et al.* (2013)	Industrial and vehicular traffic area in Kafr El-Zayat city, El-Gharbia (Egypt)	Not covered	Heavy metals (Cd, Mn, Ni, Pb, Zn and Cu)	Gastropods *Helix aspersa* (Müller)	Oxidative stress in the digestive gland of the land snail

Basile *et al.* (2013)	Botanical Gardens of University of Naples (Italy)	June 30th, 2012	Heavy metals (Cd, Pb, Cu and Zn).	Conocephalaceae - Marchantiales - Bryophyta *Conocephalum conicum*	Specimens showed mis-shaped, shrunk and small cells. Also, many vesicles were visible in the cytoplasm
Kolon *et al.* (2013)	Oława (Poland)	August, unspecified year	Airborne chemical elements: Cd, Co, Cr, Cu, Fe, Li, Mg, Mn, Ni, Pb and Zn	Mistletoe (*Viscum album* L.) and rowan tree or mountain-ash (*Sorbus aucuparia*)	Sites with lower and higher pollution levels were distinguished The ratio of metals in *V. album* to *S. aucuparia* was different depending on the pollution level
Majumder *et al.* (2013)	High traffic volume sites in Kolkata Metropolis, (India), viz. 'Salt Lake, Karunamoyee'	December–January, unspecified year	Airborne chemical elements: K, Ca, Ti, V, Mn, Cr, Fe, Cu, Zn, and Pb	Epiphytic lichen, *Flavoparmelia caperata* (L.) Hale, previously known as *Parmelia caperata*	Decrease chlorophyll a and increase in lipid peroxidation and electrolyte conductivity

(*Continued*)

Table 8.1 (*Continued*)

Reference	Place under study	Period of study	Pollutant	Biomarkers of pollution	Changes registered
	and 'Indian Botanic Garden'				
Gerdol *et al.* (2014)	Ferrara city (Italy)	September–October 2006 and after five years, September–October 2011	23 trace metals (Al, Ag, As, Ba, Bi, Cd, Co, Cr, Cu, Fe, Ga, Li, Mn, Mo, Ni, Pb, Rb, Sb, Se, Sr, U, V, Zn); total N concentration and δ^{15} N nitrogen isotopic abundance	Mosses (*Tortula muralis* Hedw.) and lichens	Diversity of epiphytic lichens increases from 2006 to 2011
Herrera-Dueñas *et al.* (2014)	Urban polluted areas of Madrid, (Spain), viz. 'Fuenlabrada' and 'centre of Madrid city'	October 2011 – February 2012	CO, NO_x, and SO_x	Birds House Sparrow (*Passer domesticus*)	Oxidative damage: decrease in haemoglobin and total antioxidant capacity, two oxidative stress markers

8.2.2 *Algae and cyanobacteria*

Algae and cyanobacteria have been used as biomarkers of air pollution to a lesser extent than lichens and mosses. Despite the well-known relevance of sulfur nutrition in cyanobacterial physiology, the relationship of algae and cyanobacteria to airborne pollutants, as well as their presence in polluted areas, remains unclear. Some authors suggest that these organisms, in particular cyanobacteria, are resistant to pollutants under favorable climatic conditions as pollution reduces diversity but not total biomass (Caneva *et al.*, 1995). However, Gladis-Schmacka *et al.* (2014), using clay roof tiles as test substrates, showed that phototrophic growth was inhibited in areas in the immediate vicinity of highways, even though the dry climate could also have played a role. On the other hand, in countries where urban air quality has improved over the past years, namely Germany and neighboring countries, the occurrence and intensity of green biofilms on building materials has increased (Gladis-Schmacka *et al.*, 2014). Along the same lines, Smith *et al.* (2011) and McCabe *et al.* (2011) documented an increase in green biofilms in some urban areas of the British Isles as a consequence of the reduction of atmospheric sulfur and an increase of atmospheric nitrogen and air humidity.

In a study conducted twelve years ago on the relationship between epiphytic biodiversity and SO_2 and NO_x levels in London, Davies *et al.* (2007) concluded that three epiphytic free-living algae (*Apatacoccus* spp., *Prasiola crispa* and *Trentepohlia* spp.) and one green algae (*Trebouxia* spp.) were among the most sensitive organisms to these two pollutants.

8.2.3 *Fungi*

Being heterotrophic, fungi are unable to use the inorganic carbonate substrate to support their growth; but they use waste products or dead cells from former microbial communities, organic nitrogenous matter from birds, decayed leaves and aerosols (Sterflinger, 2000), as well as airborne organic compounds deposited on the urban surfaces (e.g. Saiz-Jimenez, 1995; 1997). Saiz-Jimenez (1997) demonstrated that biological activity, underlining the fungal activity, impacts the fate of organic compounds deposited on building stones located in polluted environments. This is supported, according to the author, by the microbial degradation of phenanthrene that was found to be a common process in the weathered stones of European cathedrals.

Several studies have shown the capacity of fungi to accumulate pollutants and their ability to reflect air pollution. Some authors have found that fungi are more efficient accumulators than other organisms, such as moss, vascular plants and even lichens (Morre *et al.*, 1995). In a 1970 publication about the relationship of air pollution with lichens and fungi, Saunders (1970) pointed that high atmospheric concentrations of NH_3 in urban sites can prevent the establishment of blackspot on roses; a disease caused by the fungus *Diplocarpon rosae.* The author also mentioned other fungi that are absent from highly polluted areas, namely *Microsphaera alphitoides,* coniferous rusts, e.g. *Melampsora cerastii* and *Phragmidium* spp., and *Hysterium pulicare.* More recently, Mitchell and Gu (2000) when analysing fungi from an urban area in Cambridge, Massachusetts (USA), found that regardless of a suppression of culturable populations of chemolithotrophic bacteria, heterotrophic bacteria, and fungi by atmospheric pollutants, a specific population of microorganisms forming biofilms capable of utilising sulfur and hydrocarbon compounds was stimulated.

8.2.4 *Lichens*

Lichens are among the pioneer organisms that inhabited stone surfaces (Fig. 8.2). They are very well studied (Ferry *et al.*, 1973; Falla *et al.*, 2000) and highly effective biomarkers of atmospheric pollution because of their morphological and physiological characteristics. Since tree bark has a low buffering capacity in comparison with rock, soil or concrete, epiphytic lichens are particularly sensitive to air pollution and are mainly used for recording. Lichens also respond rapidly to air pollution, facilitating the detection of environmental changes on an annual basis.

Lichens can be used as bioindicators and/or biomonitors by mapping all species present in a specific area, which allows an index of

Fig. 8.2 Nitrophilic lichen *Xanthoria parietina* on an urban structure (Cathedral of Santiago de Compostela, Spain).

atmospheric purity to be established (LeBlanc and De Sloover, 1970) that estimates the levels of atmospheric pollutants (SO_2, NO_x, Cl, Cd, Pb and Zn) (Amman *et al.*, 1987). This index can be used through an individual sampling of lichen species or by transplanting lichens from an uncontaminated area to a contaminated one. In both cases, the morphological changes and/or amount of pollutants in the lichen thallus are measured (Conti and Cecchetti, 2001). Several examples have been reported for trace elements (e.g. Puckett *et al.*, 1988; Freitas *et al.*, 1999). As an example, in urban areas of Italy, different biomonitoring studies using lichens have shown that Pb is still present and very widespread, despite the introduction of lead-free petrol (Monaci *et al.*, 1997). Also, the analysis of lichens from the southwest region of France during the early twentieth century showed that the atmosphere was mainly affected by As, Pb and Cd contamination, while other type of pollutants, namely Sb, Sn, Pb and Cu have emerged more recently (Agnan *et al.*, 2013).

Different lichen species react to different pollutants in distinct ways (Nimis and Tretiach, 1995; Conti and Cecchetti, 2001). In urban environments, it is possible to find pollution-tolerant species, such as *Xanthoria parietina* (Nimis, 1987, Fig. 8.2) that showed a superior response to oxidative stress when compared to other lichen species that were analysed in London (UK) by Davies *et al.* (2007). Pollution-tolerant species also include *Pheophyscia orbicularis* and *Candelaria concolor*; the latter being present at the limits of 'lichen desert' areas, meaning areas that are unsuitable for lichen survival because of high pollution concentrations (Conti and Cecchetti, 2001). *Hypogymnia physodes* is one of the most suitable bioindicators used in the study of trace elements bioaccumulation because of its high-tolerance capacities (Jeran *et al.*, 1996). *Cladonia portentosa* is an excellent bioindicator of nitrogen and phosphorous, while *Hypogymnia physodes* is suitable for analysing nitrogen total deposition, nitrogen and sulfur. *Parmelia saxatilis* can be used in the study of nitrogen oxides (Batty *et al.*, 2003). Also, *Usnea* species have been applied to evaluate heavy metal deposition patterns in the Antarctic (Poblet *et al.*, 1997).

The effect of sulfur compounds on lichens has probably been the most extensively studied. These studies generally evaluate the effects of sulfur compounds on the physiology of lichen thalli and/or on the degradation of chlorophyll into phaeopigments. Other indicators of the effect of those compounds are increases in the production of endogenous ethylene due to the exposure to sulfur in an acid environment or the tendency to lose moisture provoked by sulfur oxides in the lichen *Ramalina ecklonii* (Spreng.) (Levin and Pignata, 1995). High levels of SO_2 and NO_x can also cause a reduction of pH of lichen thalli and lead to extinction of species, namely *Lobaria pulmonaria* and *Ramalina farinacea.*

The presence of air pollutants such as O_3 and NO_2 may damage cell membrane proteins. In 1994, Egger *et al.* reported oxidative damage to membrane and reduction in net photosynthesis in *Hypogymnia physodes* (L.) Nyl., after its transplantation to highly polluted areas with high O_3 concentrations. Moreover, under heavy metal stress, degradation of chlorophyll and alteration of malondialdehyde (membrane lipid degradation product) in lichens has been widely reported. Majumder *et al.* (2013) gave a detailed description in the epiphytic lichen *Flavoparmelia caperata* (L.) Hale (previously known as *Parmelia caperata*) of the presence of Ti, V, Mn, Cr, Fe, Cu, Zn, and Pb from high traffic volume sites in Kolkata Metropolis (India). In that study, the chlorophyll a/b ratio was successfully used to quantify the impact of pollution stress in this lichen species. Stress symptoms in thalli from the urban sites, revealed as decreased chlorophyll a (indicating less photosynthetic efficiency) and an increase in lipid peroxidation and electrolyte conductivity (indicating cell membrane injuries) were also reported.

Lichens are also excellent bioindicators of atmospheric pollution from geothermal power stations (As, B, Hg, and H_2S) (Conti and Cecchetti, 2001). They are also good indicators of radionuclides, as has been demonstrated in studies conducted in several areas of Norway, after the Chernobyl incident in April 26, 1986. Levels of ^{134}Cs and ^{137}Cs (radioactive isotopes of cesium) found in lichens from those areas were two orders of magnitude higher than those found in vascular plants (Bretten *et al.*, 1992). Also, *Parmelia sulcata* has been used as a bioindicator for the presence of radionuclides in areas close to Chernobyl, where ^{129}I and ^{36}Cl (the most important radionuclides due to their mobility and longevity) have been measured. Equally relevant, *Xanthoria parietina* has shown to be the best bioindicator of radioactive fallout.

8.2.5 *Bryophytes: mosses and liverworts*

Mosses and liverworts are bryophytes that grow on surfaces rich in humus deposits. These deposits mostly result from an accumulation of atmospheric particles and dead microorganisms (algae). Bryophyta effectively accumulate metals because of their high surface/volume ratio and the presence of a thin cuticle. The ability of the cell wall to bind with heavy metals and the accumulation of trace metals by cation exchange and chelation are well known. Schröder *et al.* (2008) suggested that these organisms can be powerful and sensitive biomarkers of metal pollution. Although less well known, bryophytes also have the ability to accumulate organic compounds. This capacity was recently documented by Harmens *et al.* (2013) in a review article about terrestrial mosses as biomonitors of atmospheric persistent organic pollutants (POPs). The list of pollutants includes polycyclic aromatic hydrocarbons (PAHs), which have been the focus of the majority of studies reported, polychlorobiphenyls (PCBs), dioxins and furans (PCDD/Fs), and polybrominated diphenyl ethers (PBDEs).

More recently, Gerdol *et al.* (2014), using mosses as bioaccumulators and lichens as bioindicators, argue for a general negative correlation between lichen abundance and trace metal concentrations in mosses and/or lichens. In Italy, Basile *et al.* (2013) observed that urban specimens of *Conocephalum conicum* (liverwort) showed smaller and mis-shaped cells, as well as an increase in cytoplasmic vesicles when compared to those from non-polluted sites. The authors also suggested a significant role of *C. conicum* cell wall in the binding of Cd, which is more toxic than other metals.

Large-scale surveys in Europe have shown a general decrease in trace metal accumulation in moss tissue over time (Gerdol *et al.*, 2014). However, there are exceptions. In a study carried out in the urban area of Ferrara (Italy), from 2006 to 2011, the levels of Zn in *Tortula muralis* (moss) remained almost unchanged (Gerdol *et al.*, 2014). Another study that analysed moss tissues collected from industrial regions in Austria has reported an increase in Mo and Pb levels from 1995 to 2005 (Zechmeister *et al.*, 2008).

8.2.6 *Vascular or higher plants*

As trees and shrubs have more efficient systems for protection against pollutants (Caneva *et al.*, 2005), they became the most common plant type found in highly polluted urban areas (Sawidis *et al.*, 1995). Despite its predominance, the relative resistance of vascular plants to air pollutants makes them less suitable to be used as biomarkers and/or biomonitors of air pollution.

Nonetheless, examples of the use of vascular plants as bioindicators have been reported. Studies have pointed out that *Viscum album*, very common in urban areas, could be used as a bioindicator for metals in large city environments throughout the year (Alessio *et al.*, 2002; Kolon *et al.*, 2013). Another example is *Sorbus aucuparia*, that grows fast even in poor soils, becomes severely contaminated by industrial emissions and waste (Malá *et al.*, 2007). In 2013, Kolon *et al.* did a comprehensive study to show that the leaves of white berry mistletoe *Viscum album* L. and rowan tree *Sorbus aucuparia* accumulated atmospheric pollutants (Cd, Co, Cr, Cu, Fe, Li, Mg, Mn, Ni, Pb and Zn) in the urban area of Oława (SW Poland). However, the capacity of *V. album* and *S. aucuparia* to accumulate such compounds varied among different pollution levels. *V. album* appeared to be a better bioaccumulator of Cd, Ni, Pb and Zn and a weaker accumulator of Co in less polluted sites of Oława (SW Poland). *S. aucuparia* was a better bioaccumulator of Cd, Ni, Pb and Zn and a weaker accumulator of Co than *V. album* in more polluted sites of the same region.

8.2.7 *Small animals*

Assessing air pollution by using small animals as biomarkers has received increasing attention in the recent few years. In Egypt, Abdel-Halim *et al.* (2014) used *Helix aspersa* (gastropod) as a sentinel organism for assessing heavy metals of atmospheric pollution in urban areas. Previously, Radwan *et al.* (2010) used the land snail *Theba pisana* for the same purpose in Alexandria (Egypt).

Several markers of oxidative stress were evaluated in the blood of house sparrows (*Passer domesticus*) by non-invasive methods (Herrera-Dueñas *et al.*, 2014). Results showed significant differences in haemoglobin and total antioxidant capacity, with lower levels in urban populations. A similar result was shown in a previous study conducted between October 2003 and June 2005 in Milan (Italy), where air pollutants (CO, NO_x, SO_x and other fuel combustion products) induced an increase of methaemoglobinaemia in pigeons (*Columba livia*) (Sicolo *et al.*, 2009). The house sparrow, which is distributed worldwide and closely associated with urban environments, had also been effectively used to monitor levels of urban heavy metals in the southern of Finland (Kekkonen *et al.*, 2012) and two cities of Palestinian territory (Swaileh and Sansur, 2006). According to the authors, this wild bird is one of the most suitable candidates for urban biomonitoring of atmosphere pollutants. This approach is consistent with other authors who agree that the analysis of oxidative stress status in urban birds can be a good biomarker of environmental quality (e.g. Koivula and Eeva, 2010).

8.3 Changes in Urban Fabric Biodiversity Due to Air Pollution

Biodiversity is the short term used for 'biological diversity', which stands for 'the total variety of all living plants and animals, and the habitats in which they live'. The definition of term also involves the interactions between living organisms. It is well known that pollution reduces the biodiversity of flora, especially lichens, but it hardly reduces the total biological cover (Caneva *et al.*, 1995). In Rome, an increase in urban pollution in 70s and 80s was considered the main factor responsible for the reduction of biological growth on monuments (Caneva *et al.*, 1995). However, climatic factors could play a more relevant role in the biodiversity than pollution. Under favorable climatic conditions (presence of water and no extreme temperatures), some organisms can resist pollution, even in very polluted environments, namely cyanobacteria, chlorophytes and some lichens (Caneva *et al.*, 1995).

Regarding the presence of cyanobacteria in polluted areas, a study carried out by Gaylarde and Gaylarde (2000) in five countries in Latin America (Argentina, Bolivia, Brazil, Mexico and Peru) showed that the biodiversity of cyanobacteria on painted building surfaces was higher in urban areas compared to residential and rural counterparts. The number of algal species, on the contrary was lower in urban sites. Gladis-Schmacka *et al.* (2014) also hypothesised that algae and cyanobacteria colonisation could increase within the next years/decades in European coastal regions. According to the authors, an opposite trend could be seen in phototrophic biofilms coverage in European continental regions, as an increase in temperature due to pollution (concomitant effect) may affect their composition and promote cyanobacteria growth. As increasing concentrations of CO_2 can promote the growth of oceanic algae (Riebesell *et al.*, 2007), it could potentially have the same effect in biofilm-forming aeroterrestrial algae (Gladis-Schmacka *et al.*, 2014). Nevertheless, such changes have not yet been observed.

SO_2 is widely considered as the pollutant with the most detrimental impact on lichen populations, causing a reduction in its frequency and biodiversity (Hawksworth and Rose, 1970; Nash, 1973). For many years, SO_2 was the main pollutant responsible for the emergence of 'lichen deserts', *i.e.* urban areas devoid of epiphytic lichens. In Europe, and mainly in Italy, the correlation between high concentrations of SO_2 in the atmosphere followed by lichen decline has been well documented. However, the reduction of atmospheric SO_2 and the progressive improvement in air quality has allowed a subsequent recolonisation as well as a significant improvement of its diversity in urban areas (e.g. Laundon, 1970; Letrouit-Galinou *et al.*, 1992; Loppi *et al.*, 1997b; Palmieri *et al.*, 1997; Khalil and Asta, 1998; Isocrono *et al.*, 2007; Lisowska, 2011). Recent recolonising species of lichens are typically characterised by very small thalli and are most often present in the lower parts of tree trunks. The colonisation improvement was slightly offset by a general and current increase in atmospheric nitrogen deposition rates. Such increases are therefore reflected on the rise of nitrophilic lichen, such as *Xanthoria parietina*, and a decrease of acidophilic species, such as *Lecanora conizaeoides*. In London (UK), during 2002 and 2003, an assemblage of 74 epiphytic lichens mainly consisting of pollution tolerant and nitrogen tolerant species were recorded (Davies *et al.*, 2007). Among them were *Hypogymnia physodes, Lepraria incana, Xanthoria parietina, Phaeophyscia orbicularis* and *Evernia prunastri.* Many of these species remained in London with higher SO_2 and NO_x during the 70s, although their presence was mostly confined to stone (Laundon, 1970), suggesting that substratum pH was a key factor for survival.

Purvis *et al.* (2003) suggested that NO_x and calcareous particles deposited on urban surfaces appear to play the major role in influencing lichen growth in the city of London (UK). This hypothesis was later on confirmed by Isocrono *et al.* (2007) in Turin (Italy). Furthermore, Gerdol *et al.* (2014), by dividing lichen species from Ferrara (Italy) in two groups, acidophilic and nitrophilic, shows that nitrophilic species, favored by pollutants, significantly contribute to lichen diversity and have been dominant between 2006 to 2011.

SO_2 is not the only pollutant that impairs lichen proliferation and diversity. A similar trend has also been seen for Pb in urban polluted areas (Isocrono *et al.*, 2007), and for NO_2 derived from traffic emissions in Seville (Spain) (Fuentes and Rowe, 1998) and in Tuscany (Italy) (Lorenzini *et al.*, 2003). In Pistoia (Italy), although lichen recolonisation improved over the past years, a lichen desert persists because of high levels of NO_x (Loppi and Corsini, 2003).

In 2007, Davies *et al.* indicated that lichen species found in areas with high PM in London were generally the same pollution-tolerant species associated with NO_x and SO_2. Furthermore, the diversity of epiphytic species in London has increased from 1970s to 2000s, mainly due to an increase in abundance of pollution-tolerant lichen species, associated with low rainfall and minimum temperature. In Ferrara (Italy), the diversity of epiphytic lichens has also increased from 2006 to 2011, along with a decrease in nitrogen concentrations in moss tissues and no damage signs were observed in epiphytic lichen communities (Gerdol *et al.*, 2014).

Air pollutant concentrations also impact the proliferation of fungal species and bacteria. Aiming at studying the combined effect of microbial growth and atmospheric pollutants in the weathering of limestone, Pitzurra *et al.* (2003) observed in Perugia (Italy) different outcomes in fungal communities between polluted and non-polluted areas. The same trend was also shown by Mitchell and Gu (2000) in chemolithotrophic bacteria, heterotrophic bacteria, and fungi inhabiting tombstones in Cambridge (Massachusetts, USA) and its surrounding area. Despite a reduction of the diversity of microflora in polluted areas, the authors documented an enrichment of both sulfur-oxidising and hydrocarbon-utilising microflora in biofilms exposed to air pollution. This phenomenon might be explained by the dominance of one population over another, while community biomass is largely stable (Saylor *et al.*, 1982). In the present year, Villa *et al.* (2014) followed up on the study of Mitchell and Gu (2000) characterising the active SAB (subaerial biofilm) communities from the tombstones and their correlation to the amount of sulfur compounds. The investigation revealed that in addition to phototrophic sulfur bacteria belonging to the genera *Thiocapsa, Halochromatium, Allochromatium, Thiococcus* and *Thermochromatium,* other sulfate-oxidising prokaryotes (e.g. the genus *Thiobacillus*) as well as sequences of *Deltaproteobacteria* from the genus *Desulfovibrio* occurred at the polluted urban site.

8.4 Air Pollution and Biodeterioration

The environment has a fundamental role in the type and extent of deterioration of urban fabric structures. Atmospheric pollution is the cause of accelerated deterioration of the exposed materials (Brimblecombe, 2003). Although the likely impact of changes in atmospheric composition on biodeterioration of engineering structures is largely unknown, it is supposed to enhance the biodeterioration hazard (Pitzurra *et al.*, 2003; Zanardini *et al.*, 2000; Herrera and Videla, 2004; Nuhoglu 2004; Nuhoglu *et al.*, 2006). Viles and Cutler (2012) indicated that i) atmospheric CO_2 enrichment could increase the productivity of microbes (specifically cyanobacteria and green algae) and lower plants growing on heritage structures, ii) reduction in the atmospheric SO_2 concentrations may reduce direct damage to heritage structures although deteriogenic lichens may become more common and iii) atmospheric nitrous oxides (NO_x) enrichment could greatly increase growth of heritage biotas and associated soiling, biodeteriorative and bioprotective effects.

The acceleration and retardation of biodeterioration is mainly controlled by chemical processes, or biologically induced and controlled physical or mechanical phenomena (Nuhoglu, 2004). Various authors have demonstrated the combined effect of microbial colonisation and atmospheric pollutants in promoting weathering. In 2003, Pitzurra *et al.* carried out assays on limestone samples exposed to different degrees of urban air pollution in Perugia (Italy). They found a combined effect of microbial colonisation and atmospheric pollutants in promoting the weathering of limestone through acid attack within the film of water present on the surface of the exposed material, and through the oxidation of metal sulfide particulate pollutant to sulfate. These results show that, over a period of a year, biological effects more than environmental ones exacerbate stone degradation.

The contribution of air pollutants to the biodeteriorative effects of the microorganisms on the stone decay historical buildings under continental-cold climatic region in Erzurum (Turkey) was studied by Nuhoglu (2004) and Nuhoglu *et al.* (2006). Their analysis demonstrated that heterotrophic microflora, composed of fungi, actinomycetes and bacteria, dominated the stone surfaces in all stone monuments during the winter months, as the metabolism of autotrophic microorganism is restricted by the cold weather. The development of heterotrophic colonisation was favoured by airborne organic compounds arising from the incomplete combustion of fossil fuels that settled on the stone surfaces during the winter. It was evident that some of these microorganisms could deteriorate historic stone facades, using as energy resource elemental components of the stone (Si, Fe, Na, Mn, Ca, S, Al) and air pollutants (sulfur dioxide, nitrogen oxides and particles on the stone surfaces), even in wintertime under continental-cold climatic conditions.

Moroni *et al.* (2004), using laboratory and outdoor exposure tests, studied the combined effect of fungal colonisation and atmospheric pollutants in the weathering of carbonate rocks, as well as the influence of the physico-chemical features of stone on the mode and extent of weathering attained. They demonstrated a significant influence of fungal growth in promoting sulfation through the oxidation of metal sulfide particulate matter deposited on the stone surface. Moreover, they pointed out the role of the lithic substratum in controlling the extent of sulfation as the chemical and textural features of stone tend to enhance chemical weathering and fungal colonisation in the short term, and to preserve the material in the long term. Moroni and Pitzurra (2008) continued this research by analysing the rock–fungi interaction in polluted and unpolluted environments to highlight the combined action of fungal colonisation and atmospheric pollutants in the weathering of limestone. Results of their experiments show the ability of fungi to proliferate on lithic substrata, fed by aerosol particulate pollutants as nutrients, and to survive on contaminated substrata for a long time after the removal of the source of pollutants. The consequence is a general acidification of the exposed surfaces, leading to the persistence of sulfation processes even in the absence of gaseous pollutants. This point emphasises the synergistic effect of microbial contaminants and atmospheric pollutants in the deteriorating exposed materials.

Biodeterioration of stone decay depends on the degree of pollution. Michell and Gu (2000) analysed historic limestone gravestones in two different locations in Massachusetts (USA) to investigate the effects of air pollution on biofilm microflora. The authors found that sulfur-oxidising and hydrocarbon-utilising microflora were enriched in the biofilms exposed to air pollution and that the enrichment of sulfur and hydrocarbon-utilising bacteria in the biofilm may contribute to dissolution of the stone. They studied the acid production and the stone penetration by microflora growing at different concentrations of sulfur and hydrocarbon and found that even at low concentrations of sulfur and hydrocarbon, the biofilm microflora was capable of producing acid. Moreover, they were able to observe biofilm penetration in limestone pores, even though it remains unclear if deterioration is caused by the bacterial acid. Nevertheless, the presence of a large population of these bacteria in limestone biofilms in habitats contaminated by atmospheric sulfur pollution is a strong indicator of the involvement of metabolic products of the sulfur bacteria in the deterioration of historic limestone.

Along the same lines, Herrera and Videla (2004) compared the relationship between biological and atmospheric effects on the structural materials in four different environments in Latin American cultural heritage: rural (Uxmal), marine (Tulum), urban (San Ignacio) and urban-industrial (Cathedral of La Plata). The authors observed a synergistic relationship between biological and atmospheric effects on the deterioration of structural materials, although in marine and in industrial/urban atmospheres with a higher degree of pollution, weathering effects predominate.

One of the consequences of pollution is the darkening of the exposed surfaces related to the deposition of airbone particulate matter. In most cases, organisms collaborate with particulates to produce soiling as their slimy extracellular products can trap airborne particles, which stimulate microbial growth (Fig. 8.3). Viles and Gorbushina (2003) analysed the spatial and temporal pattern of soiling

Fig. 8.3 Blackening on the Galician Centre of Contemporary Art (Santiago de Compostela, Spain) chiefly due to airborne particles and the buildup of microorganisms or their remains. (Photograph taken in 1997).

and microbial colonisation on exposed stone adjacent to roads in Oxford (UK) with variable traffic flows. They found that the degree of soiling has a clear relationship with traffic and NO_2 concentrations and that fungi and other microorganisms are a key element since soiling increases as fungal numbers increase. Moreover, fungal growth was associated with soiling especially within hollows on the stone surface. This study was updated in 2006 to further investigate (1) whether there are clear differences in soiling and microbial colonisation on rock samples exposed before and after restricting the access of vehicles into the historical core of Oxford; (2) if microbes are a major component of the soiling with the potential to damage underlying stone, and (3) whether air pollution from vehicular emissions, such as NO_2, and climate influence the weathering process. The authors concluded that there is a complex relationship between air pollution, soiling, and microfloral growths as, in some areas, soiling became more pronounced with lower microbial cover, and the reverse occurring in other locations.

Urzi *et al.* (1999) argued that formation of patinas and crusts are a consequence of interactions among rock surfaces, pollutants, climate, and microorganisms. They analysed multilayer patinas developed on rocks and monument surfaces and classified its layer sequence as substratum (no differences evident according to origin and composition), calcitic layer, gypsum layer, and the outer layer. They also found that the accumulation of atmospheric dust and the bioactivity are mostly confined to the outer layer and bacterial colonisation is accompanied by decay phenomena. Both dissolution and precipitation patterns are recognisable in decay phenomena, as well as the formation of niches rich in water and nutrients that stimulate microflora proliferation. On the other hand, Prieto *et al.* (2007) stated that the formation of patinas in urban environment might not always be related to biodeterioration. By comparing dark patinas taken from granitic outcrops located in unpolluted areas and from granitic buildings located in slightly polluted areas, they observed a biological and an anthropogenic difference in origin between the two.

The effect on biodeterioration of the reduction in sulfur dioxide levels has been studied by several authors. Smith *et al.* (2011) and McCabe *et al.* (2011) documented a stronger aesthetic biodeterioration of infrastructures in London in the last ten years, due the reduction in atmospheric sulfur dioxide levels followed by an increase in biocolonisation. Inkpen *et al.* (2012) recorded limestone erosion rates of a balaustrade of St Paul's Cathedral in London (UK) and analysed its connection to pollution. They also hypothesised that as air quality has continued to improve in London, by the reduction of sulfur dioxide, the increasing colonisation of microflora, between the period of 2000 to 2010, resulted in a less pronounced microtopography due to a possible reduction in erosion rates. If so, biological colonisation of monuments can have a bioprotective impact more than a biodeteriorative one.

8.5 Conclusions

Most urban habitats are affected by elevated levels of pollutants, which interact with organisms. The interaction between air pollutants and organisms is highly variable and complex with a reduction in biodiversity and biomass as the main effects. The effect of pollution on the biology of urban fabric should account for the prevailing climatic conditions as some organisms, particularly cyanobacteria, are resistant to pollutants under favorable climatic conditions. Thus, biological growth may be enhanced in high humidity conditions in spite of changes in air pollution.

SO_2, NO_x and PM are the main pollutants that impair biological proliferation and diversity in urban environments. Because of the recent atmospheric reduction of some of these pollutants, some authors have reported an increase in phototrophic biomass in urban areas.

As well as changes in biodiversity and biomass, air pollution alters biodeterioration. In situations of atmospheric pollution, microorganisms can use pollutants as energy source and nutrients, these will enhance the biodeterioration hazard. Comparison between polluted and unpolluted areas highlighted a larger proliferation of deteriogenic organisms on materials exposed to polluted environment and in some cases dissolution of the material has been observed.

Acknowledgements

Dr. Patricia Sanmartín is supported by a postdoctoral contract within the framework of the 2011–2015 Galician Plan for Research, Innovation and Growth (Plan I2C) for 2012.

References

Abdel-Halim, K.Y., Abo El-Saad, A.M., Talha, M.M., Hussein, A.A. and Bakry, N.M. (2013). Oxidative stress on land snail *Helix aspersa* as a sentinel organism for ecotoxicological effects of urban pollution with heavy metals, *Chemosphere*, 93, pp. 1131–1138.

Agnan, Y., Séjalon-Delmas, N. and Probst, A. (2013). Comparing early twentieth century and present-day atmospheric pollution in SW France: A story of lichens, *Environ. Pollut.*, 172, pp. 139–148.

Amman, K., Herzig, R., Liebendoérfer, L. and Urech, M. (1987). Multivariate correlation of deposition data of 8 different air pollutants to lichen data in a small town in Switzerland, *Adv. Aerobiol.*, 87, pp. 401–406.

Alessio, M., Anselmi, S., Confortoc, L., Improta, S., Manesd, F. and Manfra, L. (2002). Radiocarbon as a biomarker of urban pollution in leaves of evergreen species sampled in Rome and in rural areas (Lazio — Central Italy), *Atmos. Environ.*, 36, pp. 5405–5416.

Basile, A., Sorbo, S., Conte, B., Cardi, M. and Esposito, S. (2013). Ultrastructural changes and heat shock proteins 70 induced by atmospheric pollution are similar to the effects observed under in vitro heavy metals stress in *Conocephalum conicum* (Marchantiales e Bryophyta), *Environ. Pollut.*, 182, pp. 209–216.

Batty, K., Bates, J.W. and Bell, J.N.B. (2003). A transplant experiment on the factors preventing lichen colonization of oak bark in southeast England under declining SO_2 pollution, *Can. J. Botany*, 81, pp. 439–451.

Bovallius, A., Bucht, B., Roffey, R. and Anas, P. (1978). 3-Year investigation of natural airborne bacterial-flora at 4 localities in Sweden, *Appl. Environ. Microb.*, 35, pp. 847–852.

Bretten, S., Gaare, E., Skogland, T. and Steinnes, E. (1992). Investigations of radiocesium in the natural terrestrial environment in Norway following the Chernobyl accident, *Analyst.*, 117, pp. 501–503.

Brimblecombe, P. (2003) (Ed.). *The Effects of Air Pollution on the Built Environment*, Air Pollution Reviews, Vol. 2, Imperial College Press, London.

Caneva, G., Gori, E. and Montefinale, T. (1995). Biodeterioration of monuments in relation to climatic changes in Rome between 19–20th centuries, *Sci. Total Environ.*, 167, pp. 205–214.

Caneva, G., Pacini, A., Cutini, M. and Merante, A. (2005). The Colosseum floras as bioindicators of the climatic changes in Rome, *Clim. Change*, 70, pp. 431–443.

Cicek, A., Aslan, A., Yazici, K. and Koparal, A.S. (2009). Effects of environmental conditions on historical buildings: lichens and NO_x gases, *Environ. Monit. Assess.*, 154, pp. 187–195.

Conti, M.E. and Cecchetti, G. (2001). Biological monitoring: lichens as bioindicators of air pollution assessment — a review, *Environ. Pollut.*, 114, pp. 471–492.

Davies, L., Bates, J.W., Bell, J.N.B., James, P.W. and Purvis, O.W. (2007). Diversity and sensitivity of epiphytes to oxides of nitrogen in London, *Environ. Pollut.*, 146, pp. 299–310.

Egger, R., Schlee, D. and Tuérk, R. (1994). Changes of physiological and biochemical parameters in the lichen *Hypogymnia physodes* (L.) Nyl. due to the action of air pollutants — a field study, *Phyton.*, 35, pp. 229–242.

Falla, J., Laval-Gilly, P., Henryon, M., Morlot, D. and Ferard, J.F. (2000). Biological air quality monitoring: a review, *Environ. Monit. Assess.*, 64, pp. 627–644.

Ferry, B.W., Baddeley, M.S. and Hawksworth, D.L. (1973). *Air Pollution and Lichens*, The Athlone Press, London.

Freitas, M.C., Reis, M.A., Alves, L.C. and Wolterbeek, H.T. (1999). Distribution in Portugal of some pollutants in the lichen *Parmelia sulcata*, *Environ. Pollut.*, 106, pp. 229–235.

Fuentes, J.M.C. and Rowe, J.G. (1998). The effect of air pollution from nitrogen dioxide (NO_2) on epiphytic lichens in Seville, Spain, *Aerobiologia,* 14, pp. 241–247.

Gaylarde, P.M. and Gaylarde, C.C. (2000). Algae and cyanobacteria on painted buildings in Latin America, *Int. Biodeter. Biodegr.*, 46, pp. 93–97.

Gerdol, R., Marchesini, R., Iacumin, P. and Brancaleoni, L. (2014). Monitoring temporal trends of air pollution in an urban area using mosses and lichens as biomonitors, *Chemosphere,* 108, pp. 388–395.

Gladis-Schmacka, F., Glatzel, S., Karsten, U., Böttcher, H. and Schumann, R. (2014). Influence of local climate and climate change on aeroterrestrial phototrophic biofilms, *Biofouling,* 30, pp. 401–414.

Harmens, H., Foan, L., Simon, V. and Mills, G. (2013). Terrestrial mosses as biomonitors of atmospheric POPs pollution: a review, *Environ. Pollut.*, 173, pp. 245–254.

Hawksworth, D.L. and Rose, F. (1976). *Lichens as Pollution Monitors,* Institute of Biology, Studies in Biology, Vol. 66, Edward Arnold, London.

Herrera, L.K. and Videla, H.A. (2004). The importance of atmospheric effects on biodeterioration of cultural heritage constructional materials, *Int. Biodeter. Biodegr.*, 54, pp. 125–134.

Herrera-Dueñas, A., Pineda, J., Antonio, M.T. and Aguirre, J.I. (2014). Oxidative stress of House Sparrow as bioindicator of urban pollution, *Ecol. Indic.*, 42, pp. 6–9.

Inkpen, R.J., Viles, H.A., Moses, C., Baily, B., Collier, P., Trudgill, S.T. and Cooke, R.U. (2012). Thirty years of erosion and declining atmospheric pollution at St Paul's Cathedral, London, *Atmos. Environ.*, 62, pp. 521–529.

IPCC (2013). *Climate Change 2013: The Physical Science Basis,* Intergovernmental Panel on Climate Change, Working Group I Contribution to the IPCC Fifth Assessment Report (AR5), Cambridge University Press, New York.

Isocrono, D., Matteuci, E., Ferrarese, A., Pensi, E. and Piervittori, R. (2007). Lichen colonization in the city of Turin (N Italy) based on current and historical data, *Environ. Pollut.*, 145, pp. 258–265.

Jeran, Z., Jacimov, R., Batic, F., Smodis, B. and Wolterbeek, H.T. (1996). Atmospheric heavy metal pollution in Slovenia derived from results for epiphytic lichens, *Fresen. J. Anal. Chem.*, 354, pp. 681–687.

Kekkonen, J., Hanski, I.K., Väisänen, R.A. and Brommer, J.E. (2012). Levels of heavy metals in House Sparrows (*Passer domesticus*) from urban and rural habitats of southern Finland, *Ornis. Fenn.*, 89, pp. 91–98.

Kessler, N., Schauer, J.J., Yagur-Kroll, S., Melamed, S., Tirosh, O., Belkin, S. and Erel, Y. (2012). A bacterial bioreporter panel to assay the cytotoxicity of atmospheric particulate matter, *Atmos. Environ.*, 63, pp. 94–101.

Khalil, K. and Asta, J. (1998). Les lichens, bioindicateurs de pollution atmosphérique dans la région Lyonnaise, *Ecologie,* 29, pp. 467–472 (In French).

Koivula, M.J. and Eeva, T. (2010). Metal-related oxidative stress in birds, *Environ. Pollut.*, 158, pp. 2359–2370.

Kolon, K., Samecka-Cymerman, A., Klink, A. and Kempers, A.J. (2013). *Viscum album* versus host (*Sorbus aucuparia*) as bioindicators of urban areas with various levels of pollution, *J. Environ. Sci. Heal. A*, 48, pp. 205–210.

Laundon, J.R. (1970). London's lichens, *London Naturalist*, 49, pp. 20–66.

LeBlanc, F. and De Sloover, J. (1970). Relation between industrialization and the distribution and growth of epiphytic lichens and mosses in Montreal, *Can. J. Bot.*, 48, pp. 1485–1496.

Letrouit-Galinou, M.A., Seaward, M.R.D. and Deruelle, S. (1992). On the return of epiphytic lichens to the Jardin du Luxembourg, *Bull. Soc. Bot. Fr. Lett. Bot.*, 139, pp. 115–126.

Levin, A.G. and Pignata, M.L. (1995). *Ramalina ecklonii* (Spreng) as bioindicator of atmospheric pollution in Argentina, *Canadian J. Bot.*, 73, pp. 1196–1202.

Lisowska, M. (2011). Lichen recolonisation in an urban-industrial area of southern Poland as a result of air quality improvement, *Environ. Monit. Assess.*, 179, pp. 177–190.

Loppi, S., Nelli, L., Ancora, S. and Bargagli, R. (1997a). Passive monitoring of trace elements by means of tree leaves, epiphytic lichens and bark substrate, *Environ. Monit. Assess.*, 45, pp. 81–88.

Loppi, S., Giovannelli, L., Pirintsos, S.A., Putorti, E. and Corsini, A. (1997b). Lichen as bioindicators of recent changes in air quality (Montecatini Terme, Italy), *Ecol. Mediterr.*, 23, pp. 53–56.

Loppi, S. and Corsini, A. (2003). Diversity of epiphytic lichens and metal contents of Parmelia caperata thalli as monitors of air pollution in the town of Pistoia (c Italy), *Environ. Monit. Assess.*, 86, pp. 289–301.

Lorenzini, G., Landi, U., Loppi, S. and Nali, C. (2003). Lichen distribution and bioindicator tobacco plants give discordant response: A case study from Italy, *Environ. Monit. Assess.*, 82, pp. 243–264.

McCabe, S., Smith, B., Adamson, C., Mullan, D. and McAllister, D. (2011). The 'greening' of natural stone buildings: quartz sandstone performance as a secondary indicator of climate change in the British Isles?, *Atmos. Clim. Sci.*, 1, pp. 165–171.

Majumder, S., Mishra, D., Ram, S.S., Jana, N.K., Santra, S., Sudarshan, M. and Chakraborty, A. (2013). Physiological and chemical response of the lichen, *Flavoparmelia caperata* (L.) Hale, to the urban environment of Kolkata, India, *Environ. Sci. Pollut. Res.*, 20, pp. 3077–3085.

Malá, J., Máchová, P., Cvrčková, H. and Vaněk, T. (2007). Heavy metals up-take by the hybrid aspen and rowan-tree clones, *J. Forest Sci.*, 53, pp. 491–497.

Mitchell, R. and Gu, J.D. (2000). Changes in the biofilm microflora of limestone caused by atmospheric pollutants, *Int. Biodeter. Biodegr.*, 46, pp. 299–303.

Monaci, F., Bargagli, R. and Gasparo, D. (1997). Air pollution monitoring by lichens in a small medieval town of central Italy, *Acta. Bot. Neerl.*, 46, pp. 403–412.

Moriarty, F. (1999). *Ecotoxicology: The Study of Pollutants in Ecosystems*, Academic Press, London.

Moroni, B. and Pitzurra, L. (2008). Biodegradation of atmospheric pollutants by fungi: a crucial point in the corrosion of carbonate building stone, *Int. Biodeter. Biodegr*, 62, pp. 391–396.

Moroni, B., Pitzurra, L. and Poli, G. (2004). Microbial growth and air pollutants in the corrosion of carbonate building stone: results of laboratory and outdoor experimental tests, *Environ. Geol.*, 46, pp. 436–447.

Morre, T.R., Bublier, J.L., Heyes, A. and Flett, R. J. (1995). Methyl and total mercury in boreal wetland plants, experimental lakes area, Northern Ontario, *J. Environ. Qual.*, 24, pp. 845–850.

Nash III, T.H. (1973). Sensitivity of lichens to sulfur dioxide, *Bryologist*, 76, pp. 333–339.

Nimis, P.L. (1987). I macrolicheni d'Italia, chiavi analitiche per la determinazione. Atti del Museo Friulano di Storia Naturale, *Udine*, 8, pp. 101–220 (In Italian).

Nimis, P.L. and Tretiach, M. (1995). The lichens of Italy — a phytoclimatical outline, *Crypt. Bot.*, 5, pp. 199–208.

Nuhoglu, Y. (2004). The biodeteriorative action of microorganisms and the effects on stone monuments under air pollution and continental-cold climatic condition in Erzurum, Turkey, *Fresen. Environ. Bull.*, 13, pp. 591–599.

Nuhoglu, Y., Oguz, E., Uslu, H., Ozbek, A., Ipekoglu, B., Ocak, I. and Hasenekoglu. Ī. (2006). The accelerating effects of the microorganisms on biodeterioration of stone monuments under air pollution and continental-cold climatic conditions in Erzurum, Turkey, *Sci. Total Environ.*, 364, pp. 272–283.

Nylander, W. (1866). Les lichens du Jardin du Luxembourg. *Bull. Soc. Bot. France*, 13, pp. 364–372 (In French).

Palmieri, F., Neri, R., Benco, C. and Serracca, L. (1997). Lichens and moss as bioindicators and bioaccumulators in air pollution monitoring, *J. Environ. Pathol. Toxicol. Oncol.*, 16, pp. 175–190.

Pitzurra, L., Moroni, B., Nocentini, A., Sbaraglia, G., Poli, G. and Bistoni, F. (2003). Microbial growth and air pollution in carbonate rock weathering, *Int. Biodeter. Biodegr.*, 52, pp. 63–68.

Poblet, A., Andrade, S., Scagliola, M., Vodopivez, C., Curtosi, A., Pucci, A. and Marcovecchio, J. (1997). The use of epiphytic Antartic lichens (Usnea aurantiacoatra and Usnea antartica) to determine deposition patterns of heavy metals in the Shetland Islands, Antarctica, *Sci. Total Environ.*, 207, pp. 187–194.

Prieto, B., Aira, N. and Silva, B. (2007). Comparative study of dark patinas on granitic outcrops and buildings, *Sci. Total Environ.*, 381, pp. 280–289.

Puckett, K. J. (1988). 'Bryophytes and lichens as monitors of metal deposition' in: Nash III, T.H., Wirth, W. (Eds.), *Lichens, Bryophytes and Air Quality*, Bibl. Lichenol., Vol. 30, Cramer in der Gebruder Borntraeger Verlagsbuchhandlung, Berlin, pp. 231–267.

Purvis, O.W., Chimonides, J., Erotokritou, L., Jeffries, T., Jones, G.C., Louwhoff, S., Read, H. and Spiro, B. (2003). Which factors are responsible for the changing lichen floras of London?, *Sci. Total Environ.*, 310, pp. 179–189.

Radwan, M.A., El-Gendy, K.S. and Gad, A.F. (2010). Biomarkers of oxidative stress in the land snail, *Theba pisana* for assessing ecotoxicological effects of urban metal pollution, *Chemosphere*, 79, pp. 40–46.

Riebesell, U., Schulz, K.G. and Bellerby, R.G.J. (2007). Enhanced biological carbon consumption in a high CO_2 ocean, *Nature*, 450, pp. 545–548.

Saiz-Jimenez, C. (1995). Deposition of anthropogenic compounds on monuments and their effect on airborne microorganisms, *Aerobiologia*, 11, pp. 161–175.

Saiz-Jimenez, C. (1997). Biodeterioration vs biodegradation: the role of microorganisms in the removal of pollutants deposited onto historic buildings, *Int. Biodeter. Biodegr.*, 40, pp. 225–232.

Saunders, P.J.W. (1970). Air pollution in relation to lichens and fungi, *Lichenologist*, 4, pp. 337–349.

Sawidis, T., Marnasid, A., Zachariadis, G. and Stratis, J.A. (1995). A study of air pollution with heavy metals in Thessaloniki City (Greece) using trees as biological indicators, *Arch. Environ. Con. Tox.*, 28, pp. 118–124.

Saylor, G.S., Perkins, T.W., Mallory, R.E., Shiaris, L.M. and Petersen, M.O. (1982). Impacts of coal-coking effluent microbial communities: a multivariate approach, *Appl. Environ. Microbiol.*, 44, pp. 1118–1129.

Schröder, W., Pesch, R., Englert, C., Harmens, H., Suchara, I., Zechmeister, H.G., Thöni, L., Mankovská, B., Jeran, Z., Grodzinska, K. and Alber, R. (2008). Metal accumulation in mosses across national boundaries: uncovering and ranking causes of spatial variation, *Environ. Pollut.*, 151, pp. 377–388.

Sicolo, M., Tringali, M., Orsi, F. and Santagostino, A. (2009). Porphyrin pattern and methemoglobin levels in *Columba livia* applied to assess toxicological risk by air pollution in urban areas, *Arch. Environ. Contam. Toxicol.*, 57, pp. 732–740.

Smith, B.J., McCabe, S., McAllister, D., Adamson, C., Viles, H.A. and Curran, J.M. (2011). A commentary on climate change, stone decay dynamics and the 'greening' of natural stone buildings: new perspectives on 'deep wetting', *Environ. Earth. Sci.*, 63, pp. 1691–1700.

Sterflinger, K. (2000). Fungi as geologic agents, *Geomicrobiol. J.*, 17, pp. 97–124.

Swaileh, K.M. and Sansur, R. (2006). Monitoring urban heavy metal pollution using the House Sparrow (*Passer domesticus*), *J. Environ. Monit.*, 8, pp. 209–213.

Urzì, C., Garcia-Valles, M., Vendrell, M. and Pernice, A. (1999). Biomineralization processes on rock and monument surfaces observed in field and in laboratory conditions, *Geomicrobiol. J.*, 16, pp. 39–54.

Viles, H.A. and Cutler, N.A. (2012). Global environmental change and the biology of heritage structures, *Glob. Chang. Biol.*, 18, pp. 2406–2418.

Viles, H.A. and Gorbushina, A.A. (2003). Soiling and microbial colonisation on urban roadside limestone: a three year study in Oxford, England, *Build. Environ.*, 38, pp. 1217–1224.

Villa, F., Vasanthakumar, A., Mitchell, R. and Cappitelli, F. (2015). RNA-based molecular survey of biodiversity of limestone tombstone microbiota in response to atmospheric sulfur pollution, *Lett. Appl. Microbiol.*, 60, pp. 92–102.

Zanardini, E., Abbruscato, P., Ghedini, N., Realini, M. and Sorlini, C. (2000). Influence of atmospheric pollutants on the biodeterioration of stone, *Int. Biodeter. Biodegr.*, 45, pp. 35–42.

Zechmeister, H.G., Richter, A., Smidt, S., Hohenwallner, D., Roder, I., Maringer, S. and Wanek, W. (2008). Total nitrogen content and d15N signatures in moss tissue: indicative value for nitrogen deposition patterns and source allocation on a nationwide scale, *Environ. Sci. Technol.*, 42, pp. 8661–8667.

CHAPTER 9

The Whole Building and Patterns of Degradation

Rob Inkpen

Department of Geography,
University of Portsmouth, United Kingdom

9.1 Introduction

Degradation of natural stone surfaces occurs at specific locations on buildings, each set within a complicated arrangement of architectural elements. Each building, is a unique outcome of its past history of conservation treatments and a myriad of local and distant decisions that impact upon its local pollution and climatic environment. This complex context can retard or enhance the nature and rate of degradation of a particular surface relative to what might be expected from focusing research on the individual surface alone. Buildings may be simple monuments, such as standing stones, or complex configurations such as Norman cathedrals whose apparently static nature belies an evolution to their degradational character through changing conservation practices. Any surface, any building subject to degradation can not be divorced from the socio-economic framework, itself a changing phenomenon, that establishes and maintains that building and surface.

Academic research often fails to consider the complexity of the building as a whole instead focusing on building stone as an abstracted, context-less entity transferring the results from one study to another without any analysis of the significance of context on the results. Laboratory-based and small-scale studies have outlined a range of degradation mechanisms for different building materials. From these studies a whole host of results have tried to, among other things, construct indices of degradation, assess the effectiveness of different stone types and conservation treatments and identification and quantification of key erosional processes. All these results have the underlying assumption that the degradation processes and material responses identified at these small-scales operate in the same manner at the large-scale; scaling up is just a simple matter of a scaling up in the quantity rather than a qualitative change. Likewise, the environmental system and its variation that determines the occurrence and rates of change of degradational processes is assumed to be a mirror-image of the conditions constructed and strictly controlled in the laboratory. In both cases the building scale is seen merely as an extrapolation of process/form relationships and their associated environmental parameters identified at the small-scale. This chapter begins by exploring the role and limitations of small-scale studies for identifying and quantifying the processes and relationships causing stone degradation at the scale of the whole building.

Based partly on the knowledge obtained from these small-scale studies and partly from experience and field observation, as well as the iterative process of field and laboratory studies confirming each other, some researchers have tried to classify the nature of degradation on buildings into a typology of types. Using such classification schemes, the spatial variability of degradation forms thought to be associated with different erosional agents can be identified and mapped across the building (for example in Inkpen *et al.*, 2008). Similarly, the spatial variability of environmental parameters such as moisture (e.g. Hall *et al.*, 2011; McCabe *et al.*, 2013) or chemical and mineralogical composition (Cultrone *et al.*, 2012; Alves *et al.*, 2013; La Russa *et al.*, 2013) can be sampled and mapped over a surface. Combining the two distributions a correlation can be established between different degradational forms and agents with the implicit assumption of a causal relationship being present between the two. Some of these classification systems could be seen as predictive, highlighting the potential developmental pathways that degradation will take as well as where specific degradational agents will be dominant on a building. Although such classification schemes begin to tackle the complexity of degradation, they tend to look at the forms produced rather than the context of their production (Andre *et al.*, 2014) and so do not tend to consider the interaction between the production context on the building and the form identified. This chapter considers the potential and limitations of these classification schemes.

Context is not an easy term or concept to pin down. It is multidimensional as well as spatially and temporally differentiated. Context could include the general degradational environment that lies beyond the scope of the building itself (although the building itself can influence at least the nearby environment, e.g. Roekens *et al.*, 1988). The composition of the building materials, a vital factor in the complexity of interactions on a building, could also be considered, but the detailed decisions for the selection of particular building materials and particular architectural forms lie deep within the social and cultural milieux of past and present. The degradational system could be viewed hierarchically with the nature of the degradational environment setting the constraints on the type of degradational agents present and the potential for their interactions and so, indirectly, the range of degradational forms and processes found on a building. The realisation of this potential depends upon how the potential is mediated and expressed through the actual complex of interactions that occur on the building itself.

At the scale of the building itself degradation can be studied using the concept of the *catchment area* as the fundamental unit on a building. The concept, borrowed from fluvial research (Chorley, 1969), allows an area to be identified based on the pattern of water flow. Specifically, a catchment is an area that drains water falling onto it to a common outlet. Degradational process and agents will commonly be related to the distribution of water flow and the basis of the areal interactions of degradational processes and agents will depend on the channelling of water and also the characteristics of the catchment area. This chapter considers some of the conceptual issues of complex and non-liner relationships that can result from the interactions of water flows within and between catchments.

A historic building is not just a physical entity; a building has a socio-economic as well as a cultural existence all of which form the final focus of the chapter. Analysis of decay at the scale of a small surface neglects the tendency for buildings to be viewed holistically from a cultural perspective. The cultural context of a building and its possible importance as a cultural icon relate to its whole fabric rather than just a specific surface, but the relationship between cultural conservation and decay can often be seen to impact surface through policies and practices enmeshed in specific cultural contexts and their varied interpretation. The mediation of cultural values is through the network of actors such as architects, conservators, and stone masons, who collectively make decisions and interact with the stone surface. Recent research into the use of the Rock Art Stability Index (Dorn *et al.*, 2008; Allen and Lukinbeal, 2011) and its application has highlighted the importance of considering the cultural context of analysis as well as the human aspects of applying specific classifications. Unless the analysis system is considered as a whole, issues about the consistency of application are likely to arise.

Decisions on the financing of construction and conservation work can influence how decay agents interact with a surface. The socio-economic milieu also relates to wider issues over which the various actors in the network have no direct control such as global environmental change and sustainability. These larger-scale trends form the framework within which conservation is undertaken and set constraints upon the scope of activities and responses available for maintaining the fabric of the building.

9.2 Limitations of Small-Scale Studies of Stone Degradation

Large-scale international surveys of stone degradation such as the NATO/CCMS project in Europe and the National Materials Exposure Programme (NMEP) in the UK have tended to use small samples to collect data (Butlin *et al.*, 1993., Yates and Butlin, 1996). In these studies standard cut samples of stone of different types (e.g. Portland Stone Limestone, Monk's Park Limestone and White Mansfield in the case of NMEP), roughly of dimensions 50x50x10mm or similar, are exposed to environmental conditions for prolonged periods of time, in the case of the NMEP survey periods of

6 months. Stone tablets are exposed, often as unsheltered and sheltered replicas to enable the assessment of the differential impact of wet and dry deposition, and then the characteristics of the tablets, such as weight change, are assessed non-destructively, while a subset of tablets are analysed destructively to analyse changes in chemical composition. Usually weight change is the key erosional index used to characterise the decay behaviour of the stone with the weight change of all the tablets being averaged to derive the weight change for a particular location. Weight change is sometimes converted, via a set of assumptions concerning standard dimensions and homogeneous loss over a surface, to produce the average erosional height loss for a location.

The index of change for the stone is often plotted against indices representing the nature of the general erosional environment for that location such as sulfur dioxide levels, nitrogen oxides levels and precipitation. These environmental parameters are averaged or expressed as cumulative values over the exposure period for the tablets. The resultant curve representing the relationship, often referred to as the damage function, tends to be expressed as a linear relationship and can have potential implications for policy-makers as the relationship hints at a simple way of altering environmental variables to produce a clear impact upon decay rates. For a given decrease in sulfur dioxide, for example, the regression equation predicts how much erosional loss will decrease by. Such an analysis has to be viewed within the general move towards analysing critical loads for the environment as a policy tool (Bull, 1993; Bull, 1995; Cowell and ApSimon, 1995, Tidblad *et al.*, 2012) as well as for an index that can be transferred to predictive models of change with relative ease (Brimblecombe and Grossi, 2008, 2009). Some studies have used runoff slabs to derive the chemical mass balance equation for stone loss (e.g. Livingston, 1986; Reddy, 1998; Moses, 1996; see Blocken *et al.*, 2013 for review of recent studies and of impact of buildings on fluid flow) and related their finding to similar or the same indices of environmental conditions. Although such runoff slabs tend to have larger dimensions than the tablets (100 or 150 mm) they are still relatively simple and small representations of erosional and degradational surfaces.

Laboratory-based studies of degradation tend to focus on the impact of a single, or at best a small number of erosional processes and agents (Goudie, 1974, 1999; Robinson and Williams, 1982; Smith and McGreevy, 1983). These studies have tended to focus on illustrating the potential power of an agent to cause degradation and then advance by trying to make the parameters of the simulation more 'realistic'. An example of this is the quasi-standard test for durability in the UK, the sodium sulfate crystallisation test (Ross and Butlin, 1989; also prEN12370; Inkpen *et al.*, 2004). The test uses a saturated salt solution and a heating and cooling cycle to induce 'unnatural' weathering cycles. Weight loss of standard 40mm cubes under this weathering regime are then used to assess the general durability of stone to degradation relative to a standard set of limestone cubes. As noted by Inkpen *et al.* (2004), the test was originally designed to simulate frost weathering and so was meant to illustrate the potential for decay from this process. Even within the terms of its original design the test subjected samples to large variations in temperature as well as rapid temperature changes and unrealistic supply of salt solution. Use of the test as an indicator of general durability rests on the assumption that the degradational forms observed in the original trails in the mid-1920s matched the degradational forms observed on buildings. A further assumption was then made that these forms could be simulated at a smaller scale and were directly related across scales.

More complex experimental work has tried to model reality in greater detail either through manipulating the salt supply and its nature or by combining more variables at magnitudes measured in the field (e.g. Goudie, 1974; Smith *et al.*, 1987; Smith and McGreevy, 1988; Ruedrich *et al.*, 2011, Smith *et al.*, 2011, Salvadori, *et al.*, 2014). Such studies may seem to advance towards the goal of simulating 'real' degradation, but they are still restricted in the complexities and scale of degradation they can produce. These are still laboratory-based studies and so are still simplifications of reality. The variables considered and controlled are only those already identified as being important for degradation from other laboratory-based studies and field investigations that may assume form-process links. The degradational system is, however, defined by the researcher and so only those variables identified, probably those capable of measurement, and which are capable of simulation are features of the system that will be analysed in the laboratory. This can severely restrict the nature of degradational forms and processes simulated. Salt supply, for example, can vary both in terms of its mode of delivery, the spatial and temporal characteristics of its magnitude and composition (McGreevy, 1982). Altering any of these features, either singularly or in combination, could alter the nature and rate of degradation induced in the experimental samples. Variables that are difficult to measure or model may be excluded from analysis as will those not identified or considered inappropriate by the researcher such as the second to second variability of sulfur dioxide levels. Similarly, the experimental sample is analysed in isolation form its context on the building.

An analysis of 30 years of erosion rates at St Paul's Cathedral, London highlighted the potential differences between small tablet studies and analysis of change on the stone surface itself (Inkpen *et al.*, 2012). Using damage or dose-response functions derived from small tablet studies (Lipfert, 1989; Tidbald *et al.*, 2001) erosion rates of 12 to 15 microns per year are predicted for the environmental conditions experienced at St Paul's Cathedral from 1980–2010. Measurement of erosion from the actual stone surface using a micro-erosion meter (MEM) gives an erosion rate of 49 microns per year in the 1980s down to 35 microns per year by the early 2000s. The ratio between the erosion rates derived from dose-response functions as opposed to measured rates is about 1:3 over the three decades although this ratio drops to 1:2.5 if the erosion rates determined using the lead plugs in the stone are used as a baseline. These lead plugs have been in place since the cathedral was constructed so this ratio represents the difference over more than 300 years of erosion and suggests the contemporary ratio is consistent with historic differences. The MEM and dose-response function based on small tablet studies are different measurement systems (Rhoads and Thorn, 1996; Inkpen, 2005) which although converging on a common indicator of erosion, amount of loss, use different methods and underlying concepts to derive the seemingly same indicator. The MEM measures change repeated on the same 120 points within an area of $45cm^2$, whereas the tablets have a surface area of $70cm^2$ although not all that area is exposed to rainfall all the time. Unlike 'real' stone surfaces, the tablets have 'edge effects' where the initial sharp edges may be preferentially eroded. Attaching and reattaching the tablets to an exposure frame will also produce erosion that surface do not experience (Inkpen, 1995). Importantly, the first 18–24 months of exposure of the tablets is likely to be needed to overcome the smoothness of the surface caused by the cutting of each tablet (Moses, 1994).

In terms of erosional processes, there could be major differences between the action of decay agents on tablets and surfaces that could explain the difference in erosion rates. Changing scale from a small tablet to a surface changes the active relations that cause decay (Inkpen, 2011) particularly providing opportunities for decay forms that are larger than a tablet (Moses, 1996). Stone surfaces can have different combinations of histories and reflect different past environmental conditions and hence produce a mix of localised areas of high erosion (e.g. Smith and Prikryl, 2007; Smith *et al.*, 2008; Inkpen *et al.*, 2008). On a flat surface, for example, distinct runoff zones and channels can develop as there is sufficient runoff distance for them to form. Running across a surface there may also be sufficient time for runoff to become saturated with decay products and so have an ineffective impact upon eroding the stone surface. Moses (1994) found that tablets and large blocks exposed for six months followed different developmental paths, with blocks developing visible black crusts and biological and salt crusts. The presence of established microbial communities on the surface as opposed to tablets may also impact upon erosion rates (Viles, 2012; Viles *et al.*, 2008). Likewise, the thin tablets are likely to respond differently to environmental stresses than a surface. The more likely rapid response of the tablets to temperature changes, for example, may induce thermal stresses (Hall and Andre, 2001; Hall, 2004) and so accelerate erosion relatively. The potentially complicated impact of salt stored in the stone, the 'memory' effect (Cooke, 1989; Inkpen, 1991; Smith *et al.*, 2010) could also result in increased erosion from the exposed surface as opposed to the tablet.

The above discussion highlights the restrictions of small-scale field trails and laboratory-based studies in explaining degradation at the scale of the real building and even at the scale of a surface on a real building. This is not to imply, however, that such studies do not serve a useful purpose. In the St Paul's paper mentioned above (Inkpen *et al.*, 2012), although the erosion rates for the exposed surfaces were higher, the pattern of change in response to declining pollution levels was similar for both measurement systems. Likewise, controlled examination of the specific weathering agents and associated mechanisms under a set of limited conditions can help to identify the potential power of agents and combination of agents to induce degradation. Such studies are not able to cope with or illuminate the complexity of interactions that produce the patchwork mosaic of degradational forms on a building.

9.3 Classification of Building Degradation

A vital aspect in identifying and quantifying the complexity of degradation on a building is the satisfactory classification of decay forms and their patterns. Most classification system try to link process to form and thereby imply a causality between the two. This tends to mean that correlation is equated with causation, sometimes with only tenuous evidential basis from laboratory experiments or traditional explanations. Such linkages may focus on only identifying one weathering agent as significant when in reality it is the juxtaposition of contingent factors, such a the state of the stone surface, the continued supply of salt or removal of weathering products, and other factors that produce the specific pattern of forms.

Camuffo *et al.* (1982) developed a relatively simple classification based upon surface colour. White surfaces are those washed by rainwater and represent areas of surface loss while black areas are areas sheltered from rainfall and runoff and are zones of deposition where crusts, such as sulfation crusts, can form. Grey areas are areas of dust deposition rather than transitional zones between the two. Location and its relation to the degree of exposure to rainfall and runoff is central to this classification scheme. Other classification schemes have focused on identifying and mapping a selected range of features. Robinson and Williams (1996), for example, identified and mapped areas of active and inactive cavernous weathering on sandstone blocks on churches of different ages in West Sussex. They suggested that cavernous decay peaked in the eighteenth and nineteenth centuries coincide with high levels of industrial activity in the region, implying a causal link between the two. Antill and Viles (1998) use a classification scheme based on colour and visual identification of surface disruption to classify over 700 individual stone blocks in an external wall of Worcester College, Oxford. Maps were produced by sketching the wall and colouring each block according to its class. Patchy and distinct patterns of specific decay forms could be identified and related to the location of blocks in the wall. This suggested that processes and agents of decay had a locational component to their operation. Inkpen *et al.* (2001) and Inkpen *et al.* (2008) carried out similar studies, the latter using historic photography, but integrated the data into a geographical information system (GIS). In their study a predefined unit of classification, the block, was avoided by using a block map as the base upon which decay forms were recorded. This meant that researcher was not restricted to defining percentage covers for blocks or the singular allocation of blocks to a specific decay class.

Central to any scheme, however, is the development of an appropriate and standardised classification for decay forms. Fitzner (1990) developed a system based on a hierarchy of feature classification (Fitzner *et al.*, 1992; Fitzner *et al.*, 1996). The classification scheme has four groups of weathering forms at the top level of the hierarchy: loss of stone material, deposits, stone material detachment and fissures/deformation. Within each of these groupings there are further subdivisions into 29 weathering forms with finer divisions down to individual weathering features. Further differentiation is possible by adding an intensity parameter to each individual feature. Although comprehensive, the application of this hierarchical classification scheme relies on an experienced researcher capable of identifying each form and its position within the scheme.

The ICOMOS classification scheme provides an illustrated version of a classification scheme that is designed to improve communication between practitioners by simplifying and demystifying the scientific information used for classification by setting out a common language for form description. The glossary is divided into six families (general terms, cracks and deformation, detachment, features induced by material loss, discolouration and deposit and biological colonisation), which are divided into between 2 and 11 terms. This scheme retains and builds upon the hierarchical structure of other classification schemes, retaining the central idea that forms must be detectable or visible to the naked eye. Classification scheme have been derived and applied to specific conservation projects and used to provide a permanent record of the state of decay of a structure, e.g. York Minster (Holton, 2010).

Although classification schemes have become increasingly complex and increasingly computer-based, they still tend to focus on specific features of the building or individual surfaces rather than encompass the building as a whole. Even where this is the case, explanation still tends to focus on the building rather than include the context as part of the explanation of decay forms. Context, like durability, is an elusive yet vital term for decay. At the large scale, context can include factors such as the pollution environment or climate (Sabbioni *et al.*, 2010) which determines the types and quantities of pollutants available for reaction with the stone as well as the parameters that determine their rates of reaction such as temperature and precipitation (Brimblecombe, 2009; Sabbioni *et al.*, 2009; Grossi *et al.*, 2011). These parameters and their impact are likely to alter as climate changes. A simple, first approximation, for example could divide pollution environments into urban and rural (Cooke, 1989) with subtler divisions based on the variability of specific components. Recent trends in Western Europe at least, suggest that this simple division in urban and rural may have been valid in the recent past (Attewell and Taylor, 1988; Kupper and Pissart, 1976; Cooke *et al.*, 1995), but there has been a narrowing, and in some areas parity, in pollution levels (Inkpen, 1989).

Although the large scale context may set the limits on the nature of interactions possible, changing the scale of analysis changes the nature of the context relevant to building degradation. The building itself becomes an active rather than a passive creator of its own environment. In this manner, it is analogous to the relationship between an organism and its environment in evolution as outlined by Levins and Lewontin (1985). The organism only obtains information about the environment mediated by its senses. This means that environmental fluctuations may be reduced or enhanced in magnitude by these senses and the organism responds accordingly. The organism can create its own environment, via feather and fur for example, to alter the context creating a specific, beneficial micro-environment. The building could be seen in the same manner, but without the sensory coordinating aspects of an organism. Fluctuations in variables such as sulfur dioxide can be measured over small temporal and spatial scales, but the effect of these scales on the decay of the stone is unclear. These changes are mediated throughout the surface of the stone, which may have differential responses to different scales of change and to different types of pollutant. There is little literature to suggest that appropriate scales of change have been determined for the response of a building or even a single surface.

There does, however, appear to be a tendency for forms to develop at particular spatial and temporal dimensions as originally noted for marine environments by Stommel (1963). Inkpen (2011) noted the tendency for 'clumpiness' in the spatiotemporal scales at which geomorphic forms emerged and decay forms seem to exhibit a similar tendency to form at either sub-block and inter-block scales rather than at the scale a whole surface. The mediation of the context or general environment through the building itself may partly explain this clumpiness. The structure of any building results in the creation of a numerous micro-environments that either enhance or reduce both the magnitude and frequency of environmental conditions as measured away from the building. Such smaller scale microenvironments are the contexts within which decay occurs and imply the appropriate scale for form identification and process analysis. Creation of micro-environments is not a random process as any building is designed for a purpose that dictates its general structure and layout. Conservation practices add another layer to building design and also to micro-environment creation. The addition of these human design elements to a physical system results in unintentional, but structured micro-environments that are common to buildings of that type. This means that it may be possible to map and predict the nature of micro-environments between buildings and the likely scale of forms and process operation.

Although a micro-environment may have been identified, this does not mean that distinct units of study have also been identified. Micro-environments may merge and interact across a building surface, so analysis can be better served if it focuses on a distinct, functional unit of study. Defining sections of the building as catchment areas may help in this as the catchment is an established and fundamental unit of analysis in hydrology (Chorley, 1969). The particular characteristic that makes it such a useful analytical unit for hydrology are also transferrable to building studies, i.e., ease of demarcation, nested structured, ordered and interconnected, topologically and geometrically integrated and the basic unit for erosional/depositional and input/output analysis for the catchment as a system. A building catchment represents the area contributing to a runoff outlet or set of outlets, which may be linear or point features, e.g. a drainage channel or a number of drainage points across an area. Definition of the catchment area and characterisation of it using a few simple parameters indicates how water and material is transported about and between catchments. In particular, such descriptions could help to clarify the hierarchical and interconnected nature of the drainage system and the location and effectiveness of storage and removal networks across the building. Inputs and outputs are now tied to a defined, functional unit and any surface studied can be positioned within the context of this unit.

Building catchments, unlike their 'natural' counterparts in hydrology, have not developed in response to tectonic and erosional processes; rather they are designed with specific purposes in mind. Within each catchment there will be elements that can be regarded as either unconnected micro-catchments or flow barriers depending on the scale of study. These are usually isolated features which tend to deflect or concentrate runoff. Their presence can, therefore, be important in determining patterns of material movement and so impact upon the nature of the surface being eroded. Seemingly similar surfaces in terms of their slope material composition and conservation histories may, in fact, have differing degrees of connectivity to other micro-catchments and so erode differently. The catchment framework provides a conveniently scale-independent framework for incorporating and interpreting how the human dimension is likely to impact upon stone degradation.

The importance of the building and its collection of micro-catchments as mediators of the environment is an important consideration when analysing the potential impacts of environmental change. Any impact from changing climates will be mediated by the building itself. An increase in external temperature, for example, could result in a decrease in the amount of heating required in some buildings. This will alter the thermal gradient through the building materials and so reduce thermal stress. Decreasing the steepness of the thermal gradient could also influence particle deposition onto the building surface by altering the flow of air and so affecting settling velocities. Increased temperatures may, however, also enhance the urban heat island and so increase the use of air conditioning units which could locally alter thermal regimes as well, thereby cooling buildings. Environmental change is also likely to increase the frequency of extreme conditions and hence increase the episodes of high thermal stress which can be vital in causing stress induced degradation (e.g. Hall, 2004; Hall and Thorn, 2011). A similar complex-mediated sequence of alternatives can also be put forward for changes in rainfall amounts. A decrease in the rainfall amount can result in increased periods of sulfation for limestone surfaces. Increases in the frequency of extreme events, however, will result in the generation of runoff events of higher velocity and greater potential erosivity. The pathways such flows take will still depend, however, upon the specific configuration of micro-catchments on any individual building. The unique conservation history of each building will mean that the past histories of each building will be interacting with the mediated impacts of environmental change, making detailed prediction difficult. There may, however, be key pathways of degradation that such interactions will generally follow. It is the identification of these potential pathways, maybe through the analysis of past changes, such as in Inkpen *et al.* (2008) or Smith *et al.* (2010), which might help in making informed management decisions concerning conservation treatments.

9.4 Building and Complex Interactions

Buildings are composed of a set of surfaces arranged in a series of configurations. These configurations can impact upon the flow and movement of fluids across and between surfaces. To this must be added the complication that such configurations may involve different materials and there is the potential for a whole array of complex interactions. Figure 9.1 outlines some simple initial types of surfaces and their interactions. A surface needs to be viewed as part of a hierarchy of surfaces that interact to either, deflect, concentrate or retard/retain fluid movement. Each of these actions can affect the nature and rate of interaction between the fluid and the surface and so affects the nature of decay and its rate. The action of the surface hierarchy need not be constant however. Changing the rate of flow through, for example, different rainfall intensities or initial saturated conditions can change the dynamics of fluid flow. This can produce a hierarchy of surfaces and so distinct zones of decay upon a building. Surface A, for example, could be a relatively flat surface at the base of a steep slope that is the focal point for runoff from a sub-catchment of the building. Surface A will receive a high volume of flow in major

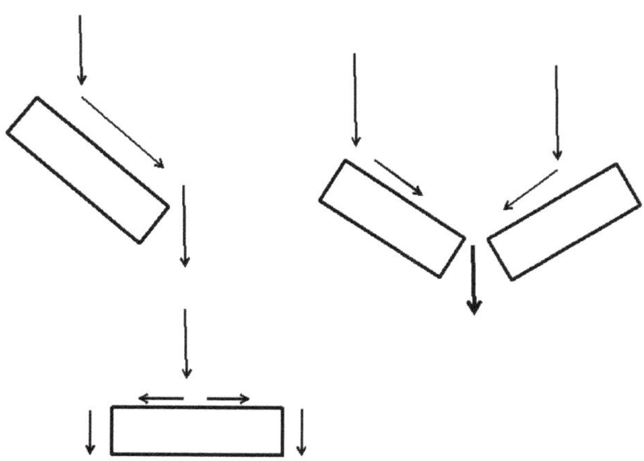

Fig. 9.1 Impact of surfaces on flow patterns: arrows represent path of water flow, its volume and velocity whilst the blocks are basic architectural features on the surface. Flow impact are deflect (top left), concentrate (top right) and retard (bottom left).

rain events, but this flow may have a high velocity over the surface so the possibility of interaction is limited. Under low rainfall events, the catchment may not be sufficiently saturated to produce over-surface flow and so surface A does not become wetted and so dissolution is limited. It is only under a range of specific flow conditions generated by rainfall events of a specific range of intensities that flow conditions result in a concentration of effective erosional flows across surface A, i.e. flow of sufficient volume and velocity to induce chemical reactions across the surface. This hypothetical example illustrates how a particularly configuration of surfaces will have a particular set of flow characteristics associated with it that maximise potential erosion.

Figures 9.2 and 9.3 represent classification of decay by colour using repeat imagery of the front of the Judges Lodgings in Oxford taken in 1892 and 1932 respectively. The building is constructed of Headington stone, a local oolithic limestone and suffered severe erosion in this 40 year period that coincided with levels of major atmospheric pollution in Oxford (Viles, 1996). The important thing to note is that there is a distinct pattern to this erosion that is a function of the interaction of the building surfaces and hence the relations between agents of erosion and the stone surface. The relatively clean and smooth surface in the 1892 image is replaced by a blackened

Fig. 9.2 Classified image of decay on front of Judges Lodgings, Oxford, 1892.

Fig. 9.3 Classified image of decay on front of Judges Lodgings, Oxford, 1932.

set of surface with severe flaking and blistering present in specific sites on the front of the building. The blackened areas are located in surfaces that tend to be sheltered from rainfall and runoff and so decay products such as sulfation crusts can accumulate. These areas are located under the eaves of the roof. In addition, the sides of the building appear to be blackened, suggesting that these are also sheltered from rainfall and runoff. Production of blackened zones reflects the action of decay processes that operate across whole areas of the building, implying they are controlled by the general configuration of surfaces, effectively the building architecture, rather than by the detailed interaction between processes and the characteristics of specific surfaces.

Areas of flaking and blistering are located on vertical exposed areas between the windows as well as under the windows. These are areas that have been exposed to rainfall and runoff as well as dry depositional processes. A closer examination of Figure 9.3 suggests that here is a distinct pattern to these flaked and blistered areas that are related to the patterns of moisture movement within the blocks of stone. This concentration of decay within blocks due to the differential strength of the mortar and the stone and is related to moisture movement within the block. The concentration of moisture near the surface, away from the harder mortar means that decay has a distinct micro-spatial pattern to its initiation. Decay then spreads from the initial zone across the stone block. The distinct spatial distribution of decay is matched by a distinct temporal pattern. Catastrophic decay as noted by Viles (2005) and Smith and Viles (2006) can be produced by this localisation of decay processes and can result in the rapid attainment of thresholds of internal strength within a stone block, producing catastrophic decay forms such as blisters. The internal stresses created within the confined stone block are likely to produce these patterns of catastrophic failure repeated across a building although the timing of such failures will vary with the individual stress conditions generated within the blocks, which will partially be a result of the severity of decay processes across the building. In addition, the development of case hardening on such block surfaces creates a cycle of catastrophic erosion that proceeds at different rates to different parts of the cycle across a building.

The specific histories of surfaces on a building can also have a dramatic impact upon the nature and rate of decay. Application of specific conservation treatments may provide a general historic context for decay, although treatments may have been applied to the building in a sporadic manner in the past both spatially and temporally. Recent research has focused on the impact of single events, such as major fires (McCabe et al., 2010; Ozguven and Ozcelik, 2013; Ugar *et al.*, 2014) and how these produce localised changes in stress histories and produce weaknesses in the fabric of the stone that decay agents can exploit. Such weaknesses may also be produced by processes such as thermal fatigue (Hall *et al.*, 2011), processes which are a function of the exposure of surfaces to solar radiation and so are dependent on the configuration of surfaces for their effectiveness. Although individual events such as fires may alter the stress histories of block or surfaces, such changes are not important until there are decay agents to act upon these alterations. The location of a surface is therefore vital in determining the extent to which weakening of a stone will be exploited by decay processes and so the location of a surface within a configuration of surfaces is essential in understanding the decay histories.

Photographic analysis of stone surfaces over the short and long term has developed as technology and techniques have developed over the last decade or so (e.g. Thornbush and Viles, 2004; Thornbush, 2013; Thornbush, 2014). Long-term analysis have integrated historic photography with archival photography to produce long-term records of form change, particularly colour change over decades (Thornbush and Viles, 2005). The development of detailed digital elevation models of building surfaces using laser scanning technology at a range of scale (Bryan and Blake, 2000; Barber *et al.*, 2002) has enabled detailed reconstruction and monitoring of stone surfaces to become a by-product of building surveys. The exploitation of this potentially vast resource of data is still, however, in its infancy. The potential for this data source to provide detailed and relatively short-term analysis of stone decay is illustrated in attempts to integrate different methodologies (e.g. Murphy *et al.*, 2013).

9.5 Buildings as Cultural Items

A building stone is not merely a physical part of the building. Its degraded state and conservation history also represent the underlying socio-economic and current currents of the society (or societies of which the building is a part. What is defined as an acceptable level of degradation, who decides this and what should be done about degradation are all questions that require a human agency to make a decision (or series of decisions). Any decisions will take place in the context of a particular historical and sociocultural setting (Inkpen, 1999). Decisions cannot be divorced from the circumstances or context of the questions posed. In this sense, any building stone or surface by its presence and subsequent decay and treatment represents the integration of socio-economic and cultural values of society over time.

The significance of social and cultural value for what is seemingly a purely physical phenomenon of degradation can be illustrated by looking at how durability and durability criteria have been used in stone degradation. Durability itself is recognised as a complex, vague and ambiguous term (Turkington, 1996). The BRE (1975) view of durability is based on the physical characteristics of the stone, resulting from a combination of the internal structure of the stone and the nature of the cementing material. The 1997 BRE digest explicitly recognises the complex nature of durability and its dependence upon the location in which the material is placed and the relationships between the material and the rest of the building environment.

> *'Durability only needs to be assessed if stone intended for longer or harsher exposure . . . or severe environments. Because such assessment must take into account both the stone's position in the building and the weathering to which it will be exposed, a single durability test cannot be of universal application.'*
>
> (BRE Digest, 420, 1997, p. 3)

The digest regards the normal service life of a building of 60 years as being too short a time for stone durability to be an issue. This illustrates the impact of defining durability relative to economic concerns rather than a purely physically determined entity.

Porosity and, in particular micro-porosity, continue to be put forward as the key physical characteristic in controlling durability (e.g. Knofel, 1991; Alves *et al.*, 2013; Molina *et al.*, 2011; Molina *et al.*, 2013) despite field observations that shows stone with similar structures and mineralogical properties exhibit differing decay behaviour. Garden (1980) viewed durability as a fluid property being the outcome of the interaction of material and the environment, including the in-service expectations. Nireki (1980) viewed performance over time as a vital element of durability. Performance as criteria, implies that the stone is continually being assessed in relation to the fulfilment of some designed function. On this basis the expected performance forms the criteria for assessing durability and so durability is not an absolute property of the stone. Duffy and O'Brien (1996) highlight that performance itself is different for different interest groups such as architects, engineers and conservators.

A key issue that none of the above authors seem to address is why a concept of durability is required in the first place. Durability is defined by architects, testing agencies and legislators in relation to a product, be it stone or any other building material. The drive to define durability via some standardised and repeatable index is part of the view of the building and building stone as a commodity (Inkpen *et al.*,

1995). In the case of natural building stone, a highly variable natural material is being characterised by a set of measureable parameters defined as important within a scientific and technological framework. It is the behaviour of these parameters within a set of controlled conditions that become the descriptors of the material and the means of presenting the material to other parties interested in purchasing stone. In this way, the socio-economic framework constrains the types of parameters considered to be valid for characterising building materials.

Putting a value onto stone also defines the temporal dimension of durability. Expectations of durability vary with the type of construction for which the stone is used, an issue recognised by Lewry and Crewdson (1994). They divide durability into four types; benchmark tests, reference or comparative tests, environmental or stress tests and site tests. Within this framework, they emphasise the importance of standardisation of testing, so that a series of standard models and common damage functions can be constructed. The context of the building and material within it is replaced by a series of experimentally derived damage functions developed and assessed for their validity over a time scale significantly less than the expected lifetime of the material and building. The imperative of standardisation within a market and the confidence in a product influence the time scale over which durability is determined. In this case, the time scale of assessment is reduced and formalised as a damage function applicable to a given material in given environment. Grossi *et al.* (2008) continue this damage function approach in relation to the impact of environmental change in decay rates, but how this is translated or rather mediated through individual building surfaces is unclear.

In what could loosely be described as cultural monuments, value lies in the continuance of original material or its replacement by 'identical' material. Durability in this context has a long time frame with material expected to survive over hundreds of years. The value of the stone lies not only within itself, but also in the context into which it is placed. Stone used for conservation work has value imparted to it by the cultural significance of the monument and also takes on expectations associated with such a context. Within a modern building, stone can be used as a facing material and this affects both its expected life and the properties that are important for its decay as it is cut thinly (4–5mm thickness). This multiplicity of valuation can become an issue when techniques such as life-cycle analysis are applied to stone and its degradation. If such techniques are then extended to the valuation of conservation methods such as lime-washing or application of epoxy resin, then the complicated relationships between the time and space frames of each interact so as to blur the appropriate spatiotemporal scale for analysing degradation.

Despite the prevalence of cost-benefit analysis to value competing demands, it is also recognised that valuation of cultural heritage is difficult (Lowenthal, 1994). In particular, the purpose of valuation for its outcome is a key concern. Mohr and Schmidt (1997) argue that the introduction of a concept of non-valuation may be of use. Non-use valuation assumes that when confronted with a change in use or characteristics of an object, a person can be compensated for those changes. The individual decline in well-being experienced can be translated into an acceptable range of alternatives with monetary value for that individual. In economic terms, there is an assumption of substitutability in the valuation of cultural heritage, i.e. individuals are willing to make trade-offs. This requires that the non-use value of cultural heritage be related to a marketable good or service. Viewed as a social good, this would require the trade-off to be expressed in terms of other social goods.

Valuation of cultural heritage can also be undertaken by techniques such as contingent valuation. This method uses different hypothetical situations, such as the current state of decay of the building and its characteristics after conservation, which it converts to a monetary valuation that reflects the willingness of individual to accept or pay for the different situations. Importantly, any economic valuation technique can not be divorced from the context of its use.

Culturally, the stone surface and building also need to be viewed as part of wider, larger 'architectural ecosystem'. The valuation of a single building needs to be viewed in relation to its position within this ecosystem and the devaluation to the ecosystem as a whole its decay produces. Maintaining the diversity of this architectural ecosystem may not only be important for maintaining the economics of tourism to the urban area. Snickars (1997) also suggests that the general attractiveness of an urban area will be affected by the amount of well-maintained cultural heritage it contains. If culturally valued buildings are allowed to degrade beyond a certain threshold of recovery, then this can spark a reaction from residents and the wider society. The attractiveness of the area declines and out-migration results with a perception or public image of the urban area suffering as a result. A process of cumulative negative causation can result and magnify the initial impact of the degradation of the cultural heritage. This illustration highlights the need to value culturally significant buildings not only in terms of their components and upkeep, but also in terms of their relationship to their milieu.

9.6 Conclusions

Studies of degradation have tended to concentrate on extrapolating small-scale field and laboratory-based analysis of process-form relationships to the scale of the whole building and even beyond. Extrapolation in this fashion assumes that the different spatial and temporal scales of processes and forms associated with the whole building have no impact upon the operation of decay agents and processes while field observations suggest that spatially and temporally distinct clusters of decay forms are produced on buildings. Although these small scale studies aid in developing a detailed understanding of the mechanisms of degradation and the limitations of their operations, they do not necessarily help in understanding why degradation occurs with a particular magnitude, frequency and form at particular locations on a building. Mapping patterns of degradation is an important step in recognising that order and pattern are present at and resultant from the form of the building itself. Such patterns cannot, however, be simply reduced to explanation by reference to the presence or absence of particular agents of degradation. The questions of 'why there and why now?' still remain.

Expressing stone decay within the wider context of the environment and how its influence is mediated to a particular surface is essential in understanding how and why degradation occurs. The catchment framework is a useful analytical tool for such an analysis as it provides a physical, functional unit for simple and complex systems analysis.

From the above discussion and examples, however, it is also clear that the role of socio-economic factors in stone degradation should not be under valued or be regarded as mere extensions or add-ons to understanding the physical basis of stone decay. Changes in socio-economic and cultural conditions will increasingly become the key drivers of the environments of stone decay, determining the nature of future conditions of stone decay. Focusing on the specific issue of degradation tends to limit consideration of appropriate and effective treatments and policies to those that focus on the surface itself. The catchment framework reminds us that there is a wider context to consider even within just the physical environment. Recognition of the socio-economic and cultural basis of conservation work widens the scope of analysis even further.

References

Allen, C.D. and Lukinbeal, C. (2011). Practicing physical geography: an actor-network view of physical geography exemplified by the Rock Art Stability Index, *Prog. Phys. Geogr.*, 35, pp. 227–248.

Alves, C., Figueiredo, C., Mauricio, A., Braga, M.A.S., Aires-Barros, L. (2011). Limestones under salt decay tests: assessment of pore network-dependent durability predictors, *Environ. Earth Sci.*, 63, pp. 1511–1527.

Alves, C., Figueiredo, C., Mauricio, A. and Aire-Barros, L. (2013). Susceptibility of limestone petrographic features to salt weathering: A scanning electron microscopy study, *Microsc. Microanal.*, 19, pp. 1231–1240.

Andre, M.F., Vautier, F., Voldoire, O., Roussel, E. (2014). Accelerated stone deterioration induced by forest clearance around the Angkor temples, *Sci. Total Environ.*, 493, pp. 98–108.

Antill, S.J. and Viles, H.A. (1998). 'Deciphering the impacts of traffic on stone decay in Oxford: some preliminary observations from old limestone walls' in Jones, M.S. and Wakefield, R.D. (Eds.), *Aspects of Stone Weathering, Decay and Conservation*, Imperial College Press, London, pp. 28–42.

Attewell, P.B. and Taylor, D. (1988). 'Time dependent atmospheric degradation of building stone in a polluting environment' in Marinos, G. and Koukis, G. (Eds.), *Engineering Geology of Ancient Works and Historical Sites*, Balkema, Rotterdam, pp. 739–753.

Barber, D.M., Mills, J.P. and Bryan, P.G. (2002). Experiences of laser scanning for close range structural recording, *Proceedings, International Workshop on Scanning for Cultural Heritage Recording*, Corfu, Greece, pp. 121–126.

Blocken, B., Derome, D. and Carmeliet, J. (2013). Rainwater runoff from building facades: A review, *Build. Environ.*, 60, pp. 339–361.

BRE (1975). *Decay and Conservation of Stone Masonry*, Building Research Establishment, BRE Digest, 177.

BRE (1997). *Selecting Natural Building Stone*, Building Research Establishment, BRE Digest, 420.

Brimblecombe, P. and Grossi, C.M. (2008). Millennium-long recession of limestone facades in London, *Environ. Geol.*, 56, pp. 463–471.

Brimblecombe, P. and Grossi, C.M. (2009). Millennium-long damage to building materials in London, *Sci. Total Environ.*, 407, pp. 1354–1361.

Brimblecombe, P. (2010). 'Mapping heritage climatologies' in Bunnik, T., de Clerq, H., van Hees, R., Schellen, H., Schueremans, L. (Eds.), *Effect of Climate Change on Built Heritage*, WTA Publications, Pfaffenhofen, pp. 18–30.

Bryan, P.G. and Blake, B. (2000). *Metric survey specification for English Heritage*, English Heritage, Swindon.

Bull, K.R. (1993). 'Development of critical loads concept and UN-ECE mapping programme' in Harnung, M. and Skeffington, R.A. (Eds.), *Critical Loads: Concepts and Applications, ITE Symposium*, No. 28, eds. pp. 8–10.

Butlin, R.N., Yates, T.J.S., Coote, A.T., Lloyd, G.O. and Massey, S.W. (1993). *The First Phase of the National Materials Exposure Programme 1987–1991*, BRE Report CR253/93.

Camuffo, D., Del Monte, M., Sabbioni, C. and Vittori, O. (1982). Wetting, deterioration and visual features of stone surfaces in an urban area, *Atmos. Environ.*, 16, pp. 2253–2259.

Chorley, R.J. (1969). 'The drainage basin as the fundamental geomorphic unit' in Chorley, R.J. (Ed.), *Water, Earth and Man*, Methuen, London, pp. 77–100.

Cooke, R.U. (1989). II. Geomorphological contributions to acid rain research: studies of stone weathering, *Geogr. J.*, 155, pp. 361–366.

Cooke, R.U., Inkpen, R.J. and Wiggs, G.F.S. (1995). Using gravestones to assess changing rates of weathering in the United Kingdom, *Earth Surf. Proc. Land.*, 20, pp. 531–546.

Cowell, D.A. and ApSimon, H.M. (1995). 'Estimating the cost of damage to buildings by acidifying atmospheric pollution in Europe', in Lee, D.S. and McMullen, T.A. (Eds.), *The Effects of Air Pollutants on Materials, Proceedings of a Workshop Held in London, 27–28 July 1994*, AEA Technology, pp. 53–62.

Cultrone, G., Luque, A. and Sebastian, E. (2012). Petrophysical and durability tests on sedimentary stones to evaluate their quality as building materials, *Q. J. Eng. Geol. Hydrogeol.*, 45, pp. 415–422.

Dorn, R.I., Whitley, D.S., Cerveny, N.V., Gordon, S.J., Allen, C.D. and Gutbrod, E. (2008). The Rock Art Stability Index: A new strategy for maximising the sustainability of rock art, *Cultural Heritage*, 1, pp. 37–70.

Duffy, A.P. and O'Brien, P.F. (1996). 'A basis for evaluating the durability of new building stone' in Smith, B.J. and Warke, P.A. (Eds.), *Processes of Urban Stone Decay*, Donhead, London, pp. 253–260.

Fitzner, B. (1990). 'Mapping of natural stone monuments — documentation of lithotypes and weathering forms' in Venilae, F. and Zezza, U. (Eds.), *Analytical Methodologies for the Investigation of Damaged Stone, Proceedings of the Advanced Workshop, Pavia, 14–21 October 1990*, La Goliardica Pavese.

Fitzner, B., Heinrichs, K. and Kownatzki, R. (1992). 'Classification and mapping of weathering forms' in *Proceedings of the 7th International Congress on Deterioration and Conservation of Stone, Lisbon 15–18 October 1992*, Laboratorio Nacional de Engenharia Civil, pp. 957–968.

Fitzner, B., Heinrichs, K. and Volker, M. (1996). 'Monument mapping — a contribution to stone preservation' in Zezza, F. (Ed.), *Origins, Mechanisms and Effects of Salts on Degradation of Monuments in Marine and Continental Environments, Proceedings European Commission Research Workshop, 25–27 March 1996*, Bari, Italy, Protection and conservation of European Cultural Heritage Research report No. 4, pp. 345–355.

Garden, G.K. (1980). 'Design determines durability' in Sereda, P.J. and Litvan, G.G. (Eds.), *Durability of Building Materials and Components*, ASTM STP 691, pp. 31–37.

Goudie, A.S. (1974). Further experimental investigation of rock weathering by salt and mechanical processes, *Zeitschrift fur Geomorphologie*, Suppl. Bd. 21, pp. 1–12.

Goudie, A.S. (1999). Experimental salt weathering of limestones in relation to rock properties, *Earth Surf. Proc. Land.*, 24, pp. 715–724.

Grossi, C.M., Bonazza, A., Brimblecombe, P., Harris, I., and Sabbioni, C. (2008). Predicting twenty-first century recession of architectural limestone in European cities, *Environmental Geology*, 56, pp. 455–461.

Grossi, C.M., Brimblecombe, P., Mendez, D., Benavente, D., Harris, I. and Deque, M. (2011). Climatology of salt transitions and implications for stone weathering, *Sci. Total Environ.*, 409, pp. 2577–2585.

Hall, K. (2004). Evidence for freeze-thaw events and their implications for rock weathering in northern Canada, *Earth Surf. Proc. Land.*, 29, pp. 43–57.

Hall, C., Hamilton, A., Hoff, W.D., Viles, H.A. and Eklund, J.A. (2011). Moisture dynamics in walls: response to micro-environment and climate change, *Proc. R. Soc. A.*, 467, pp. 194–211.

Hall, K. and Andre, M.F. (2001). New insights into rock weathering as deduced from high-frequency rock temperature data: an Antarctic study, *Geomorphology*, 41, pp. 23–35.

Hall, K. and Thorn, C. (2011). The historical legacy of spatial scales in freeze-thaw weathering: misrepresentation and resulting misdirection, *Geomorphology*, 130, pp. 83–90.

Holton, A. (2010). *The Archaeology and Conservation of the East Front of York Minster*, unpublished PhD thesis, University of York. Available at: http://etheses.whiterose.ac.uk/1506/ (Accessed 24th September 2014).

ICOMOS Illustrated Glossary on Stone Deterioration Patterns: http://www.icomos.org/publications/monuments_and_sites/15/pdf/Monuments_and_Sites_15_ISCS_Glossary_Stone.pdf (Accessed 24th September 2014).

Inkpen, R.J. (1989). *Stone Decay and Atmospheric Pollution in a Transect Across Southern Britain*, unpublished PhD thesis, University of London.

Inkpen, R.J. (1991). *Stone Weathering and the 'Memory' Effect*, Department of Geography, Portsmouth Polytechnic, Working Paper 17.

Inkpen, R. (1995). Errors in measuring the percentage dry weight change of stone tablets, *Earth Surf. Proc. Land.*, 20, pp. 783–793.

Inkpen, R.J. (1999). Atmospheric pollution and stone degradation in nineteenth century Exeter, *Environ. Hist.*, 5, pp. 209–220.

Inkpen, R.J. (2005). *Science, Philosophy and Physical Geography*, Routledge, London.

Inkpen, R.J. (2011). Scale as a relational network in weathering studies, *Geomorphology*, 130, pp. 10–16.

Inkpen, R.J., Petley, D. and Murphy, W. (2004). 'Durability and rock properties' in Smith, B.J. and Turkington, A.V. (Eds.), *Stone Decay: Its Causes and Controls*, Donhead, London, pp. 33–52.

Inkpen, R.J., Duane, W., Burdett, J. and Yates, T. (2008). Assessing stone degradation using an integrated database and geographical information system (GIS), *Environ. Geol.*, 56, pp. 789–801.

Inkpen, R., Viles, H., Moses, C. and Baily, B. (2012). Modelling the impact of changing atmospheric pollution levels on limestone erosion rates in central London, 1980–2010, *Atmos. Environ.*, 61, pp. 476–481.

Knofel, D. (1991). 'Causes, mechanisms and measurements of damage in cultural heritage materials: the state of the art: concrete structures', in Baer, N.S., Sabboni, C. and Sores, A.I. (Eds.), *Science, Technology and European Cultural Heritage: Proceedings of the European Symposium, Bologna, Italy, 13–16 June 1989*, Butterworth-Heinemann, pp. 138–147.

Kupper, M. and Pissart, A. (1976). Vitesse d'erosion en Belique de calcaires d'age primarie exposes a l'air libre ou soumis a location de l'eau courante, *Report of the Commission on Present-Day Geomorphological Processes (International Geographical Union)*, pp. 39–50.

La Russa, M.F., Belfiore, C.M., Comite, V., Barca, D., Bonazza, A., Ruffolo, S.A., Crisci, G.M. and Pezzino, A. (2013). Geochemical study of black crusts as a diagnostic tool in cultural heritage, *Appl. Phys. A*, 113, pp. 1151–1162.

Levins, R. and Lewontin, R. (1985). *The Dialectical Biologist*, Harvard University Press, Cambridge, Massachusetts.

Lewry, A.J. and Crewdson, F.E. (1994). Approaches to testing the durability of materials used in the construction and maintenance of buildings, *Construction Building Materials*, 8, pp. 211–222.

Lipfert, F.W. (1989). Atmospheric damage to calcareous stones: comparison and reconciliation of recent experimental findings, *Atmos. Environ.*, 23, pp. 415–429.

Livingston, R.A. (1986). 'Evaluation of building deterioration by water runoff', in Davis, G. (Ed.), *Building Performance: Function, Preservation and Rehabilitation*, ASTM STP 901, American Society for Testing and Materials, pp. 181–189.

Lowenthal, D. (1994). 'The value of age and decay' in Krumbein, W.E., Brimblecombe, P., Cosgrove, D.E. and Staniforth, S. (Eds.), *Durability and Change: The Science, Responsibility, and Cost of Sustaining Cultural Heritage*, Wiley and Sons, Chichester, pp. 39–66.

McCabe, S., Brimblecombe, P., Smith, B.J., McAllister, D., Srinivasan, S. and Basheer, P.A.M. (2013). The use and meanings of 'time of wetness' in understanding building stone decay, *Q. J. Eng. Geol. Hydrogeol.*, 46, pp. 469–476.

McGreevy, J.P. (1982). Frost and salt weathering: further experimental results, *Earth Surf. Proc. Land.*, 7, pp. 475–488.

Mohr, E. and Schmidt, J. (1997). 'Aspects of economic valuation of cultural heritage' in Baer, N.S. and Snethlage, R. (Eds.), *Saving Our Architectural Heritage: The Conservation of Historic Stone Structures*, John Wiley and Sons Ltd., Chichester, pp. 333–348.

Molina, E., Cultrone, G., Sebastian, E., Alonso, F.J., Carrizo, L., Gisbert, J. and Buj, O. (2011). The pore system of sedimentary rocks as a key factor in the durability of building materials, *Eng. Geol.*, 118, pp. 110–121.

Molina, E., Cultrone, G., Sebastian, E. and Alonso, F.J. (2013). Evaluation of stone durability using a combination of ultrasound, mechanical and accelerated aging tests, *J. Geophys. Eng.*, 10, pp. 1–18.

Moses, C.A. (1994). *The Origin and Implications of Microsolutional Features on the Surface of Limestones*, unpublished PhD thesis, Queen's University Belfast.

Moses, C.A. (1996). 'Methods for investigating stone decay mechanisms in polluted and 'clean' environments, Northern Ireland' in Smith, B.J. and Warke, P.A. (Eds.), *Processes of Urban Stone Decay*, Donhead, London, pp. 212–227.

Murphy, M., McGovern, E. and Pavia, S. (2013). Historic building information modelling — adding intelligence to laser and image based surveys of European classical architecture, *ISPRS J. Photogramm. Remote Sens.*, 76, pp. 89–102.

Nireki, T. (1980). 'Examination of durability test methods for building materials based on performance evaluation' in Sereda, P.J. and Litvan, G.G. (Eds.), *Durability of Building Materials and Components*, ASTM STP 691, pp. 119–130.

Ozguven, A. and Ozcelik, Y. (2013). Investigation of some property changes of natural building stones exposed to fire and high heat, *Constr. Build. Mater.*, 38, pp. 813–821.

prEN 12370. *Tests on Natural Stone — Determination of Resistance by Crystallization Cycles.*

Reddy, M. (1988). Acid rain damage to carbonate stone: a quantitative assessment based on the aqueous geochemistry of rainfall runoff from stone, *Earth Surf. Proc. Land.*, 13, pp. 335–354.

Rhoads, B.L. and Thorn, C.E. (1996). 'Toward a philosophy of geomorphology' in Rhoads, B.L. and Thorn, C.E. (Eds.), *The Scientific Nature of Geomorphology*, Wiley, Chichester, pp. 115–143.

Robinson, D.A. and Williams, R.B.G. (1982). Salt weathering of rock specimens of varying shapes, *Area*, 14, pp. 293–299.

Robinson, D.A. and Williams, R.B.G. (1996). 'An analysis of the weathering of Wealden sandstone churches' in Smith, B.J. and Warke, P.A. (Eds.), *Processes of Urban Stone Decay*, Donhead, London, pp. 133–149.

Roekens, E., Komy, Z., Leysen, L., Veny, R. and Van Grieken, R. (1988). Chemistry of precipitation near a limestone building, *Water Air Soil Pollut.*, 38, pp. 273–282.

Ross, K.D. and Butlin, R.N. (1989). *Durability Tests for Building Stone*, BRE Report 141.

Ruedrich, J., Kirchner, D. and Siegesmund, S. (2011). Physical weathering of building stones induced by freeze-thaw action: a laboratory long-term study, *Environ. Earth Sci.*, 63, pp. 1573–1586.

Sabbioni, C., Cassar, M., and Brimblecombe, P. (2009). Vulnerability of cultural heritage to climate change, *Pollut. Atmos.*, 203, pp. 285–298.

Sabbioni, C., Brimblecombe, P. and Cassar, M. (2010). *The Atlas of Climate Change Impact on European Cultural Heritage*, Anthem Press, London.

Salvadori, B., Pinna, D. and Porcinai, S. (2014). Performance evaluation of two protective treatments on salt-laden limestones and marble after natural and artificial weathering, *Environ. Sci. Pollut. R.*, 21, pp. 1884–1896.

Smith, B.J. and McGreevy, J.P. (1983). Simulation study of salt weathering in hot deserts, *Geograf. Ann.*, 65A, pp. 127–133.

Smith, B.J., McGreevy, J.P. and Whalley, W.B. (1987). Silt production by weathering of a sandstone under hot arid conditions: an experimental study, *J. Arid Environ.*, 12, pp. 199–214.

Smith, B.J. and McGreevy, J.P. (1988). Contour scaling of a sandstone by salt weathering under simulated hot desert conditions, *Earth Surf. Proc. Land.*, 13, pp. 697–705.

Smith, B.J., Srinivasan, S., Gomez-Heras, M., Basheer, P.A.M. and Viles, H.A. (2011). Near-surface temperature cycling of stone and its implications for scales of surface deterioration, *Geomorphology*, 130, pp. 76–82.

Smith, B.J. and Viles, H.A. (2006). 'Rapid, catastrophic decay of building limestones: Thoughts on causes, effects and consequences' in Fort, R., Alvarez Debuergo, M., Gomez-Heras, M. and Vazquez-Calvo, C. (Eds.), *Heritage, Weathering and Conservation*, Taylor and Francis, London, pp. 191–198.

Smith, B.J., Gomez-Heras, M. and Viles, H. (2010). Underlying issues in the selection, use and conservation of building limestone, *Geological Society of London, Special Publications*, 331, pp. 1–111.

Snickars, F. (1997). 'How to assess and assert the value of cultural heritage in planning negotiations' in Baer, N.S. and Snethlage, R. (Eds.), *Saving Our Architectural Heritage: The Conservation of Historic Stone Structures*, John Wiley and Sons Ltd., Chichester, pp. 349–370.

Stommel, H. (1963). Varieties of oceanographic experience, *Science*, 139, pp. 572–576.

Tidblad, J., Kucera, V., Mikailov, A.A., Henrikson, J., Kreislova, K. and Yates, T. (2001). UN ECE ICP materials; dose–response functions on dry and wet acid deposition effects over 8 years of exposure, *Water Air Soil Pollut.*, 130, pp. 1457–1462.

Tidblad, J., Kucera, V., Ferm, M., Kreislova, K., Bruggerhoff, S., Doytchinov, S., Screpanti, A., Grontoft, T., Yates, T., de la Fuente, D., Roots, O., Lombardo, T., Simon, S., Faller, M., Kwiatkowski, L., Kobus, J., Varotsos, C., Tzanis, C., Krage, L., Schreiner, M., Melcher, M., Grancharov, I. and Karmanova, N. (2012). Effects of air pollution on materials and cultural heritage: ICP Materials celebrates 25 years of research, *International Journal of Corrosion*, 2012, Article ID 496321, 16 pages.

Thornbush, M. (2013). Photogeomorphological studies of Oxford stone — a review, *Landform Analysis*, 22, pp. 111–116.

Thornbush, M. (2014). A soiling index based on quantitative photography at Balliol College in Central Oxford, UK, *Journal of Earth, Ocean and Atmospheric Sciences*, 1, pp. 1–15.

Thornbush, M. and Viles, H. (2004). Integrated digital photography and image processing for the quantification of colouration on soiled limestone surfaces, Oxford, England, *J. Cult. Herit.*, 5, pp. 285–290.

Thornbush, M. and Viles, H. (2005). The changing face of Magdalen College, Oxford: reconstructing long-term soiling patterns from archival photographs and traffic records, *Journal of Architectural Conservation*, 11, pp. 40–57.

Turkington, A.V. (1996). 'Stone Durability' in Smith, B.J. and Warke, P.A. (Eds.), *Processes of Urban Stone Decay*, Donhead, London, pp. 19–31.

Ugur, I., Sengun, N., Demirdag, S. and Altindag, R. (2014). Analysis of the alterations in porosity features of some natural stones due to thermal effects, *Ultrasonics*, 54, pp. 1332–1336.

Viles, H.A. (1996). 'Unswept stone, besmeer'd by sluttish time': Air pollution and building stone decay in Oxford, 1790–1960, *Environment and History*, 2, pp. 359–372.

Viles, H.A. (2005). 'Can stone decay be chaotic?' in Turkington, A.V. (Ed.), *Stone Decay in the Architectural Environment*, Geological Society of America, Special Papers, 390, pp. 11–16.

Viles, H.A. (2012). Microbial geomorphology: A neglected link between life and landscape, *Geomorphology*, 157, pp. 6–16.

Viles, H.A., Naylor, L.A., Carter, N.E. and Chaput, D. (2008). Biogeomorphological disturbance regimes: progress linking ecological and geomorphological systems, *Earth Surf. Proc. Land.*, 33, pp. 1419–1435.

Yates, T.J.S. and Butlin, R. (1996). 'Predicting the weathering of Portland limestone buildings' in Smith, B.J. and Warke, P.A. (Eds.), *Processes of Urban Stone Decay*, Donhead, London, pp. 194–204.

CHAPTER 10

The Historic and Archaeological Heritage: Pollution and Non-Urban Sites

JoAnn Cassar

Department of Conservation and Built Heritage,
Faculty for the Built Environment,
University of Malta, Malta

10.1 Introduction

When considering air pollution and its effects on cultural heritage, the mind instantly goes towards urban pollution, primarily from energy generation, industry and traffic. And indeed, most studies available today centre around pollution generated in, and affecting, cultural heritage in the city (e.g. Brimblecombe, 2003; Sabboni, 2003; Inkpen, 2003; Yates, 2003; Nava *et al.*, 2010; Tidblad *et al.*, 2012).

However, cultural heritage, and here the focus will be on built heritage including historic buildings and archaeological sites, is not exclusively located in urban centres. There is a wealth of heritage located outside main city centres often in rural areas or the periphery of cities, which one does not commonly associate with air pollution.

This chapter will look at the various sources of air pollution which can affect built heritage located in non-urban areas. It will first look at the wider air pollution problem — including primary and secondary pollutants associated with the main sources of such emissions, and their effects on building materials, including how in developed countries (including Europe and North America) concentrations of pollutants have been falling over recent decades. Next, this chapter will discuss non-urban sources of pollution and their emissions. These will include both natural and anthropogenic sources such as volcanoes and forest fires, biomass burning, agricultural activity, brick making and waste incineration — and the pollutants they produce, with special emphasis on pollutants in relation to the Asian continent.

The great variety of built heritage types which can be affected by pollutants will be introduced, with pertinent examples. The effects of global pollutants will be briefly mentioned as these are similar to the effects on the same materials in cities (though of course, through dilution factors, the concentrations and impact will generally be less). Local pollution sources will be introduced which will include also tourism and the associated transport that moves tourists, often in very large numbers, to the heritage site itself. A specific thrust will be directed towards semi-confined, confined and underground sites, including decorated caves, grottoes and tombs in various parts of the world, specific pollutants and the issues they raise. A further section will deal with the acidification of soil through acid rain and the potential effects on both buried heritage and on the context or "setting" of historic and archaeological sites.

10.2 Sources and Types of Air Pollutants

10.2.1 *Urban sources*

Primary urban generated pollutants include acidic gases, organic carbon compounds and particulate matter, including carbon and heavy metals (Sabbioni, 2003). The main sources of these pollutants are industries including the petrochemical and metals industries, power generation and traffic (EPA, 2002). The main gases and particulates (primary pollutants) emitted by the 28 EU countries (as an example of an industrialised area) are nitrogen oxides (NO_x), non-methane volatile organic compounds (NMVOCs), sulfur oxides (SO_x), ammonia (NH_3), and carbon monoxide (CO). Also emitted are particulate matter (PM), including fine particulate matter ($PM_{2.5}$) and coarse particulate matter (PM_{10}), often included together as total suspended particulates (TSPs), heavy metals and persistent organic pollutants (EEA, 2014).

It is pertinent to note at the outset that developed countries, particularly in Europe, as well as North America, have been showing trends in pollution reduction in recent years, with the worst conditions (peak SO_2 emissions) occurring in the industrialised countries of the west in the 1970s and 1980s (Goudie, 2013). In Europe, there has been a general decrease, being about 27% for SO_2 emissions from 1980 onwards (EUROSTAT *et al.*, 1999). Between 1974 and 1990, SO_2 emissions in the UK decreased by 35% (Goudie, 2013). In less developed countries, the burning of coal and oil, and wood for household use has however resulted in rising levels of SO_2 as well as PMs (Goudie, 2013). In the case of rapidly developing economies, emissions are rising rapidly. China is responsible for 64–71% of total Asian SO_2 emissions, with rapid increases being registered in the 1980s–1990s, as well as in 2000–2006. Sulfur emissions in China in fact increased by 53% from 2000 to 2006, i.e. 7.3% annually (Lu *et al.*, 2010, quoted in Goudie, 2013). Goudie (2013, quoting Smith *et al.*, 2001) also reports that whereas in 1980 the US, Canada and Europe were responsible for 60% of global emissions, these countries produced only 38% of global emissions by 1995.

The emissions of NO_x also decreased across the EU but only by 6% since 1980 (EUROSTAT *et al.*, 1999). In Japan, where S emissions have in fact also decreased, N gases have not decreased (Seto *et al.* 2002, quoted in Goudie, 2013). In the US, while most air pollutants are declining, "concentrations of nitrogen dioxide in the atmosphere have remained basically steady over the past decade as emissions of NO_x have increased by two percent" (EPA, 2002). The EPA (2002) report identifies "large combustion units (including power generation and other heavy industry) and transportation-related sources (dominated by cars and trucks, but also including trains, ships, airplanes, and other non-road vehicles)" as the main sources of NO_x emissions. An EEA report (2010) also identifies heavy-duty vehicles as important sources of NO_x. Other more localised sources include the burning of wood and coal by households, which also produce considerable $PM_{2.5}$ emissions, which can become widely dispersed over very large areas.

Ozone is a secondary pollutant which can affect cultural heritage, directly or indirectly. The formation of ozone (O_3) can be due to both natural and anthropological activity. Photochemical oxidation is a natural mechanism; ozone precursors (e.g. NO_x and hydrocarbon emissions, especially from traffic) are important, making ozone widely considered as an important secondary pollutant (EPA, 2002; Karaca and Öztürk, 2012) which is thought to enhance the formation of acids from SO_2 and NO_x (Brimblecombe and Grossi, 2007). In addition, "photochemical oxidants such as ozone (O_3) are also capable of damaging certain materials" including "culturally important materials, such as a number of artists' pigments and dyes, (which) may also be damaged (EPA 1996, quoted in World Bank and State Environmental Protection Administration, P. R. China, 2007). Carbon dioxide (CO_2) is also emitted in large concentrations, important sources being the combustion of fossil fuels as well as cement production (Goudie, 2013). An EEA report (2010) identifies the energy sector as "a large source of air pollution, accounting for around 70% of Europe's suphur oxide (SO_x) emissions and 21% of NO_x output despite significant reductions since 1990".

Acidic gas emissions can lead to the production of damaging acid rain. The term "acid rain" was first thought of by R.A. Smith, the first of a number of industrial air pollution inspectors appointed in the UK following the Alkali Act in 1863, which was "designed to control acid emissions from alkali works" (Woodin, 1989). In 1872, Smith wrote a book on chemical climatology (Smith, 1872, quoted in Woodin, 1989) where he "described the polluted state of air and rain in detail" (Woodin, 1989). Acid rain can be defined as "rain which has a pH of less than 5.65, this being the pH which is produced by carbonic acid in equilibrium with atmospheric CO_2" (Goudie, 2013). The presence of other acidic gases such as SO_2 and NO_x can lower the pH of rain and snow down to even 2.1; values between 4 and 4.5 can be quite frequent (e.g. eastern USA) (Goudie, 2013). Commonly referred to as "acid rain", wet deposition mechanisms can transfer gaseous pollutants (usually sulfur and nitrogen compounds) from the atmosphere to the ground by means of rain, snow or fog. This type of deposition is intermittent. Also possible is dry deposition which transfers pollutants to exposed surfaces without rain, snow, fog or sleet as transfer vectors, thus constituting a more constant source of pollutant deposition. On the other hand, cloud deposition at high altitudes transfers acidic droplets to exposed surfaces by direct contact (EPA, 2002).

Pollutants can travel hundreds of kilometers from their origin to reach very distant areas, including also the Arctic where polluted air arrives from mid-latitude regions, including from Europe and Eurasia. These can include emissions such as from Russian and Siberian forest fires (Quennehen *et al.*, 2012). As an example, in the Greenland ice cap, the lead content rose substantially from the industrial revolution onwards, though located far from the source of emission (usually industry or traffic). It has however decreased greatly since the 1970s (Bindler, 2011, quoted in Goudie, 2013). When considering the transport of pollutants away from their sources of emission, it is acid rain (wet deposition) which can transport them over long distances (depending on climatological conditions); the concentrations of gases (dry deposition) tend to decrease with increasing distance from where they are first emitted (Woodin, 1989). It is paradoxical that one of the measures taken to reduce air pollution, that of increasing the height of chimneys, resulted in decreasing local pollution, but also produced a regional acid rain problem, as dispersal occurred over wide areas (Goudie, 2013).

10.2.2 *Non-urban sources of pollution*

Global sources of pollution include natural events such as lightening which can generate nitrogen oxides (EPA, 2002). Other natural emissions include SO_2 released during volcanic eruptions which also emit CO_2 and smaller amounts of others gases, including: hydrogen sulfide (H_2S), hydrogen (H_2), carbon monoxide (CO), hydrogen chloride (HCl), hydrogen fluoride (HF), and helium (He) (USGS, 2010). Local and regional sources of pollution on the other hand can have great impact on their surroundings and any archaeological or historic sites which are situated in the vicinity. Anthropological activities include incineration, forest fires and other biomass burning, crude oil burning, coal burning also used for brick making, and waste incineration, as well as other activities such as agriculture and mining. Transportation, sometimes even heavy traffic in the vicinity of heritage sites carries tourists, often in huge numbers, to view the sites; this also needs to be considered.

10.2.2.1 *Agriculture*

The main pollutants released by agricultural activity include methane (CH_4), ammonia (NH_3), nitrous oxide (N_2O) and particulate materials (Agricultural Change & Environment Observatory Programme, 2006; EPA, 2002). The main source of Europe's NH_3 emissions has been identified as agriculture, an important contributor to the increase of N levels in the atmosphere (EPA, 2002). This includes ammonia containing fertilisers. Besides contributing 80% of ammonia emissions (mainly from cattle), agriculture also produces over 60% of nitrous oxide emissions (much coming from animal waste), and ca 40% of all UK methane emissions, again mostly from livestock (Agricultural Change & Environment Observatory Programme, 2006). The average duration of gaseous NH_3 in the atmosphere is of 14–36 hours, this being weather dependent (Jacobson *et al.*, 2003). Here NH_3 and NH_x combinations contribute towards acidification (together with SO_x, NO_x and VOCs) (Groot Koerkamp, 1994, in Jacobson *et al.*, 2003). Low concentrations of H_2S are also released in and around buildings housing animals (Jacobson *et al.*, 2003). NMVOCs and dust particulates are also in part agriculturally derived (Jacobson *et al.*, 2003).

10.2.2.2 *Incineration*

Forest fires can generate high concentrations of a great variety of different emissions, including carbon dioxide, methane, nitrogen oxides, ammonia, particulate matter and non-methane hydrocarbons, and can lead to the formation of ozone and secondary organic aerosols (SOAs) (Martins *et al.*, 2012). In their study on forest fires in Portugal during 2003–2005, focusing on PM and ozone levels, Martins *et al.* (2012) conclude that the forest fires did affect PM_{10} values, though the influence on ozone formation was not evident. In a similar study on fires in Greece (in July 2000), it was concluded that emissions from these fires increased the concentration of PM_{10} as well as that of gases such as CO, NO_x and NH_3 (Lazaridis *et al.*, 2008).

Burning of vegetation also produces a large amount of emissions, again including CO_2, CO, NO_x, NH_3, PMs and VOCs (Bossioli *et al.*, 2012, quoting Andreae and Merlet, 2001). In their study on small-scale dry-grass fires in Russia, concentrating on the spring of 2006, and their effects on ozone production, Bossioli *et al.* (2012) investigated in particular the effects of the burning material and its chemical properties on the O_3–NO_x–VOC sensitivity in the areas where the fires occurred. The authors concluded that, given the meteorological conditions existing at the time of the burning events ("low wind speeds limiting the dilution of anthropogenic pollutants over Central and Eastern Europe"), the fires proved to be a determinant in the formation of ozone (Bossioli *et al.*, 2012).

Biomass is also burnt as a fuel. Hata *et al.* (2014) carried out laboratory experiments to characterise particles from the burning of various biomass fuels, including rubber wood, whole wood, wood pellets and rice husks. They found that different particles were emitted during the burning process. Woodsmoke also releases NH_3, CO, NO_2 and SO_2 as well as fine particulate matter (Naeher *et al.*, 2005).

10.2.2.3 *Waste incineration/waste management*

Damgaard *et al.* (2010) discuss the energy recovered from municipal solid waste incineration, and emissions released to the atmosphere when waste is burnt. Besides the gases and particulate matter already mentioned by other authors discussing forest fires and biomass burning (see above), SO_2, HCl and dioxin are also mentioned, as well as the release of heavy metals. The same authors refer to the difference that exists among countries in their attitude towards waste incineration. In some countries, over 65% of urban waste is burnt e.g. Switzerland, Japan and Denmark. A number of such plants are being planned or constructed also in other countries such as the UK and the USA where previously this option was not preferred for waste management. The authors go on to discuss eight different technologies which can be used to control air pollution (Damgaard *et al.*, 2010). Waste incineration has been used for waste management also in developing countries. Buenos Aires, Mexico City, New Delhi, and São Paulo, among other cities, have had to shut down incinerators due to high costs or environmental considerations (UNEP, 1996). Anaerobic bacteria living off decomposing garbage, both uncollected and in unmanaged landfills, release methane. CH_4 concentrations can make up as much as 50% of the landfill gas at maximum anaerobic decomposition (Cointreau-Levine, 1996 in Zerbock 2003).

10.2.2.4 *Crude oil, brick making and coal burning*

The accidental combustion of crude oil, considered as a disaster when it occurs, can lead to great impacts both in the immediate vicinity of the incident, and beyond. Bonazza *et al.* (2007) discuss the possible effects of oil well fires in Kuwait following the Gulf War, when in 1991 Iraqi forces set fire to half of Kuwait's 1000 oil wells which fires burned on for months. An estimate is that, on a daily basis, 5–6 million barrels of crude oil and 70–100 million m^3 of natural gas were burned. The fires burned with great intensity such that "about 96% of the burned carbon emitted as CO_2" (Bonazza *et al.*, 2007). One possible consequence could have been on monuments in the vicinity; it was even feared that there could be an impact on monuments as far away as southern Iran. However, a detailed study on several monuments in Iran by Bonazza *et al.* (2007), including also the Tomb of Cyrus in the Pasargadae Complex (Fars province) (Fanood, 2012) concluded that there was no evidence of the accumulation of PMs produced by these oil fires, and that the gypsum found probably originated from gypsum mining. Iran is one of the largest producers of gypsum. The blackening was attributed to the presence of cyanobacteria and was not found to be anthropogenic in origin (Bonazza *et al.*, 2007).

Brick making is a significant consumer of energy, often produced by the burning of coal, especially in developing countries. The main other energy sources for brick making are other solid fuels (Pet-coke), electricity, and diesel. These can result in a number of negative environmental impacts due to acidification, when low-grade fuel with a high sulfur content is used. Main gases released are large amounts of CO_2, as well as SO_2 and NO_x (Koroneos and Dompros, 2007). In Asia and Africa, where fired brick and wood are mostly used for construction purposes, bricks are often locally produced in small factories. Here, the fuel used for this energy-intensive process includes wood, as well as crop residues and animal dung (Alam and Starr, 2009).

Fuel oil is used in more modern kilns. Here, the main gas released is CO_2. Alam and Starr (2009) also mention that in Asia the amount of incomplete combustion is so great that emitted gases such as CO, CH_4 and non-methane hydrocarbons "account for 4.5% of total biomass fuel carbon emissions". In Sudan, the brick making industry only contributes 2% of the country's overall emissions of carbon dioxide; but it must also be noted that fuelwood is also heavily consumed to produce energy in households (Alam and Starr, 2009). The two main brick producers in the world are China, followed by India. Khan and Vyas (2008) argue that in India, where 140 billion bricks are produced annually, their production is not by large-scale industries but is rather carried out as small, traditional production. Coal is usually the fuel consumed here, in a process which requires one month of continual firing for the process to be finalised (Khan and Vyas, 2008). In their study on a village in India, it was concluded that the presence of these industries results in increasing concentrations of SO_x, NO_x and PM.

The environmental impacts of the use of coal include not only emissions from coal burning (acidic gases, mainly SO_2 and NO_x, PMs, hydrocarbons and heavy metals) but also dust and methane released during mining operations and pollutants released during transportation, as well as emissions from coal fires in abandoned coal mines and coal waste piles. In the United States, CO_2 from coal fired power plants equals a third of all of the country's carbon dioxide emissions, and is equivalent to the amount of CO_2 released by all means of transport combined (including land, sea and air transport) (Source Watch, 2013). Coal production in China "trebled between 2000 and 2010" (Goudie, 2013).

10.2.2.5 *Traffic and the tourism industry*

Road transport is responsible for large scale pollutant emissions. Heavy vehicles such as those employed in goods transport are responsible for great NO_x emissions, while lighter (passenger) vehicles release mainly CO, NO_x, $PM_{2.5}$ and NMVOCs. NO_x is primarily produced by high temperature combustion as occurs in motor vehicle engines, as well as through power generation and other heavy industry (EPA, 2002). When it comes to the release and transformation of N compounds, Tidblad *et al.* (2012), in their review article on air pollution and cultural heritage, discuss recent trends, including the formation, transportation and deposition of HNO_3, NO_2 and particulate nitrate. The authors explain that, when primary pollutants such as NO and NO_2 are transformed in the atmosphere, the most stable forms which result are HNO_3 and particulate nitrate. However, urban air is very rarely analysed for these two pollutants, although they can be very damaging. Nitric acid is a strong acid which is rapidly absorbed on most surfaces, and the salts it forms are very hygroscopic (Tidblad *et al.*, 2012) and soluble.

A UNEP report (n.d.) on the impact of tourism on the environment cites pollution as one of the main negative impacts, with air, road and rail transport being important sources of air emissions. International air passengers are reported to have increased from 88 million in 1972 to 344 million in 1994. Tourism now accounts for more than 60% of air travel. The World Bank gives data on international tourism and quotes some astounding figures: from 2009 to 2012, tourists to Cambodia increased from 2.162 million to 3.584 million; to China the increase was from 50.875 million to 57.725 million; Indonesia from 6.324 million to 8.044 million; Malaysia from 23.646 million to 25.033 million; Thailand from 14.150 million to 22.354 million; and Vietnam from 3.747 million to 6.848 million (World Bank, 2014).

Besides contributing to rising global levels of air pollutants, tourism is identified as an important source of local pollution, with some impacts being specific to tourism. An example is when tour buses leave their engines running for hours so that tourists on an excursion can return to a comfortable temperature in their air-conditioned bus (UNEP, n.d.). An associated problem is described in relation to the Angkor Wat ruins in Cambodia. An article in *GlobalPost* (Hurby, 2014) points to “decades-old tour buses that belch dark, toxic fumes”, prompting UNESCO country director to comment to the same journalist that “pollution and acid rain will affect the stone”. The situation at Angkor Wat and other iconic sites in various countries will be addressed in the sections below.

10.3 The Effects of Air Pollutants on Historic and Archaeological Sites, and Their Contexts

10.3.1 *Exposed sites*

10.3.1.1 *General trends and examples*

In 1968, Hiroshi Daifuku, then Head of the Section for the Development of Cultural Heritage, UNESCO, wrote a leading article published in The UNESCO Courier entitled "Saving our Heritage in Stone" (Daifuku, 1968). He then stated: "Sites and monuments, as with all material things, are subject to change and degradation. Objects which are recovered from archaeological excavation may have rested for centuries in a very stable environment. Upon exposure and perhaps transport to a new environment, the object or the monument is once more subject to diurnal and seasonal changes. A new dimension has been added with industrialization, however. The combustion of fossil fuels has added a host of other compounds which are not only deleterious to human health, but also attack stones, wood, glass and other building materials ... Limestone and marble, sandstone and other soft stones, wood and other organic material used for building purposes reveal the results of atmospheric contamination. With time even the more resistant materials are subject to attack" (Daifuku, 1968). The situation still applies today, although the "stable environment" mentioned needs to be revisited in the light of the potential acidification of soils i.e. buried environments (see below).

In a 2001 working paper for the STOA Unit of the European Parliament (Cassar, 2001) it was emphasised that although SO_2 concentrations in Europe had decreased (see also above), "levels of smoke and particles are still high, and continue to damage urban monuments". The large contribution by cars and other vehicles to the deterioration of monuments was also highlighted. One case study refers to research into the effects of environmental pollution on ancient and modern hydraulic mortars — a study which can be highly relevant to Roman archaeological sites, where hydraulic lime mixes were widely used.

A 2007 publication on the situation as regards pollution in China states that, in general, "Acid rain, caused by SO_2 reacting in the atmosphere with water, oxygen, and other substances, can reduce crop and timber yields and forest canopy and damage buildings and monuments, as can SO_2 in gaseous form" (World Bank and State Environmental Protection Administration, P. R. China, 2007).

Air pollution effects on non-urban buildings and sites can generally be deduced by extrapolating from existing information related to the same types of building materials in an urban environment, where most studies are concentrated. Of the pollutants emitted from both natural and anthropogenic sources, SO_2 and NO_x can be identified as the main corrosive gases in the deterioration of building materials in outdoor environments. In addition, particulate matter is responsible for the blackening of surfaces and the catalysis of SO_2 oxidation on stone surfaces (Sabbioni, 2003). NO_2, in the presence of heavy metals and carbon particles, also enhances the oxidation of SO_2 (Sabbioni, 2003). The mechanisms of the sulfation of carbonate stones, which include limestone, marble and lime mortars, is discussed in detail by Sabbioni (2003). This leads to the formation of gypsum ($CaSO_4.2H_2O$), which can also bind carbon and other particles in what are known as "black crusts" (Sabbioni, 2003). Carbonate stone can undergo dissolution even in the absence of air pollution, known as the karst effect. Rain water containing CO_2, producing weak carbonic acid, can dissolve $CaCO_3$, resulting in mobilisation of the calcium carbonate, and its deposition in other areas (Inkpen, 2003). This action can affect limestone, marbles and dolomites (Goudie, 2013). Acid rain will increase the rate of this dissolution (Sabbioni, 2003). In the case of brick and concrete structures, Yates (2003) gives a comprehensive overview.

In the case of non-urban emissions, including agriculture, forest fires, landfills and even volcanoes, the gases emitted, mostly acidic gases, have been mentioned. The acidity of rainfall (wet deposition) as well as dry deposition and cloud deposition (as mentioned above) will enhance the dissolution of carbonate containing materials, and, where SO_2 emissions are especially concentrated, e.g. close to volcanic eruptions, sulfation will be promoted. This has special relevance for archaeological sites close to volcanoes, such as e.g. Pompeii and Herculaneum situated close to Mt Vesuvius in Italy. Experiments have shown that exposure to SO_2 can convert the red pigment haematite (Fe_2O_3) to magnetite (Fe_3O_4), which causes a darkening of the pigment (Maguregui *et al.*, 2013). In the case of ammonia, on the other hand, interactions with acidic aerosols (sulfuric acid (H_2SO_4) and nitric acid (HNO_3)) leads to formation of secondary particulate matter ($PM_{2.5}$) (Aneja *et al.*, 2008; Behera and Sharma, 2010a, quoted by Behera and Sharma, 2011). Salts of ammonium (NH_4^+) including nitrates (NO_3^-), sulfates (SO_4^{2-}) and chlorides (Cl^-) are thus formed (Russell *et al.*, 2004; Behera and Sharma, 2010b quoted by Behera and Sharma, 2011) which formation depends on concentrations of NH_3 as well as NO_2 and SO_2, temperature and relative humidity, amongst other conditions (McMurry *et al.*, 1983; Harrison and Kitto, 1992, quoted by Behera and Sharma, 2011). If deposited on porous stone surfaces, these can lead to salt damage given the right environmental conditions (see also Steiger, 2003).

10.3.1.2 *Some sites*

The most iconic sites which come to mind when thinking of the detrimental effects of air pollution on monuments are the Parthenon in Athens, the Taj Mahal in India, and the Pyramids in Egypt often mentioned in publications on the effects of air pollution from traffic and industry. Heritage materials, including building materials are negatively affected, particularly marble and limestone. What about monuments and sites located far from urban centres? And what about the effects of local sources? Some specific examples are given below.

A paper by Ullén *et al.* (2004), quoting reports by Lindborg (1990) and Bertilsson and Löfvendahl (1992) mentions that "anthropogenic pollutants are responsible for the accelerated deterioration of rune stones, rock carvings and rock paintings in various parts of Scandinavia." Once again, however, decreasing pollution levels in recent years in these countries must be highlighted. Another study, in Sicily, the southernmost part of Europe, on the surface layers of the stones of the Agrigento Temples, confirmed that PAHs (polycyclic aromatic hydrocarbons) detected had originated primarily from anthropogenic combustion processes. The same sources as for the PAH emissions was confirmed by good correlations existing between total PAH concentrations and the content of sulfates and nitrates of the analysed samples (Orecchio, 2010).

On the Asian continent, the example of Angkor Wat (Cambodia), and the effects of tourism-related pollution have already been mentioned. A study by Furuuchi *et al.* (2007) was carried out on PAHs in the Angkor monument area, sampling the air in and around the site and analysing concentrations of total suspended particulates (TSP), PAHs, and heavy metals. It was concluded that during the rainy season there was significant air pollutant transport from the city to the monument and PAH concentrations in Angkor Wat were similar to those measured in Bangkok indicating a serious air pollution situation (Furuuchi *et al.*, 2007). This would thus be adding to the local pollution generated by the ever-increasing tourists to the area and their aged means of transport (see above).

In addition, other regional sources must also be taken into consideration. The emissions by volcanoes, a threat to areas where active volcanoes exist, including Asia, have been mentioned. A current reality on the Asian continent is also the extensive forest fires which often rage in Central Asia. For example, as reported in *The Observer* in March 2014 (Vidal, 2014) "more than 3,000 separate fires have been recorded across Indonesia, Thailand and Malaysia since mid-January, more than in June 2013 when the pollution spiked to dangerous levels", fires attributed to the burning of peat soils and vegetation. The same article reports that "Asia is now the centre of global air pollution", attributed to not only fires but also traffic, industries, power generation and "emissions from millions of inefficient cookers burning wood and cow dung" and adds that "Scientists now fear that the Asian haze will intensify and become an annual event" (Vidal, 2014). The greatest SO_2 emitter in the world is now China. "Recent trends in energy consumption, particularly increased coal use, provide a possible explanation for the increase in SO_2 emissions" (World Bank and State Environmental Protection Administration, P. R. China, 2007).

Awareness of the effects of air pollution on health, and even tourism (due to the immediate impact of dwindling tourist numbers) is widespread. But what about awareness on the effects of pollutants on cultural heritage materials? An interesting survey was carried out by Agbota *et al.* (2013) in developing countries, in over 20 sites in Africa, Asia and Latin America. This study evaluates the concern towards the presence and effects of air pollution on heritage sites, also asking whether air pollution data were being collected. It was concluded that, although many respondents were concerned about the possible effects of air pollution on their sites and collections (84% considered pollution as a possible problem), many of the answers given indicated that the respondents were not knowledgeable on which pollutants were actually damaging. As regards actual monitoring, it was interestingly noted that that whereas air temperature and humidity were often monitored, with 88% confirming this as taking place at their sites, only 24% of respondents had access to pollution data. In addition, descriptions of pollutants were not generally precise (Agbota *et al.*, 2013).

10.3.2 *Semi-confined, confined and underground sites*

10.3.2.1 *General trends and examples*

When discussing archaeological heritage and the effects of air pollution on this heritage, special attention must be given to semi-confined, confined and underground sites. There are many publications on indoor air pollutants, their sources and effects on museum collections (e.g. Brimblecombe, 1990; Blades *et al.*, 2000; Grzywacz, 2006; Rhyl-Svendsen, 2006; Brimblecombe, 2012; Cavicchioli *et al.*, 2012; Smolik *et al.*, 2012; Anaf *et al.*, 2014) as well as international collaborative projects (e.g. ENVIArt, 2005–2010) to study these issues. These have been generally centred on buildings located in urban areas. Here the effects of external pollutants must be evaluated in relation also to the indoor pollutants emitted, to the frequency and extent of air exchanges which the building and showcases can allow, to visitor traffic, and to the building itself, including construction and finishing materials (Ryhl-Svendsen, 2006; UCL Centre for Sustainable Heritage, 2008).

An important but very fragile type of archaeological site consists of those with wall decorations, including caves in many parts of the world, such as the Palaeolithic caves of France and Spain (examples are Lascaux and Altamira), as well as the painted grottoes on the Silk Road in northwestern China. Impressive wall paintings are also found inside tombs such as those of Etruscan tombs in Tarquinia (Italy), Roman catacombs, and Egyptian tombs including those of Nefertari and Tutankhamun. These enclosed areas are severely affected by perturbations brought about by the presence of ever-increasing numbers of visitors, especially by the exhalation of CO_2 and moisture (Maekawa, 1993), the forced air exchanges brought about by visitor groups to the sites, and the influx of organic nutrients and microbiological organisms which can negatively affect the sites. Examples of these are given below. Even semi-confined spaces can be affected by air pollutants in a different way from external and from totally enclosed sites, due to the partial restriction on air exchange. An example of how such parameters can affect a site, though citing a study carried out in an urban area, can illustrate some of the issues such confined and semi-confined space can face if sources of pollution exist nearby, and can possibly be extrapolated to include sites with primarily local sources of pollution such as small scale industry or domestic fuel consumption.

A paper by Nava *et al.* (2010) assesses air quality and microclimatic conditions in the main Courtyard (Michelozzo's Courtyard) in Palazzo Vecchio in Florence, Italy. The authors explain that this is one of the first studies to investigate a semi-confined space, concluding that the air quality measured was "comparable to the urban environment" with "heavy pollution loadings typical of the city" (Nava *et al.*, 2010). The authors also mention similar studies carried out in underground environments (Yungang grottoes in China, Altamira cave (N Spain) — see also below), but emphasise that these studies in confined environments are not directly comparable with their study centered on a semi-confined environment.

A question here arises: can the semi-confined areas found in relation to sheltered archaeological remains be considered as comparable? Shelters can vary from being totally enclosed, to partially enclosed, to consisting only of a roof above a ruin. Examples are to be found in various countries, including Israel, Malta, Greece, and Italy. Other countries are now considering them (Fanood, 2012). It is widely known that totally enclosed spaces can lead to accelerated deterioration of the sheltered areas (Azlan, 1997; Rizzi, 2008; Coningham *et al.*, 2013). However, it would be of great interest if, besides the usual temperature and humidity monitoring that takes place at sheltered sites (Cassar *et al.*, 2011; Cassar *et al.*, 2013) air pollution monitoring were also to take place, leading to an understanding of the presence, pathways, and possible accumulation of pollutants in such environments. This could result in an understanding of their effects on the building materials in such confined and semi-confined spaces (Cassar, 2007). Aerosol monitoring in such cases has usually been limited to TSP and biological aerosols (ISAC-CNR, 2006).

10.3.2.2 *Specific sites*

Numerous studies have been carried out on the decorated Palaeolithic caves of France and Spain, including those of Lascaux (Dordogne, France) and Altamira (Cantabria, Spain), both World Heritage sites (e.g. Vouvé *et al.*, 1982; Brunet *et al.*, 1997; Sánchez-Moral *et al.*, 1999; Saiz-Jimenez *et al.*, 2011). It is the stability of the confined environment underground, and the natural buffering of the external environment by the cave walls which are the most important factors leading to the preservation of the rock art. This has ensured the preservation of the fragile paintings for nearly 40,000 years (Bourges *et al.*, 2014). As already mentioned, tourism can introduce pollutants, and also into confined areas. These here include not only external pollutants generated by vehicles transporting crowds to the sites, but also those generated indoors especially carbon dioxide exhaled by the visitors, which can result in the dissolution of limestone surfaces. Corrosion of the rocky substrate and paintings can thus be caused by the presence of visitors. Sánchez-Moral *et al.* (1999) carried out studies, including microclimatic monitoring, in Altamira Cave (Cantabria, Spain). These studies found that enhanced corrosion was due to the presence of visitors (increased by up to 78 times) with no significant effects from outdoor pollutants. This is a scenario which is replicated in other underground limestone sites (Cassar, 2000).

Other changes also result in degradation of caves and their environments. Saiz-Jimenez *et al.* (2011) writing on the Altamira cave, state that huge numbers of visitors to the cave before its closure to the public in 2002, as well as other changes including excavations, transformed what was essentially a nutrient-poor environment into an ecosystem with nutrients available to support the growth of microorganisms. These organisms, whose introduction into, and proliferation within, caves is caused by visitors, in these confined ecological niches can be considered as threats, bringing about severe devastation to sensitive rock art. Several papers dealing with these biological problems have been published (including: Jurado *et al.*, 2010; Butler, 2010; Saiz-Jimenez, 2012; Martin Sanchez *et al.*, 2014; Garcia-Anton *et al.*, 2014; Mulec, 2014) raising awareness of this previously little studied problem.

Painted grottoes along the Silk Road in China are also deteriorating, some of which are feeling the effects of external air pollutants. In a study of the Yungang Grottoes, another World Heritage site, Huang (2010) attributes the deterioration of some of the caves to "the high SO_2 content in the atmosphere, as a result of coal mining and burning" producing acid rain and snow. Some of the Mogao Grottoes, on the other hand, are reported to be affected primarily by internally generated pollutants; heavy soot present on the wall paintings is said to have been caused after the Russian Revolution in the 1920s, when Russians used the caves and made fires there (Li, 2010). The burning of incense during religious ceremonies in the past could have also led to the formation of soot deposit.

The Museum of Qin Terracotta Warriors and Horses in Xi'an China, which covers an area of 16,300 m^2 and is visited by over 2 million tourists a year, was built to house the over 7000 terracotta soldiers, horses, chariots and weapons discovered in three pits from 1974 to 1976 (Tourism Administration of Xi'an, 2006–2008). These artefacts were placed on the UNESCO World Heritage List in 1987. The museum is located 30 km from Xi'an, "a city of eight million people with increasing traffic, high coal consumption, and intensive building construction. The museum is also surrounded by agricultural fields in which biomass is periodically burned" (Hu *et al.*, 2010). From their study on dust deposition inside the museum, Hu *et al.* found that fly ash was present as well as gypsum particles. These have been blamed at least in part on the presence of pollutants including SO_2. In June 2005, the South China Morning Post reported the launch of a two-year campaign to study pollutants within the museum (Chan, 2005).

In Egypt, a six-year joint project by the Getty Conservation Institute and the Egyptian Antiquities Organization in the tomb of Nefertari and its magnificent wall paintings, included microclimatic monitoring, continuing beyond completion of the project. Agnew (1993) sums up the deterioration of the wall paintings, as evaluated by the Getty Conservation Institute, as follows: "Now the deterioration is accelerating as uncontrolled tourism, pollution, vandalism, and often theft are tipping the scales against the long-term survival of the artworks, monuments, and the ancient legacy of cultures".

The scale of the devastating effects of CO_2 and microorganisms on wall paintings in confined environments is a function of cave ventilation and anthropogenic disturbances. A study on the distribution of cave aerosols (Dredge *et al.*, 2013), originating internally and externally, both from natural and anthropogenic sources, concluded that the degradation of cave environments "from combustion emissions and cave visitors is becoming increasingly common and is developing as a cave management problem".

10.4 Landscapes and Archaeological Context

10.4.1 *General trends and examples*

The European Convention on the Protection of the Archaeological Heritage defines archaeological heritage as "all remains and objects and any other traces of mankind from past epochs ... the preservation and study of which help to retrace the history of mankind and its relation with the natural environment; ... for which excavations or discoveries and other methods of research into mankind and the related environment are the main sources of information ...

The archaeological heritage shall include structures, constructions, groups of buildings, developed sites, moveable objects, monuments of other kinds as well as their context, whether situated on land or under water" (Council of Europe, 1992).

Following this specific reference in 1992 to the "context" of the site as part of the archaeological heritage, awareness of the importance of protecting the surroundings of a building or site, as well as the site itself, continued to grow. In 2005 in Xi'an, China, the 15th General Assembly of ICOMOS adopted the Xi'an Declaration on the Conservation of the Setting of Heritage Structures, Sites and Areas and stated that "The setting of a heritage structure, site or area is defined as the immediate and extended environment that is part of, or contributes to, its significance and distinctive character" (ICOMOS, 2005).

When dealing with World Heritage Sites, UNESCO is clear about emphasising the importance of creating "buffer zones" around the monument or site, where development should be drastically curtailed. The "Operational guidelines for the implementation of the World Heritage Convention" describe a buffer zone as "an area surrounding the nominated property which has complementary legal and/or customary restrictions placed on its use and development to give an added layer of protection to the property. This should include the immediate setting of the nominated property, important views and other areas or attributes that are functionally important as a support to the property and its protection" (UNESCO, 2013a). An international meeting on World Heritage and Buffer Zones drew up a report in 2008 (Martin and Piatti, 2009) where it was stated that "Buffer zones are an important tool for conservation of properties inscribed on the World Heritage List. All along the history of implementation of the World Heritage Convention, the protection of the "surroundings" of the inscribed properties was considered an essential component of the conservation strategy, for cultural and natural sites alike. The World Heritage Committee valued the concept and included it into the *Operational Guidelines* as one of the elements to be considered in the preparation of a nomination proposal" (Martin and Piatti, 2009).

Following on from the above, not only do we need to protect the site itself from threats but also the surroundings of the site, especially where, as in this chapter, we are dealing with air pollution in a non-urban environment. The damage to a site or monument made of limestone or marble can be great if uncontrolled pollution transfers to the vicinity of the site, or if local industries, or heavy traffic, or the transport of tourists to the site, release damaging acidic gases and particulate matter, resulting in the dissolution, sulfation and/or blackening of surfaces. This may lead to enormous and irreversible damage to a unique site. And as already stated, local and regional sources, such as fires, volcanoes and agriculture also contribute their share of emissions.

The "context", "landscape", "setting" or "surroundings" of a building or site, which are often a natural ecosystem in a rural area, are equally priceless and must equally be protected. Damage to plants and trees by pollution and related changes in ecology through the loss of certain species and possibly the intrusion of others will not only be to the detriment of local flora but can also lead to changes in the fauna through the destruction of their habitat, with consequences to the site's "setting".

Damage to vegetation can be either direct, due to exposure to air pollutants, or indirect, mainly through soil acidification or other changes taking place in the soil. "Direct damage from SO_2 emissions is very likely in some regions. Other possible causes of damage include high concentrations of ozone and other photo-oxidants and, in some areas, hydrogen fluoride in the air. In many countries, ozone is the main damaging agent" (World Bank and State Environmental Protection Administration, P. R. China, 2007).

Direct damage through acid rain can have a significant effect on trees and plants because it can affect leaves and also increases the toxicity of the soil (EPA, 2012; Goudie, 2013); biodiversity can also be affected. Acid rain can also weaken trees not only by damaging leaves or restricting the availability of nutrients, leading to injury or death, but also often in combination with one or more additional problems such as for example fungal attack. In the 1970s and the 1980s, in southern Norway and Sweden, pollutants produced by the industrialised economies of Europe including outside Scandinavian borders resulted in the formation of acid rain which crossed borders and acidified the shallow nutrient-poor soils. This phenomenon was first noticed in the 1970s in Germany from where reports started appearing of damage to forests (Woodin, 1989). This was followed in the 1980s by reports from other countries, including Poland and the former Czechoslovakia, and then in Sweden, of damage to forests. In response to this scenario, scientists in the 1980s were predicting "dramatic forest dieback", which however never materialised. Although defoliation has occurred in some areas of European forests, these are generally now growing well (World Bank and State Environmental Protection Administration, P. R. China, 2007), primarily because of measures taken in recent decades to reduce air pollutants in Europe.

This same World Bank report says of the effects of pollution on forests in China "Damages to forests could not be quantified due to lack of monitoring data in remote areas and adequate dose-response functions"; however, for crops, "It is estimated that acid rain, caused mainly by increased SO_2 emissions due to increased fossil fuel use — causes over 30 billion yuan in damages to crops, primarily vegetable crops (about 80 percent of the losses). This amounts to 1.8 percent of the value of agricultural output" (World Bank and State Environmental Protection Administration, P. R. China, 2007).

Acidic water can also wash away nutrients and minerals from the soil and release toxins such as aluminium. A large acidic input can also result in the leaching of cations from the soil, leading to nutrient deficiency (Goudie, 2013). In forests in high mountain regions surrounded by acidic clouds and fog, removal of essential nutrients from leaves and needles can make the trees more prone to damage by other environmental factors and pathogens. Other plants can be harmed in the same way, besides being damaged by other air pollutants such as ground level ozone (EPA, 2012). Other effects could result from direct damage by acidic gases (SO_2 and NO_x) alone or jointly, high ozone levels triggering the leaching of nutrients from leaves (Woodin, 1989).

One of the resulting problems which can affect flora and indirectly also fauna is acidification of soils. Areas particularly affected by "acid rain" are those which have soils with a low buffering capacity and therefore a low resistance to acidification. This will affect in particular soils which lie on non-calcareous bedrock (e.g. granite or gneiss), which can build up high levels of acidity. This does not occur when a calcareous substrate exists, as this allows for the buffering of the acidity thus introduced to the soil. Critical loads are defined as "exposures below which significant harmful effects on sensitive elements of the environment do not occur according to currently available knowledge" (Goudie, 2013). Critical loads are reported to have been exceeded in areas of high industrialisation, such as eastern USA, Europe, Russia and large parts of Asia; risks are also on the increase in parts of Brazil, Argentina and South Africa (Bouwman *et al.*, 2002, quoted in Goudie, 2013). However, in other countries, reduced pollution levels have resulted in soil acidity decreasing (RoTAP, 2012, quoted in Goudie, 2013), including for example an increase in soil pH in forests in the UK in recent years (Vanguelov *et al.*, 2010, quoted in Goudie, 2013).

An ecological change due to air pollution which is of particular interest to archaeological heritage is the sensitivity of lichens to SO_2 and nitrate pollution (Hawksworth, 1990; van Herk *et al.*, 2003; quoted in Goudie, 2013). Lichens not only colonise trees, but also monuments and archaeological sites in rural areas. While lichens can cause both physical and chemical damage to stone surfaces (St Clair and Seaward, 2004; Piervitori *et al.*, 2004), they can also give a monument or site "an air of age and romance" and make it look attractive, while removing the lichens "can leave the stone looking stark and denuded" (Doehne and Price, 2010). There are however evident signs of recovery in lichen colonisation in urban areas where SO_2 levels have been decreasing for years, as for example in Tampere (Finland) (Ranta, 2001, quoted in Goudie, 2013) and London (Larsen *et al.*, 2007, quoted in Goudie, 2013).

10.4.2 *Specific sites*

Some iconic sites owe their beauty to their natural surroundings, especially those in mountainous or forest locations. One site where the forest is as much a part of the beauty as the site itself is Angkor Wat in Cambodia. UNESCO describes this site, inscribed on the World Heritage List in 1992, as follows: "Angkor is one of the most important archaeological sites in South-East Asia. Stretching over some 400 km^2, including forested area, Angkor Archaeological Park contains the magnificent remains of the different capitals of the Khmer Empire, from the 9th to the 15th century. They include the famous Temple of Angkor Wat and, at Angkor Thom, the Bayon Temple with its countless sculptural decorations. UNESCO has set up a wide-ranging programme to safeguard this symbolic site and its surroundings" (UNESCO, 1992–2014). A study by André *et al.* (2014) has shown that the forest can also exert a beneficial contribution to the preservation of the site itself. The authors state that forest clearance has provoked on the building sandstone "an average ten-fold increase in stone loss rates (0.2 instead of 0.02% per year). This accelerated decay is assigned to the climatic stress provoked by the exposure of fragile ornamented sandstones to the harsh impact of tropical sunshine and monsoon rains" (André *et al.*, 2014).

The mountain Temple of Borobudur in Indonesia (Java) is one of the world's finest Buddhist monuments, included in the UNESCO World Heritage List in 1991. Located about 40 km from the city of Yogyakarta this magnificent site is surrounded by four active volcanoes (UNESCO, 2013b). Its surroundings, the rainforest, are here also an intrinsic part of the site, so much so that when the site was abandoned in the 15th century, the temple site was overrun by plants, trees and bushes, which protected the site from further damage (UNESCO, 2013b). Borobudur has long been affected by volcanic eruptions from nearby volcanoes, the most recent such event being in 2010 (UNESCO, 2013b), when ash from Mount Merapi covered the temple in ash. The temple is built of volcanic rocks without the use of mortar, the stone being "in composition between an andesite and a basalt, very porous and not very hard or strong" (Voûte, 1973). Studies concluded that "deterioration takes place essentially as an erosion of the surface of the stone" (Voûte, 1973) which at the time was attributed to "the composition of the material, the climatic conditions, the influence of the ambient humidity, the infiltration of water and biologic action in particular that of the cryptogamic vegetation" (Hyvert, 1973), but not to pollution effects from the nearby volcanoes, probably because of the composition of the building materials.

10.5 Buried Heritage

Air pollution, and in particular acid rain, is not only affecting exposed sites and the landscape, but can also have a negative impact on the buried heritage, which is alarming as this is the heritage which is still unknown to us. One of the main pollution-related problems here is the acidification of soil.

There is however still limited research on the effects of ground acidification through air pollution on buried heritage. Studies in Scandinavia were carried out by Ullén *et al.* (2004) and Nord *et al.* (2005a and b), centred on corrosion of buried bronze artefacts. Other studies were on buried archaeological bone (Gordon and Buikstra, 1981; Kars and Kars, 2002) and buried iron artefacts (Gerwin and Baumhauer, 2000; Neff *et al.*, 2004 and 2007).

The studies on bronze artefacts buried in soils in Sweden showed that (1) finds from the west coast, which tended to be more polluted, were more degraded than those from the east coast; (2) especially well preserved objects were usually found in calcareous regions (3) artefacts excavated more than a hundred years ago were in better condition that those excavated in the last 50–100 years and in the same areas i.e. the major corrosion in the buried environment took place in the last 50–100 years; (4) whether the objects had been buried for two thousand or one thousand years was irrelevant to the state of conservation (Nord *et al.*, 2005a,b; Ullén *et al.*, 2004). All these conclusions pointed to the fact that soils which had a low critical load (see also above) resulted in more corroded items, especially if the objects were excavated in recent years from more heavily polluted areas.

This revolutionised the thinking, prevalent at the time, that the west of the country was poor in bronze finds whereas the east was richer; in fact "many artifacts may have disappeared because of heavy pollution" (Nord *et al.*, 2005a). In their paper, Nord *et al.* (2005b) state that, whilst affirming that most of the available knowledge is based on studies centred on Europe, the problem could also exist in developing countries, like India and China, with their increasing pollution problems and their immense archaeological heritage.

One of the main principles of conservation — acceptable in professional circles for the last 25 years, that of 'preservation *in situ*' — has led to the survival of much of the archaeological record. Chitty (2007) states that this position may have to change if *in situ* environmental conditions are no longer stable, and whilst attributing her comments to the effects of climate change, one can also use the same argument to explain the changes that could be taking place in the soil and subsoil due to acidification. One now therefore has to carefully evaluate the previously held premise that buried objects "may have rested for centuries in a very stable environment" (Daifuku, 1968) and even possibly rethink when and where "rescue archaeology" is pertinent.

10.6 Conclusion

This chapter has aimed at throwing light on the specific problems linked to the survival of archaeological and historical heritage located in non-urban sites, with specific reference to air pollution. A wide range of pollution sources have been looked at, including urban emissions which can be transported very long distances from their sources, to regional and local sources, both anthropogenic and natural. The types of emissions produced by these various sources have been discussed, together with their effects on the most common building materials, especially carbonate based ones (limestone, marble, lime-based mortars and plasters and also some pigments). Extrapolating from the very large number of publications which have dealt with pollution effects in urban environments, and taking into account transport and dilution factors, a picture of what can occur to the same materials in non-urban areas has been drawn, whilst keeping in mind that other specific sources, such as volcanoes, fires, local industries such as brick-making and agricultural activity can contribute substantially to the damage seen in these cases. Specific attention has also been given to special cases of underground, confined and semi-confined environments, and the buried heritage. It however can be concluded that this field of air pollution and its effects on monuments in non-urban sites is grossly under-researched and that information in this area of study, as well as the quantification of possible related damage, are sorely lacking. As the effects of these pollutants become ever more critical, and as tourism becomes ever more affected by the consequences, such as tourists abandoning heavily polluted areas, targeted studies should be undertaken to help understand the related phenomena and, hopefully, to find means of mitigating against their deleterious effects in what is now a world-wide pressing issue.

References

Agbota, H., Young, C. and Strlič, M. (2013). Pollution monitoring at heritage sites in developing and emerging economies, *Stud. Conserv.*, 58, pp. 129–144.

Agnew, N. (1993). 'Conclusion', in Corzo M.A. and Afshar, M. (Eds.), *Art and Eternity*, The J. Paul Getty Trust, pp. 123–124.

Agricultural Change & Environment Observatory Programme (2006). *OBS 03: Quantitative Approaches to Assessment of Farm Level Changes and Implications for the Environment*, Final report, Central Science Laboratory, York and Countryside and Community Research Unit, University of Glouchester.

Alam, S.A. and Starr, M. (2009). Deforestation and greenhouse gas emissions associated with fuelwood consumption of the brick making industry in Sudan, *Sci. Total Environ.*, 407, pp. 847–852.

Anaf, W., Bencs, L., Van Grieken, R., Janssens, K. and Se Wael, K. (2014). Indoor particulate matter in four Belgian heritage sites: Case studies on the deposition of dark-colored and hygroscopic particles, *Sci. Total Environ.*, 506–507, pp. 361–368.

André, M-F., Vautier, F., Voldoire, O. and Roussel, E. (2014). Accelerated stone deterioration induced by forest clearance around the Angkor temples, *Sci. Total Environ.*, 493, pp. 98–108.

Andreae, M.O. and Merlet, P. (2001). Emissions of trace gases and aerosols from biomass burning, *Global Biogeochem. Cy.*, 15, pp. 955–966.

Aneja, V.P., Schlesinger, W.H., and Erisman, J.W. (2008). Farming pollution, *Nature Geosci.*, 1, pp. 409–411.

Azlan, Z. (1997). Protective structures for the conservation and presentation of archaeological sites, *Journal of Conservation and Museum Studies*, http://www.jcms-journal.com/articles/10.5334/jcms.3974/ (accessed November 2014).

Behera, S.N., Sharma, M. (2010a). Investigating the potential role of ammonia in ion chemistry of fine particulate matter formation for an urban environment, *Sci. Total Environ.*, 408, pp. 3569–3575.

Behera, S.N., Sharma, M. (2010b). Reconstructing primary and secondary components of $PM_{2.5}$ aerosol composition for an urban atmosphere, *Aerosol Sci. Technol.*, 44, pp. 983–992.

Behera, S.N. and Sharma, M. (2011). Degradation of SO_2, NO_2 and NH_3 leading to formation of secondary inorganic aerosols: An environmental chamber study, *Atmos. Environ.*, 45, pp. 4015–4024.

Bertilsson, U. and Löfvendahl, R. (1992). 'Rock carvings', in Gullman, J. (Ed.), *Air Pollution and the Swedish Heritage, Konserveringstekniska Studier*, vol. RIK 6, National Heritage Board of Sweden, Stockholm.

Bindler, R. (2011). Contaminated lead environments of man: reviewing the lead isotopic evidence in sediments, peat and soils for the temporal and spatial patterns of atmospheric lead pollution in Sweden, *Environ. Geochem. Health*, 33, pp. 311–329.

Blades, N.W., Bordass, W., Oreszczyn, T. and Cassar, M. (2000). *Guidelines on Pollution Control in Museum Buildings*, Museums Association, London.

Bonazza, A., Sabbioni, C., Ghedini, N., Hermosin, B., Jurado, V., Gonzalez, J.M. and Saiz-Jimenez, C. (2007). Did smoke from the Kuwait oil well fires affect Iranian archaeological heritage? *Environ. Sci. Technol.*, 41, pp. 2378–2386.

Bossioli, E., Tombrou, M., Karali, A., Dandou, A., Paronis, A. and Sofiev, M. (2012). Ozone production from the interaction of wildfire and biogenic emissions: a case study in Russia during spring 2006, *Atmos. Chem. Phys.*, 12, pp. 7931–7953.

Bourges, F., Genthon, P., Genty, D., Lorblanchet, M., Mauduit, E. and D'Hulst, D. (2014). Conservation of prehistoric caves and stability of their inner climate: lessons from Chauvet and other French caves, *Sci. Total Environ.*, 493, pp. 79–91.

Bouwnam, A.F., van Vuuren, D.P., Derwent, R.G. and Posch, M. (2002). A global analysis of acidification and eutrophication of terrestrial ecosystems, *Water Air Soil Pollut.*, 141, 349–382.

Brimblecombe, P. (1990). The composition of museum atmospheres, *Atmos. Environ.*, 24B, pp. 1–8.

Brimblecombe, P. (Ed.) (2003). *The Effects of Air Pollution on the Built Environment*, Imperial College Press, London.

Brimblecombe, P. (2012). 'Indoor air quality in the future', in *Indoor Air Quality 2012, 10th International Conference Indoor Air Quality in Heritage and Historic Environments "Standards and Guidelines"*, Book of Abstracts, p. 52.

Brimblecombe, P., Grossi, C.M. (2007). Damage to buildings from future climate and pollution, *APT Bulletin*, 38, pp. 13–18.

Brunet, J., Vouvé, J. and Malaurent, P. (1997). 'Conservation of subterranean historic and prehistoric monuments', in Agnew, N. (Ed.), *Conservation of Ancient Sites on the Silk Road*, The Getty Conservation Institute, Los Angeles, pp. 259–269.

Butler, D. (2010). French bid to save rock art, *Nature*, 467, pp. 375–376.

Cassar, J. (2000). 'Conservation studies at the Hal Saflieni Hypogeum', in Pace, A. (Ed.), *The Hal Saflieni Hypogeum 4000BC–2000AD*, National Museum of Archaeology, Museums Department, Malta, pp. 45–52.

Cassar, J. (2007). Malta's prehistoric temples: conservation issues, *STONE Newsletter on Stone Decay*, No. 1., pp. 3–7.

Cassar, J., Galea, M., Grima, R., Stroud, K. and Torpiano, A. (2011). Shelters over the Megalithic Temples of Malta: debate, design and implementation, *Environ. Earth Sci.*, 63, 1849–1860.

Cassar, J., Cefai, S., Galea, M., Grima, R., Stroud, K., Torpiano, A. (2013). *Preserving the Megalithic Temples of Malta — the Interdisciplinary Approach.* Available at: http://www.bh2013.polimi.it/papers/bh2013_paper_52.pdf (accessed November 2014).

Cassar, M. (2001). *Technological Requirements for Solutions in the Conservation and Protection of Historic Monuments and Archaeological Remains,* working paper for the STOA Unit, European Parliament Directorate-General for Research, Luxembourg.

Cavicchioli, A., de Souza, R.O.C., Ranieri Reis, G. and Fornaro, A. (2012). 'Indoor air quality in heavily polluted cities: ozone and nitrogen dioxide contamination in the indoor atmosphere of two museums of São Paulo, Brazil', in *Indoor Air Quality 2012, 10th International Conference Indoor Air Quality in Heritage and Historic Environments "Standards and Guidelines"*, Book of Abstracts, p. 14.

Chan, S.-S. (2005). *Terracotta Army in a Battle for Survival,* South China Morning Post. Available at: http://www.scmp.com/article/506151/terracotta-army-battle-survival (accessed November 2014).

Chitty, G. (2007). 'Meeting the climate change challenge', in *The Archaeologist,* 66 Institute for Archaeologists, pp. 16.

Cointreau-Levine, S. (1996). *Sanitary landfill siting and criteria,* Washington DC: World Bank Infrastructure Notes, Urban No. UE-12.

Coningham, R.A.E., Acharya, K.P., Strickland, K.M., Davis, C.E., Manuel, M.J., Simpson, I.A., Gilliland, K., Tremblay, J., Kinnaird, T.C. and Sanderson, D.C.W. (2013). The earliest Buddhist shrine: excavating the birthplace of the Buddha, Lumbini (Nepal), *Antiquity,* 87, pp. 1104–1123.

Council of Europe (1992). *European Convention on the Protection of the Archaeological Heritage (Revised),* Valletta.

Daifuku, H. (1968). Saving our heritage in stone: a worldwide challenge, *The UNESCO Courier,* June 1969, pp 4–11.

Damgaard, A., Riber, C., Fruergaard, T., Hulgaard, T. and Christensen, T. H. (2010) Life-cycle-assessement of the historical development of air pollution control and energy recovery in waste incineration, *Waste Manage.*, 30, pp. 1244–1250.

Doehne, E. and Price, C. (2010). *Stone Conservation: An Overview of Current Research,* 2nd Ed., The Getty Conservation Institute, Los Angeles.

Dredge, J., Fairchild, I.J., Harrison, R.M., Fernandez-Cortes, A., Sanchez-Moral, S., Jurado, V., Gunn, J., Smith, A., Spötl, C., Mattey, D., Wynn, P.M. and Grassineau, N. (2013). Cave aerosols: distribution and contribution to speleothem geochemistry, *Quat. Sci. Rev.*, 63, pp. 23–41.

EEA (2010). *The European Environment State and Outlook 2010 Air Pollution,* European Environment Agency, Publications Office of the European Union, Luxembourg.

EEA (2014). *European Union Emission Inventory Report 1990–2012 under the UNECE Convention on Long-Range Transboundary Air Pollution (LRTAP)* EEA Technical Report No. 12/2014, European Environment Agency, Publications Office of the European Union, Luxembourg.

ENVIArt (2005–2010). *Chemical Interactions between Cultural Artefacts and Indoor Environment.* Available at: http://www.cost.eu/COST_Actions/cmst/Actions/D42

EPA (1996). *Air Quality Criteria for Ozone and Related Photochemical Oxidants,* Vol. II of III, Report EPA/600/P-93/004bF, U.S. Environmental Protection Agency, Office of Research and Development, National Center for Environmental Assessment, Research Triangular Park, North Carolina.

EPA (2002). *Nitrogen: Multiple and Regional Impacts,* Report EPA-430-R-01-006, U.S. Environmental Protection Agency, Clean Air Markets Division, Washington, D.C.

EPA (2012) *Acid Rain.* Available at: http://www.epa.gov/acidrain/effects/forests.html#a2 (accessed November 2014).

EUROSTAT, the Agriculture DG and the Environment DG (1999). *Agriculture, Environment, Rural Development: Facts and Figures – A Challenge for Agriculture.* (European Commission) Available at: http://ec.europa.eu/agriculture/publi/reports/envir1999_en.pdf (accessed November 2014).

Fanood, M.R. (2012). The debate on the need for a protective shelter over the Mausoleum of Cyrus the Great, *Journal of Architectural Conservation,* 18, pp. 53–70.

Furuuchi, M., Murase, T., Tsukawaki, S., Hang, P., Sieng, S. and Hata, M. (2007). Characteristics of ambient particle-bound polycyclic aromatic hydrocarbons in the Angkor Monument Area of Cambodia, *Aerosol Air Qual. Res.*, 7, pp. 221–238.

Garcia-Anton, E., Cuezva, S., Jurado, V., Porca, E., Miller, A.Z., Fernandez-Cortes, A., Saiz-Jimenez, C. and Sanchez-Moral, S. (2014). Combining stable isotope ($\delta^{13}C$) of trace gases and aerobiological data to monitor the entry and dispersion of microorganisms in caves, *Environ. Sci. Pollut. Res.*, 21, pp. 473–484.

Gerwin, W. and Baumhauer, R. (2000). Effect of soil parameters on the corrosion of archaeological metal finds, *Geoderma,* 96, pp. 63–80.

Gordon, C.C. and Buikstra, J.E. (1981). Soil pH, bone preservation, and sampling bias at mortuary sites, *Am. Antiqu.*, 46, pp. 566–571.

Goudie, A. (2013). *The Human Impact on the Natural Environment* (7th ed), Wiley-Blackwell, Chichester.

Groot Koerkamp, P.W.G. (1994). Review on emissions of ammonia from housing systems for laying hens in relation to sources, processes, building design and manure handling, *J. Agr. Eng. Res.*, 59, pp. 73–87.

Grzywacz, C. (2006). *Monitoring for Gaseous Pollutants in Museum Environments,* The Getty Conservation Institute, Los Angeles.

Harrison, R.M. and Kitto, A.N. (1992). Estimation of the rate constant for the reaction of acid sulphate aerosol with NH_3 gas from atmospheric measurements, *J. Atmos. Chem.* 15, pp. 133–143.

Hata, M., Chomanee, J., Thongyen, T., Bao, L., Tekasakul, S., Tekasakul, P., Otani, Y. and Furuuchi, M. (2014). Characteristics of nanoparticles emitted from burning of biomass fuels, *J. Environ. Sci.*, pp. 1913–1920.

Hawksworth, D.L. (1990). The long-term effects of air pollutants on lichen communities in Europe and North America, in G.M. Goodwell (Ed.), *The Earth in Transition: Patterns and Processes of Biotic Impoverishment*, pp. 45–64, Cambridge University Press, New York.

Hurby, D. (2014). *In Cambodia, Toxic Air Threatens Timeless Ruins*, Globalpost. Available at: http://www.globalpost.com/dispatch/news/regions/asia-pacific/cambodia/140630/cambodia-toxic-air-threatens-timeless-ruins (accessed November 2014).

Hu, T., Cao, J., Lee, S., Li, X., Shen, Z., Rong, B. and An, Z. (2010). Microanalysis of dust deposition inside Emperor Qin's Terra-Cotta Warriors and Horses museum, *Aerosol Air Qual. Res.*, 10, pp. 59–66.

Huang, J.Z. (2010). 'The influence of water on the stone carvings of the Yungang Grottoes', in Agnew N. (Ed), *Conservation of Ancient Sites on the Silk Road*, The Getty Conservation Institute, Los Angeles, pp. 196–202.

Hyvert, G. (1973). Borobudur, les bas-reliefs materiaux — facteurs responsables des degradations — programme de conservation, *Stud. Conserv.*, 18, pp. 131–155.

ICOMOS (2005). *Xi'an Declaration on the Conservation of the Setting of Heritage Structures, Sites and Areas.* Available at: http://www.international.icomos.org/charters/xian-declaration.pdf (accessed November 2014).

ISAC-CNR (2006). *Environmental Monitoring at Hagar Qim and Mnajdra Temples Final Report*, unpublished report, CNR-ISAC, Italy.

Inkpen, R. (2003). 'The whole building and patterns of degradation', in Brimblecombe, P. (Ed.), *The Effects of Air Pollution on the Built Environment*, Imperial College Press, London, pp. 393–422.

Jacobson, L.D., Bicudo, J.R., Schmidt, D.R., Wood-Gay, S., Gates, R.S. and Hoff, S.J. (2003). *Air Emissions From Animal Production Buildings, Proceedings XI International Congress ISAH Mexico City.* Available at: http://www.isah-soc.org/index.php?action=15 (accessed November 2014).

Jurado, V., Porca, E., Cuezva, S., Fernandez-Cortes, A., Sanchez-Moral, S. and Saiz-Jimenez, C. (2010). Fungal outbreak in a show cave, *Sci. Total Environ.*, 408, pp. 3632–3638.

Karaca, F. and Öztürk, Z. (2012). Ground level ozone exposure and distribution over the historical Peninsula of Istanbul, *J. Cult. Herit.*, 13, pp. 69–76.

Kars, E.A.K. and Kars, H. (2002). *The Degradation of Bone as an Indicator in the Deterioration of the European Archaeological Property*, final report, with appendices A-H; Project ENV4-CT98-0712, internal report, European Commission, Brussels.

Khan, R. and Vyas, H. (2008) A study of impact of brick industries on environment and human health in Ujjain City (India), *Journal of Environmental Research and Development*, 2, pp. 421–425.

Koroneos, C. and Dompros, A. (2007). Environmental assessment of brick production in Greece, *Build. Environ.*, 42, pp. 2114–2123.

Larsen, R., Bell, J.N.B., James, P.W., Chimonides, P.J., Rumsey, P.J., Tremper, A. and Purvis, O.W. (2007). Lichen and bryophyte distribution on oak in London in relation to air pollution and bark acidity, *Environ. Pollut.*, 146, pp. 332–340.

Lazaridis, M., Latos, M., Aleksandropoulou, V., Hov, Ø., Papayannis, A. and Tørseth, K. (2008). Contribution of forest fire emissions to atmospheric pollution in Greece, *Air Qual. Atmos. Health*, 1, pp. 143–158.

Li, Z.X. (2010). 'Deterioration and treatment of wall paintings in grottoes along the Silk Road in China and related conservation efforts', in Agnew, N. (Ed.), *Conservation of Ancient Sites on the Silk Road*, The Getty Conservation Institute, Los Angeles, pp. 46–55.

Lindborg, U. (1990). *Air Pollution and Cultural Heritage Action Plan 90, Konserveringstekniska Studier*, vol. RIK 1, National Heritage Board of Sweden, Stockholm.

Lu, Z., Streets, D.G., Zhang, Q., Wang, S., Carmichael, G. R., Cheng, Y. F., Wei, C., Chin, M., Diehl, T. and Tan, Q. (2010). Sulfur dioxide emissions in China and sulfur trends in East Asia since 2000, *Atmos. Chem. Phys.*, 10, pp. 6311–6331.

Maguregui, M., Castro, K., Morillas, H., Trebolazabala, J., Knuutinen, U., Wiesinger, R., Schreiner, M. and Madariaga, J.M. (2013). Multianalytical approach to explain the darkening process of hematite pigment in paintings from ancient Pompeii after accelerated weathering experiments, *Anal. Methods*, 6, pp. 372–378.

Maekawa, S. (1993). 'Environmental monitoring at the tomb of Nefertari', in Corzo M.A., Afshar, M. (Eds.), *Art and Eternity*, The J. Paul Getty Trust, pp 105–121.

Martin, O. and Piatti, G. (Ed.) (2009). *World Heritage and Buffer Zones, International Expert Meeting on World Heritage and Buffer Zones*, Davos, Switzerland 11–14 March 2008, World Heritage Papers 25, UNESCO World Heritage Centre.

Martin-Sanchez, P.M., Jurado, V., Porca, E., Bastian, F., Lacanette, D., Alabouvette, C. and Saiz-Jimenez, C. (2014). Airborne microorganisms in Lascaux Cave (France), *Int. J. Spel.*, 43, pp. 295–303.

Martins, V., Miranda, A.I., Carvalho, A., Schaap, M., Borrego, C. and Sá, E. (2012). Impact of forest fires on particulate matter and ozone levels during 2003, 2004 and 2005 fire seasons in Portugal, *Sci. Total Environ.*, 414, pp. 53–62.

McMurry, P.H., Takano, H. and Anderson, G.R. (1983). Study of the ammonia (gas) sulphuric acid (aerosol) reaction rate, *Environ. Sci. Technol.*, 17, pp. 347–352.

Mulec, J. (2014). Human impact on underground cultural and natural heritage sites, biological parameters of monitoring and remediation actions for insensitive surfaces: case of Slovenian show caves, *J. Nat. Conserv.*, 22, pp. 132–141.

Naeher, L.P., Smith, K.R., Brauer, M., Chowdhury, Z., Simpson, C., Koenig, J.Q., Lipsett, M. and Zelikoff, J.T. (2005). *Critical Review of the Health Effects of Woodsmoke*, report funded by the Air Health Effects Division, Health Canada, Ottawa.

Nava, S., Becherini, F., Bernardi, A., Bonazza, A., Chiari, M., García-Orellana, I., Lucarelli, F., Ludwig, N., Migliori, A., Sabbioni, C., Udisti, R., Valli, G. and Vecchi, R. (2010). An integrated approach to assess air pollution threats to cultural heritage in a semi-confined environment: the case study of Michelozzo's Courtyard in Florence (Italy), *Sci. Total Environ.*, 408, pp. 1403–1413.

Neff, D., Reguer, S., Bellot-Gurlet, L., Dillmann, P. and Bertholon, R. (2004). Structural characterization of corrosion products on archaeological iron, an integrated analytical approach to establish corrosion forms, *J. Raman Spectrosc.*, 35, pp. 739–745.

Neff, D., Vega, E., Dillmann, P., Bellot-Gurlet, L., Descostes, M. and Béranger, G. (2007). 'Contribution of iron archaeological artefacts to the estimation of average corrosion rates and the long term corrosion mechanisms of low carbon steel buried in soil', in Dillmann, P., Béranger, G., Piccardo, P. and Matthiessen, H. (Eds.), *Corrosion of Metallic Heritage Artefacts: Investigation, Conservation and Prediction for Long Term Behaviour*, Woodhead Publishing, Cambridge, pp. 41–76.

Nord, A.G., Tronner, K., Mattson, E., Borg, G.C. and Ullén, I. (2005a). Environmental threats to buried archaeological remains, *Ambio*, 34, pp. 256–262.

Nord, A.G., Mattson, E. and Tronner, K. (2005b). Factors influencing the long-term corrosion of bronze artefacts in soil, *Protection of Metals*, 41, pp. 309–316.

Orecchio, S. (2010). Analytical method, pattern and sources of polycyclic aromatic hydrocarbons (PAHs) in the stone of the Temples of Agrigento (Italy), *J. Hazard. Mater.*, 176, pp. 339–347.

Piervitori, R., Salvadori, O. and Isocrono, D. (2004). Literature on lichens and biodeterioration of stonework IV, *The Lichenologist*, 36, pp. 145–157.

Quennehen, B., Schwarzenboeck, A., Matsuki, A., Burkhart, J.F., Stohl, A., Ancellet, G. and Law, K.S. (2012). Anthropogenic and forest fire pollution aerosol transported to the Arctic: observations from the POLARCAT-France spring campaign, *Atmos. Chem. Phys.*, 12, pp. 6437–6454.

Ranta, P. (2001). Changes in urban lichen diversity after a fall in sulphur dioxide levels in the city of Tampere, SW Finland, *Annales Botanici Fennici*, 38, pp. 295–304.

Rizzi, G. (2008). *Sheltering the Mosaics of Piazza Armerina: Issues of Conservation and Preservation.* Available at: http://www.archaeological.org/news/hca/71 (viewed November 2014).

RoTAP (2012). *Review of Transboundary Air Pollution: Acidification, Eutrophication, Ground Level Ozone and Heavy Metals in the UK*, Wallingford: Centre for Ecology and Hydrology.

Russell, M., Allen, D., Collins, D. and Fraser, M. (2004). Daily, seasonal, and spatial trends in $PM_{2.5}$ mass and composition in Southeast Texas, *Aerosol Sci. Technol.*, 38, pp. 14–26.

Ryhl-Svendsen, M. (2006). Indoor air pollution in museums: a review of prediction models and control strategies, *Reviews in Conservation*, 7, pp. 27–41.

Sabbioni, C. (2003). 'Mechanisms of air pollution damage to stone', in Brimblecombe, P. (Ed.), *The Effects of Air Pollution on the Built Environment*, Imperial College Press, London, pp. 63–106.

Saiz-Jimenez, C. (2012). Microbiological and environmental issues in show caves, *World J. Microbiol. Biotechnol.*, 28, pp. 2453–2464.

Saiz-Jimenez, C., Cuezva, S., Jurado, V., Fernandez-Cortes, A., Porca, E., Benavente, D., Cañaveras, J.C. and Sánchez-Moral, S. (2011). Palaeolithic art in peril: policy and science collide at Altamira cave, *Science*, 334, pp. 42–43.

Sánchez-Moral, S., Soler, V., Cañaveras, J.C., Sanz-Rubio, E., Van Grieken, R. and Gyesis, K. (1999). Inorganic deterioration affecting Altamira cave, N Spain: quantitative approach to wall-corrosion (solutional etching) processes induced by visitors, *Sci. Total Environ.*, 243–244, pp. 67–84.

Seto, S., Nakamura, A., Noguchi, I., Ohizumi, T., Fukuzaki, N., Toyama, S., Maeda, M., Hayashi, K., Hara, H. (2002). Annual and seasonal trends in chemical composition of precipitation in Japan during 1989–1998, *Atmos. Environ.*, 31, pp. 3505–3517.

Smith, R.A. (1872). *Air and Rain: The Beginnings of a Chemical Climatology*, Longmans, Green and Co., London.

Smith, S.J., Pitcher, H. and Wigley, T.M.L. (2001). Global and regional anthropogenic sulphur dioxide emissions, *Global Planet. Change*, 29, pp. 99–119.

Smolík, J., Ondráčková, L., Mašková, L., Ondráček, J. and Zíková, N. (2012). 'Behaviour of particulate matter in the indoor environment of the National Library in Prague' in *Indoor Air Quality 2012, 10th International Conference Indoor Air Quality in Heritage and Historic Environments "Standards and Guidelines"*, Book of Abstracts, p. 45.

SourceWatch (2013). *Environmental Impacts of Coal*, Sourcewatch. Available at: http://www.sourcewatch.org/index.php/Environmental_impacts_of_coal (accessed November 2014).

St Clair, L. and Seaward, M.R.D. (Eds.) (2004). *Biodeterioration of Stone Surfaces: Lichens and Biofilms as Weathering Agents of Rocks and Cultural Heritage*, Kluwer Academic Publishers, Dordrecht.

Steiger, M. (2003). 'Salts and crusts', in Brimblecombe, P. (Ed.), *The Effects of Air Pollution on the Built Environment*, Imperial College Press, London, pp. 133–181.

Tidblad, J., Kucera, V., Ferm, M., Kreislova, K., Brüggerhoff, S., Doytchinov, S., Screpanti, A., Grøntoft, T., Yates, T., de la Fuente, D., Roots, O., Lombardo, T., Simon, S., Faller, M., Kwiatkowski, L., Kobus, J., Varotsos, C., Tzanis, C., Krage, L., Schreiner, M., Melcher, M., Grancharov., I. and Karmanova, N. (2012). Effects of air pollution on materials and cultural heritage: ICP Materials celebrates 25 years of research, *International Journal of Corrosion*, 2012, Article ID 496321, 16 pages.

Tourism Administration of Xi'an (2006–2008). *Museum of Qin Terra-cotta Warriors and Horses*. Available at: http://en1.xian-tourism.com/xiantraveldetail.asp?listID=352 (accessed November 2014).

UCL Centre for Sustainable Heritage (2008). *IMPACT Modelling Tool*. Available at: http://www.ucl.ac.uk/sustainableheritage-save/impact/index.htm (accessed November 2014).

Ullén, I., Nord, A.G., Fjaestad, M., Mattsson, E., Borg, G.C. and Tronner, K. (2004). The degradation of archaeological bronzes underground: evidence from museum collections, *Antiquity*, 78, pp. 380–390.

UNEP (n.d.). *Tourism's Three Main Impact Areas*, United Nations Environment Programme, Environment for Development. Available at: http://www.unep.org/resourceefficiency/Business/SectoralActivities/Tourism/TheTourismandEnvironmentProgramme/FactsandFiguresaboutTourism/ImpactsofTourism/EnvironmentalImpacts/TourismsThreeMainImpactAreas/tabid/78776/Default.aspx (accessed November 2014).

UNEP (1996). *International Source Book on Environmentally Sound Technologies (ESTs) for Municipal Solid Waste Management (MSWM)*, United Nations Environment Programme, Nairobi. Available at: http://www.unep.or.jp/ietc/ESTdir/pub/MSW/sp/sp5/sp5_1.asp (accessed November 2014).

UNESCO (1992–2014). *Angkor*, UNESCO World Heritage Centre, World Heritage List. Available at: http://whc.unesco.org/en/list/668 (accessed November 2014).

UNESCO (2013a). *Operational Guidelines for the Implementation of the World Heritage Convention.* Available at: http://whc.unesco.org/archive/opguide13-en.pdf (accessed November 2014).

UNESCO (2013b). *Conserving the Borobudur Temple for the Future*, UNESCO Office in Jakarta and Ministry of Education and Culture, Indonesia.

USGS (2010). *Volcanic gases and their effects.* Available at: http://volcanoes.usgs.gov/hazards/gas/ (accessed November 2014).

Van Herk, C.M., Mathijssen-Spiekman, E.A.M. and deZwart, D. (2003). Long distance nitrogen effects on lichens in Europe, *Lichenologist*, 35, pp. 347–359.

Vanguelov, E.I. *et al.* (2010). Chemical fluxes in time through forest ecosystems in the UK — soil response to pollution recovery, *Environ. Pollut.*, 158, pp. 1857–1869.

Vidal, J. (2014). *Indonesia's Forest Fires Feed "Brown Cloud" of Pollution Choking Asia's Cities*, The Observer. Available at: http://www.theguardian.com/environment/2014/mar/23/indonesia-forest-fires-pollution-asia (accessed November 2014).

Voûte, C. (1973). The restoration and conservation project of Borobudur Temple, Indonesia, *Stud. Conserv.*, 18, pp. 113–130.

Vouvé, J., Brunet, J., Vidal, P. and Marsal, J. (1982). *Lascaux en Perigord Noir: Environnement, Art Parietal et Conservation*, P. Fanlac, Périgueux.

Woodin, S.J. (1989). Environmental effects of air pollution in Britain, *J. Appl. Ecol.*, 26, pp. 749–761.

World Bank (2014). *International Tourism, Number of Arrivals*, The World Bank. Available at: http://data.worldbank.org/indicator/ST.INT.ARVL (accessed November 2014).

World Bank and State Environmental Protection Administration, P. R. China (2007). *Cost of Pollution in China: Economic Estimates of Physical Damages*, The World Bank.

Available at: http://www-wds.worldbank.org/external/default/WDSContentServer/WDSP/IB/2007/03/30/000090341_20070330141612/Rendered/PDF/392360CHA0Cost1of1Pollution01PUBLIC1.pdf (accessed January 2015).

Yates, T. (2003). 'Mechanisms of air pollution damage to brick, concrete and mortar', in Brimblecombe, P. (Ed.), *The Effects of Air Pollution on the Built Environment*, Imperial College Press, London, pp. 107–132.

Zerbock, O. (2003). *Urban Solid Waste Management: Waste Reduction in Developing Nations*, M.S. report, School of Forest Resources & Environmental Science, Master's International Program, Michigan Technological University. Available at: http://appropriatetechnology.wikispaces.com/file/view/Waste_reduction_and_incineration_FINAL.pdf (accessed November 2014).

CHAPTER 11

Evaluation of Cleaning Methods for Graffiti Removal

Patricia Sanmartín

Laboratory of Applied Microbiology, School of Engineering and Applied Sciences, Harvard University, United States;
Departamento de Edafología y Química Agrícola, Facultad de Farmacia, Universidad de Santiago de Compostela, Spain

Ralph Mitchell

Laboratory of Applied Microbiology, School of Engineering and Applied Sciences, Harvard University, United States

Francesca Cappitelli

Dipartimento di Scienze per gli Alimenti, la Nutrizione e l'Ambiente, Università degli Studi di Milano, Italy

11.1 Introduction

Buildings and structures undergo deterioration due to various factors such as air pollution, physical and chemical weathering, microbial discolouration, and also the presence of undesirable graffiti. Indeed, graffiti represent an important form of disfigurement to the urban fabric and besides that, their application generates a relatively large amount of pollution due to the associated emissions, which are

Fig. 11.1 Spray-painted graffiti on an urban building. Source: Alberto Ramil (Department of Industrial Engineering II. University of A Coruña, Spain).

chiefly volatile organic compound (VOC)-based. Many affected sites (Fig. 11.1) are considered deteriorated and thus lose some of their economic value. According to the EU project GRAFFITAGE, around 3,500,000 protected monuments in European cities are affected by graffiti (GRAFFITAGE). In the preservation of public spaces and historic buildings, graffiti removal (viz. cleaning of aerosol spray paint, which is the most commonly used staining agent in practice) requires a huge financial outlay by local governments and agencies.

Conservators rarely remove graffiti and only if cultural heritage surfaces are affected. Graffiti-removal operations are well-developed in the maintenance industry who use cleaning methods based on abrasive mechanical action or chemical substances. Graffiti removal by means of hot water is generally suggested in the technical sheets. Chemical removers are recommended only to enhance unsatisfactory results of the cleaning process. However, the use of hot water could melt the applied anti-graffiti systems (graffiti barrier coatings, aimed at reducing paint adhesion and to facilitate removal of potential graffiti using a specific cleaning methodology), which can be potentially removed along with the stain (Goidanich *et al.*, 2010). Penetration of the staining agent into the pores of the substrate, sometimes even in large amounts, as a consequence of a high porosity of the material, hinders effective cleaning. Some cases have been reported in which repeated cleaning procedures, involving hot water, mechanical action and chemical removers, did not completely remove the stain (e.g. Lettieri and Masieri, 2014).

Protection of buildings and structures against graffiti is therefore a serious challenge in urban environments. In this situation, the market for graffiti removal systems and products, as well as anti-graffiti systems, is developing fast and many new methods are introduced every year. However, until now graffiti cleaning has always been a balance between removal of unwanted material and substrate damage, since none of the cleaning methods available at present are capable of removing the graffiti from substrates without also affecting the underlying material itself in some way (Sanmartín *et al.*, 2014a). Likewise, there are no standardised methods for the evaluation of anti-graffiti treatments on porous built materials (García and Malaga, 2012), and the efficacy of anti-graffiti systems has not been discussed in depth (Lettieri and Masieri, 2014).

Importantly, despite the high cost of cleaning and preventive operations and the damage to the image of cities, it is surprisingly difficult to find scientific publications related to graffiti-removal and anti-graffiti systems (Goidanich *et al.*, 2010). Some patents concerning novel graffiti cleaning methods and a huge number of new anti-graffiti products have been reported in the literature. However, very few papers discuss the efficacy and potentially damaging actions with respect to conservation criteria. In addition, the available literature is unfortunately not easily accessible to the scientific community.

This chapter will look at ways of evaluating current methods of graffiti removal that are accessible to the scientific community and will include an updated overview of published studies on graffiti cleaning.

Finally, current perspectives and future directions towards a method that is not harmful to substrates, and that is environmentally friendly and safe for human health, are presented. A brief revision of graffiti barrier coatings (anti-graffiti) is also reported.

11.2 Published Studies on Graffiti Cleaning (1995–2014)

Because graffiti paints and how they age are so varied, graffiti removal processes cannot be standardised. Consequently, different methods are used to remove graffiti. High-pressure hot water and chemical solvents (which typically contain methyl ethyl ketone, methylene, chloride and phenol) are capable of dissolving or breaking down stains. These methods and mechanical cleaning (with abrasives) are considered the most traditional techniques of graffiti removal (Weaver *et al.*, 1995; Sanmartín *et al.*, 2014a), although the solvents may penetrate into the substrate and cause irreversible damage. The solvents may also contain substances that are pollutants and thereby constitute serious environmental and health hazards. Traditional physical methods, such as scalpel, abrasive dust and sandblasting may be appropriate for removing graffiti from surfaces under highly controlled conditions, or occasionally from delicate masonry surfaces (e.g. Chapman, 2000), but are generally not appropriate. Novel cleaning techniques, such as ultrasonic/megasonic agitation, plasma spray, arc or thermal spray, dry ice blasting (CO_2-based) and soda blasting are currently under development for application in graffiti removal. However, their relative merits and limitations remain to be fully assessed. Laser-based techniques for the removal of graffiti have become especially important in the last ten years, and studies involving these techniques are the most prolific in the recent literature, although nowadays they are mainly applied to historic buildings and structures. The major benefits of laser cleaning include the non-contact nature of the treatment and its selectivity, gradual removal of graffiti, repeatability of the treatment, control of the area processed, and the low environmental impact (e.g. Costela *et al.*, 2003; Gómez *et al.*, 2006; Rivas *et al.*, 2012; Siano *et al.*, 2012; Ortiz *et al.*, 2013). However, these techniques also have some drawbacks. In particular, they can visibly alter the colour of the substrate (e.g. inducing yellowing), remove some grains and produce craters, and they can also transform, melt and/or fracture the minerals (Sanmartín *et al.*, 2014a and references therein).

Laser-based techniques have been used in all studies presented in this chapter. The substrate most commonly involved is stone, although other types of materials such mortar, glass, polymers, wood and metals must also be protected from graffiti. According to Goidanich *et al.* (2010), this is mainly because many structures with exposed surfaces in urban environments are made from stone, conservation of which poses a difficult challenge to conservators because of the porous nature of the material.

Using a laser method, Ye *et al.* (2014) successfully removed the graffiti ink that had appeared on the Yungang Grottoes (Chinese Buddhist sandstone temple, located in Shanxi, China), dating from the 5th and 6th centuries and designated as a World Heritage Site by UNESCO in 2001. Only one side effect was reported by the authors: a slight residual contamination on the site cleaned by dry laser cleaning (see Table 11.1, which includes information about graffiti removal methods used in key studies between 1995 and 2014). Fiorucci *et al.* (2013, 2011) and Rivas *et al.* (2012) used another laser method to clean several graffiti spray paints (see Sanmartín *et al.* (2014a) for a complete description of the graffiti paint characteristics reported in these papers and the others mentioned here) from granite rocks. The results were satisfactory, mainly for red, blue and black paint, with minimal or even no paint remaining. However, laser treatment was less efficient for silver paint. Removal of silver paint caused appearance of an opaque film that notably modified the stone appearance and the scatter of silver particles over the surface. In silver-painted samples, the Nd:YVO_4 laser (355 nm) can also alter granite minerals, causing biotite grains to merge and, in some cases, fracture of quartz crystals. Furthermore, the Nd:YVO_4 laser (355 nm) may induce changes in the degree of hydration of the iron oxides and oxyhydroxides that tend to cover granite fissures. Ortiz *et al.* (2013) used several physical, chemical and laser methods (Table 11.1) to remove spray paints, markers and black cutting-edge ink from dolomitic white marble. Cleaning with chemical products yielded satisfactory results for most of the samples considered, whereas pressurised water was not

Table 11.1 Graffiti removal methods used by several authors.

Reference	Methods used
Ye *et al.* (2014)	Nd:YAG laser, first harmonic (1064 nm, IR), pulse duration: 10 ns and repetition rate: 1 Hz. Two methods were used: (1) dry laser cleaning (DLC) and (2) steam laser cleaning (SLC) that uses a liquid (water or a water/alcohol mixture) applied prior to the laser treatment.
Fiorucci *et al.* (2013, 2011) and Rivas *et al.* (2012)	Nd:YVO_4 laser, third harmonic (355 nm, UV) at a high repetition rate (nanoseconds).
Ortiz *et al.* (2013)	(1) Pulsed laser cleaning (PLC) techniques: near-infrared (IR) and ultraviolet (UV) Nd:YAG laser pulses, (2) pressurised water: working pressures of 0.27, 1.09 and 7.67 kg/cm^2, and (3) chemical treatments: *Elephant snot*® and *Graffitipaste*® (cleaning products).
Penide *et al.* (2013)	High power diode laser (HPDL) emitting in modulated wave (MW) and continuous wave (CW) laser operation regimes.
Gómez *et al.* (2006)	(1) Nd:YAG laser, first harmonic (1064 nm, IR), energy: up to 49 mJ/pulse, pulse duration: 6 ns, repetition rate: 1.25 Hz and laser fluence: 1.8–3.3 J/cm^2, and (2) excimer laser (XeCl) (308 nm, UV), energy: up to 180 mJ/pulse, pulse duration: $\sim$ 25 ns, repetition rate: 1 Hz and laser fluence: 1.1–2.7 J/cm^2.
Costela *et al.* (2003)	(1) Nd:YAG laser, second harmonic (532 nm, VIS), energy: 300 mJ/pulse, pulse duration: $\sim$ 6.6 ns and repetition rate: 1.25 Hz, and (2) Nd:YAG laser, third harmonic (355 nm, UV), energy: 40 mJ/pulse, pulse duration: $\sim$ 5.4 ns and repetition rate: 1.25 Hz.
Chapman (2000)	(1) Nd:YAG laser, first harmonic (1064 nm, IR) (at West Kennet Avenue), (2) steam at low-pressure (at West Kennet Avenue), (3) Nd:YAG laser, first and second harmonics (1064 nm, IR, and 532 nm, VIS) (Heel Stone), and (4) chemical treatment: acetone (Heel Stone).
Liu and Garmire (1995)	(1) Nd:YAG laser, energy: up to 300 mJ/pulse, and (2) CO_2 laser, power of the cw: 500W.

efficient and induced appreciable surface damage. Pulsed laser cleaning (PLC) techniques were advantageous for rapid cleaning of small or inaccessible areas. Among them, a UV laser emitting at 266 nm was particularly efficient for cleaning graffiti of different tonalities although it caused damage to the surface at a microscopic scale (removal of grains and production of craters). An IR laser emitting at 1064 nm successfully removed black spray, but also induced yellowing and was not sufficient to remove red and green sprays and markers or black cutting-edge ink. Also with laser technique, Penide *et al.* (2013) obtained the best results with the CW (continuous wave) regime (double pass), which maintained the colour and roughness of the stone under study (ignimbrite, rhyolite igneous rock, viz. pink Morelia quarry). Visible colour changes were, however, detected with the other laser techniques under study. The MW (modulated wave) regime caused unavoidable changes to the appearance of the stone surface, and black spots were observed in the treated area where the stone melted. The CW regime (single pass) induced transformation of quartz to crystobalite and amorphisation (vitrification) of quartz and a minority phase of clay. It also melted the clay compounds in the stone. Gómez *et al.* (2006) and Costela *et al.* (2003) used different types of Nd:YAG and excimer lasers to remove spray paints from glass, steel, wood, marble and concrete. The use of an IR laser (1064 nm) yielded a more efficient process, while the use of a UV laser (308 and 355 nm) completely removed pigment and polymeric base from paint on all studied materials without any damage to the substrate. The results of VIS (visible) emission at 532 nm were very similar to those achieved with UV emission at 355 nm for the most of the paints under study.

Among the negative results, the authors highlighted the non-removal of several colours such as blue, pink, green and yellow from wood, marble and concrete, as well as damage to the substrate. The studies by Chapman (2000) and Liu and Garmire (1995) represent pioneering research on graffiti removal. The case studies performed by Chapman (2000) involved graffiti on Sarsen stones (neolithic sandstone) at the West Kennet Avenue at Avebury and the Heel Stone at Stonehenge, both in Wiltshire (UK). Liu and Garmire (1995) carried out a laboratory study with building materials, such as brick, concrete, metals, wood and glass. At West Kennet Avenue, the laser appeared to change the state of the ferrous oxides present in the stone, resulting in a change of colour to either dark grey or dark brown. In the most porous areas, a laser used in combination with methylene dichloride gave successful results, with no discolouration of the substrate or any effects on the lichen colonies. On the Heel stone, acetone successfully removed most of the paint without damaging the stone, and the laser proved to be an appropriate method for softening the impact of the graffiti, camouflaging it against the lichen-covered substrate, which was mainly unaffected by the cleaning process. Although paint residues remained, these may disappear gradually over time. In the study by Liu and Garmire (1995), Nd:YAG laser treatment successfully removed the paint from the wood substrate but was not able to remove paint from Plexiglass without damaging the substrate. Use of a CO_2 laser also failed to remove the paint from wood or Plexiglass. The laser beam burned the substrate at the same time as it removed the paint.

11.3 Methods for Evaluating Graffiti Cleaning

This section examines the main methods used in eleven studies that evaluated graffiti removal, in direct relation to the specific subject considered in this chapter. Physical and imaging techniques are used almost exclusively to evaluate the efficiency and effectiveness of graffiti removal methods. Table 11.2 summarises the most important contributions in terms of degree of cleaning achieved and damage to the substrate.

11.3.1 *Optical and electronic microscopy*

Costela *et al.* (2003) used an optical microscope coupled to a colour video camera to characterise the Nd:YAG laser-irradiated zones at 532 nm and 355 nm during the ablation process. In a study by Gómez *et al.* (2006), morphological analysis by optical microscopy and environmental scanning electron microscopy (ESEM) of the laser irradiated areas (by XeCl excimer (308 nm, UV) and Nd:YAG (1064 nm, IR) lasers) revealed remarkable features of interest for

Table 11.2 Techniques used by various authors for evaluating graffiti removal (including the degree of cleaning and damage to the substrate).

Reference	Techniques for assessing the effectiveness of the cleaning	Techniques for assessing damage to the substrate
Ye *et al.* (2014)	Laboratory experiments: — Optical microscope — 3D digital microscope — Electronic microscopy (SEM–EDS) *In-situ*: — 3D digital microscope — Colourimetric measurements	Laboratory experiments: — Optical microscope — 3D digital microscope — Electronic microscopy (SEM–EDS) *In-situ:* — 3D digital microscope — Colourimetric measurements
Ortiz *et al.* (2013)	— Optical microscope — Colourimetric measurements — X-ray microfluorescence	— Optical microscopy — XRD and XRF — Petrographic microscope with digital camera.
Penide *et al.* (2013)	— Preliminary visual analysis — Optical microscopy — Electronic microscopy (XRD and SEM)	— Preliminary visual analysis — Optical microscopy — XRD and SEM
Rivas *et al.* (2012)	— Optical microscopy — Colourimetric measurements — Polarised microscopy — Electronic microscopy (SEM–EDS) — Infrared spectroscopy (ATR-FTIR)	— Colourimetric measurements
Siano *et al.* (2012)	— Raman spectroscopy — Laser induced plasma spectroscopy (LIPS)	— Raman spectroscopy — Laser induced plasma spectroscopy (LIPS)
Fiorucci *et al.* (2011)	— Optical microscopy — Electronic microscopy (SEM–EDS) — Confocal microscopy	— Optical microscopy — Electronic microscopy (SEM–EDS) — Confocal microscopy

(*Continued*)

Table 11.2 (*Continued*)

Reference	Techniques for assessing the effectiveness of the cleaning	Techniques for assessing damage to the substrate
Gómez *et al.* (2006)	— Optical microscopy — Electronic microscopy (ESEM) — Laser-induced breakdown spectroscopy (LIBS) elemental analysis	— Optical microscopy — Electronic microscopy (ESEM) — Laser-induced breakdown spectroscopy (LIBS) elemental analysis
Costela *et al.* (2003)	— Optical microscopy — Laser-induced breakdown spectroscopy (LIBS) elemental analysis	— Optical microscopy — Laser-induced breakdown spectroscopy (LIBS) elemental analysis
Chapman (2000)	— Visual analysis	— Visual analysis

ATR-FTIR: Attenuated total reflectance-Fourier transform infrared.
ESEM: Environmental scanning electron microscope.
FTIR: Fourier transform infrared.
SEM: Scanning electron microscopy.
SEM–EDS or SEM-EDX: with energy dispersive X-ray spectroscopy.
XRD: X-ray diffractometer.
XRF: X-ray fluorescence.

understanding the dependence of ablation efficiency on wavelength. Fiorucci *et al.* (2011) used optical microscopy on graffiti-affected mortar and granite stone samples before and after laser treatment, respectively, for preliminary evaluation of the cleaning process. Fiorucci *et al.* (2011) used scanning electron microscopy with energy dispersive X-ray spectroscopy (SEM-EDX or SEM-EDS) in a more in-depth study to analyse the effectiveness of paint removal and the different morphological effects of the laser on the granite rock-forming minerals, such as biotite, which melts at low irradiation fluence.

Rivas *et al.* (2012) used optical microscopy to obtain a preliminary qualitative evaluation of the efficacy of laser treatment in removing the graffiti from granite surfaces. These authors also examined cross sections of the cleaned surfaces by polarised microscopy and SEM–EDS to characterise the efficacy of paint removal and the effects of the laser on granite-forming minerals. Similarly, Ortiz *et al.* (2013) used optical and electronic microscopy to observe the morphological changes induced on the marble substrate.

With the aim of determining whether any changes took place in either the colour or shape of ignimbrite during the high power diode laser (HPDL) treatment, Penide *et al.* (2013) examined images of the cross-sections of the heat-affected zone by means of optical microscopy and scanning electron microscopy. These images were compared with others taken from untreated stone. Scanning electron microscopy (SEM) enables observation of the difference between the treated area (in which the pores were spherically-shaped, indicating that the clay compounds had melted) and the untreated area. Finally, Ye *et al.* (2014) used optical microscopy and SEM–EDS in laboratory experiments to observe the sample surface before and after Nd:YAG (1064 nm) laser cleaning. *In-situ* examination of samples by 3D digital microscopy confirmed that laser cleaning was effective for removing graffiti ink.

11.3.2 *Infrared spectroscopy*

Fourier transform infrared (FTIR) can be used to at least partly to characterise paint and evaluate the chemical changes. A good example of this is provided in the study by Cappitelli *et al.* (2005), although these authors did not work with spray paint. Attenuated total reflectance-Fourier transform infrared (ATR-FTIR) was used to evaluate the existence of traces of paint on granite surfaces in the apparently cleaned areas, as assessed by optical microscopy (Rivas *et al.*, 2012).

11.3.3 *Colourimetric measurements*

Paint is susceptible to loss of colour through pigment fading and to breakdown of polymer in the binder (Bentley and Turner, 1998). In some studies (e.g. Chapman, 2000), the colour changes that occurred during or after the cleaning treatment were only assessed by examination with the naked eye. However, colour is a very subjective property that depends on an individual's perception, and its quantification and representation in a colour space is, therefore, a better option. For example, Rivas *et al.* (2012) used the CIELAB colour space (CIE Publication 15–2, CIE 1986) in which three values: L*a*b* (Cartesian coordinates) or $L^*C^*_{ab}h_{ab}$ (polar or cylindrical coordinates) determine each colour. L* is the lightness coordinate and moves from top (value: 100, white) to bottom (value: 0, black), a* and b* coordinates are associated with changes in redness-greenness (positive a* is red and negative a* is green) and in yellowness-blueness (positive b* is yellow and negative b* is blue). C^*_{ab} is the chroma of colour ($C^*_{ab} = [a^{*2} + b^{*2}]^{1/2}$) and h_{ab} is the hue angle or tone (h_{ab} = arctan (b*/a*)), which indicates redness, yellowness, greenness or blueness in a circular scale that starts at 0° and increases counterclockwise to 360°. Moreover, this space enables quantification of the colour difference between two samples by combining the differences in three coordinates, both Cartesian and polar. This is the basis of the CIELAB-based formulae (ΔE^*_{ab}, CMC, ΔE_{94}, ΔE_{00}) (for more details of this colour space see, e.g. Wyszecki and Stiles, 1982; Sanmartín, 2012). The authors used spectrophotometric colour measurements to assess the cleaning efficiency and the impact of the $Nd{:}YVO_4$ laser treatment (355 nm) on the original colour of *Vilachán* and *Rosa Porriño* granites. In all cases, the colour of the samples darkened and become grayish in tone after cleaning; the total colour difference (ΔE^*_{ab}) surpassed the visual threshold of 3 CIELAB units, which is perceptible to the human eye (Wyszecki and Stiles, 1982; Völz, 2001; Sanmartín, 2012). These authors pointed out that colourimetric change measurements should be used with caution for evaluating the effectiveness of laser paint removal, especially for porous rocks such as granites, because colour is greatly influenced by the amount of paint remaining below the surface.

More recently, Ye *et al.* (2014) used a spectrophotometer to assess the surface colour in the Yungang Grottoes (Shanxi, China) and reported the total colour differences (ΔE^*_{ab}) between laser-cleaned areas and areas without graffiti. Bearing in mind the known L*, a* and b* parameters, the differences ΔL^*, Δa^* and Δb^* and the total colour difference ΔE^*_{ab} in the CIELAB colour system, Ortiz *et al.*

(2013) defined a cleaning effectiveness (CE%) parameter as CE% = $(\Delta E^*_W - \Delta E^*_c)*100/\Delta E^*_W$, where ΔE^*_W is the colour difference between the native marble used in their study and the weathered surfaces thereof, and ΔE^*_c is the colour difference between the native marble and the cleaned surface. Note that when ΔE^*_c tends to zero, the cleaning effectiveness approaches 100%, so that the cleaned surface has recovered a colour similar to the original, in this case of the native marble.

The next step would be to monitor removal rate or thickness graffiti by using a digital camera converted into a non-contact colourimeter (Fig. 11.2), because with contact-type colour measuring devices (colourimeters and spectrophotometers, used until now) it is not always possible to reach the target object with the instrument. Based on a technique developed recently by Sanmartín *et al.* (2014b), involving a method for obtaining L*a*b* colour measurements from RGB digital values that has been successfully applied to measure the colour of Spanish granites, the suitability of using digital cameras to determine the efficacy of graffiti removal on stone substrates may be tested.

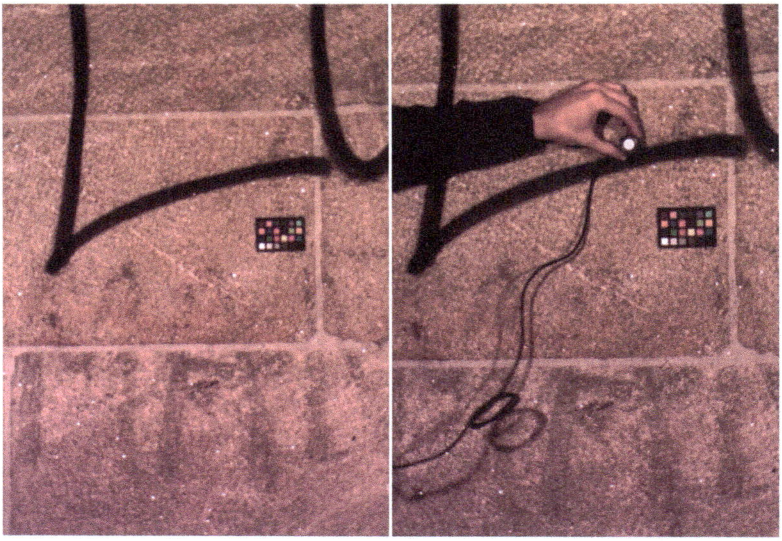

Fig. 11.2 Image acquisition system setup on a granite façade with spray-painted inscriptions. Source: Patricia Sanmartín.

11.3.4 *Miscellaneous other techniques*

During the laser-based cleaning of graffiti, Costela *et al.* (2003) and Gómez *et al.* (2006) used the laser-induced breakdown spectroscopy (LIBS) technique, which is a laser-based analytical method used in the field of artwork conservation. This technique enables elemental analysis of the ablated material from both spray paint and building material substrates and, thus, on-line control of the removal process. Costela *et al.* (2003) focused its use in preventing substrate damage and demonstrated the feasibility of use of the system by unskilled operators with minimal training.

Fiorucci *et al.* (2011) measured the topography of granite samples with an optical imaging profiler and confocal microscopy. Siano *et al.* (2012) measured laser induced plasma spectroscopy (LIPS) elemental depth profiles and Raman spectra (both *in situ*, using portable analytical devices) to assess the extent to which these diagnostic methods can integrate or replace invasive petrographic studies. Results on compositional features were obtained, providing prompt compositional answers and thereby avoiding the need for sampling.

X-ray diffraction enabled comparison of the existing crystalline phases in HPDL laser-treated and untreated surfaces, as well as determination of whether some mineral phases disappeared after laser treatment (Penide *et al.*, 2013).

Ye *et al.* (2014) used a high precision balance to measure the mass of samples. The mass loss (ML) of samples was calculated as $ML = MB - MA$, where MB and MA are the mass of sample before and after laser irradiation, respectively. ML obviously increased with damage to the sample, and the results indicated an abrupt increase in ML when a visible change began to appear on the sample surface.

11.4 Graffiti Barrier Coatings (Anti-Graffiti Coatings)

Anti-graffiti protection is becoming common practice in many urban buildings, especially in areas of social decay. Anti-graffiti or barrier coatings constitute preliminary treatments carried out to prevent graffiti from penetrating the substrate and to facilitate their removal. For this purpose, products with a low surface free energy and therefore reduced surface wettability are generally chosen (Lettieri and Masieri, 2014). Amongst others elements, these include acrylics, waxes, polysaccharides, polyurethanes, silicon resins and fluorinated polymers. Organic–inorganic hybrid materials have also recently been proposed as anti-graffiti systems. As with any other surface treatment for building materials, anti-graffiti treatments should alter the surface appearance as little as possible, and they should not prevent the exchange of water vapor between the substrate and the environment. Current anti-graffiti systems can be broadly divided into permanent and sacrificial. Sacrificial products are removed during the cleaning process and must be renewed. Permanent systems can withstand repeated cleaning cycles (in some cases more than ten). The former type of treatment is best suited to materials of low porosity, such as metals and concrete, but does not fit the requirements for porous ancient materials. The latter is sometimes used on monuments, but it cannot be considered as an appropriate solution, since the removal procedures can damage substrates. Hybrid anti-graffiti systems, i.e., a mixture of the two previous systems, can be considered. These semi-permanent products can normally withstand two or three cleaning cycles, after which re-application of the coating has to be performed.

Anti-graffiti systems have been tested on stone materials; however, very few studies have been carried out with other type of materials. As already mentioned, the most recent efforts in the anti-graffiti field have focused on the application of a product that makes the surface (independently of its nature) highly hydrophobic, thus protecting it against water, minimising adhesion of the paint and facilitating removal of the graffiti by simple manual cleaning with a cloth. These products include the organic-inorganic hybrids based on methyl-derived silanes (e.g., polydimethylsiloxane) (Oteo *et al.*, 2006) and polysiloxane modified polyurethanes (modified by addition of silica nanoparticles) (Rabea *et al.*, 2012).

Methyl-derived silanes are more effective with low-porosity materials, such as granite and marble, than with more porous materials such as calcareous stones. However, the following drawbacks have been observed:

(i) atmospheric agents (e.g. oxygen, sunlight) induce degradation and consequent loss of the protective ability of the products and optical disfigurement;
(ii) the products may cause chromatic variations on the treated stones, unacceptably altering the aesthetic appearance of monuments;
(iii) the products may be sensitive to some components (e.g., solvents) of the paint used to draw graffiti (Marzolla, 2008).

In their patent specifications, Bourdette and Amara (2012) state that the use of the latex type fluorinated polymers is well known in anti-graffiti treatments. Marzolla (2008) reported the results of a study performed on a commercialised PFPE-based polymer called Fluorolink® P56. The product was applied by water-based microdispersion to the surface of Carrara marble, in the presence and absence of an epoxysilane and of a polyaziridine (both curing agents). The variations in the stone surface properties induced by the application of the protective material were mainly studied in terms of chromatic changes and waterproofness. The European project GRAFFITAGE, conducted between 2005 and 2008, aimed at developing a coating for protecting heritage structures against graffiti (GRAFFITAGE). The product was based on complexes of polymeric amines that were modified by fluorocarbon residues, specifically developed to protect monuments. The coating was structurally similar to traditional coatings such as casein, gelatine and egg white. It displayed desirable properties in terms of transparency, low surface energy, permeability to water vapor and impermeability to liquid water.

11.5 Current Perspectives and Future Directions

Currently available techniques to remove graffiti do not fully meet the requirements for the protection of the substrate, sometimes leading to serious damage to the material surface. Therefore, a new risk-free method for the removal of graffiti, especially designed for use on cultural heritage materials and civil buildings constructed in porous materials, such as natural stone, cement, bricks and mortar lime is urgently needed.

Bioremediation provides a novel approach to graffiti removal. Bioremediation is the use of living organisms (primarily microorganisms) to remove environmental pollutants, through biodegradation, or to prevent pollutants from entering the environment, through waste treatment and the biological production of biodegradable substances (Atlas, 1994). Microorganisms are good candidates for bioremediation because of the vast metabolic diversity that allows them to use various organic/inorganic compounds for growth. Bioremediation offers several advantages over chemical and physical cleaning methods: (1) it selectively targets xenobiotic compounds; (2) clean-up costs are often lower than with the other techniques, especially if done *in situ*; (3) it is an environmental-friendly method; (4) it is relatively simple, particularly in comparison with laser-based and novel physical methods; and (5) it has greater public acceptance. Microorganisms have been proposed for the removal of natural organic matter and synthetic polymers from cultural heritage surfaces (Troiano *et al.*, 2014). However, to our knowledge, so far only one published study has focused directly on bioremediation of graffiti spray paint (Giacomucci *et al.*, 2012). According to this study, a more efficient technique for the biodegradation of nitrocellulose-based spray-painted graffiti can be achieved via an understanding of the mechanisms of nitrocellulose degradation. Ammonification was postulated as a degradation pathway, and it was also demonstrated that the sulfate-reducing bacterium *Desulfovibrio desulfuricans* ATCC 13541 is able to transform nitrocellulose-based paint binder and is, therefore, a good candidate as a nitrocellulose-degrading bacterium. More recently, a related study stated that black fungi are not suitable for bioremediation purposes as they are very harmful to stone treated with synthetic polymers (Troiano *et al.*, 2014).

The overarching scientific aim of the project *BioRemoGraf: Application of bioremediation to remove graffiti on cultural heritage and civil buildings*, in which we are involved, is the implementation and development of a biological methodology to remove graffiti from construction materials by using microorganisms, leaving the substrates undamaged. Microorganisms found in aged paint and waste paint will presumably be good candidates for graffiti removal, as they can be bought from international collections of microorganisms (ATCC, DSMZ, etc.) or isolated from graffiti paintings and related

Fig. 11.3 Sampling of microorganisms with potential biocleaning capacity. Source: Patricia Sanmartín, Archana Vasanthakumar and Alice Dearaujo (Laboratory of Applied Microbiology. Harvard University, USA.)

environments, such as car bodies found in junk yards, soils of the sampled areas, the inside of the spray paint cans (Fig. 11.3). We screened several dozen culturable bacteria and fungi for their ability to degrade spray paint to explore the metabolic diversity of microorganisms in breaking down graffiti (Sanmartín *et al.*, 2015). Several organisms efficiently degraded acrylic spray paint in liquid suspensions. The ultimate goal of the study is to establish a biological method of using culturable bacteria or fungi to remove spray-painted graffiti. To accomplish this goal, further investigation of how graffiti components are metabolised or degraded by culturable microorganisms, identification of microorganisms with bioremediation potential, and characterisation of novel degradative capabilities are required.

Acknowledgements

Dr. Patricia Sanmartín is financially supported by a postdoctoral contract within the framework of the 2011–2015 Galician Plan for Research, Innovation and Growth (Plan I2C) for 2012.

References

Atlas, R. (1994). *The Scope of Bioremediation.* Paper presented at Bioremediation: The Tokyo '94 Workshop, Tokyo, Japan.

Bentley, J. and Turner, G.P.A. (1998). *Introduction to Paint Chemistry and Principles of Paint Technology*, 4th Ed., Chapman & Hall, London.

Bourdette, A. and Amara, M. (2012). *Graffiti-Cleaning Composition*, US patent No: US 20120202729 A1. Filing date: July 30, 2010. Publication date: August 9, 2012.

Cappitelli, F., Vicini, S., Piaggio, P., Abbruscato, P., Princi, E., Casadevall, A., Nosanchuk, J.D. and Zanardini, E. (2005). Investigation of fungal deterioration of synthetic paint binders using vibrational spectroscopic techniques, *Macromol. Biosci.*, 5, pp. 49–57.

Chapman, S. (2000). Laser technology for graffiti removal, *J. Cult. Herit.*, 1, pp. 75–78.

Costela, A., Garcia-Moreno, I., Gómez, C., Caballero, O. and Sastre, R. (2003). Cleaning graffitis on urban buildings by use of second and third harmonic wavelength of a Nd:YAG laser: a comparative study, *Appl. Surf. Sci.*, 207, pp. 86–99.

Fiorucci, M.P., López, A.J., Ramil, A., Pozo, S. and Rivas, T. (2013). Optimization of graffiti removal on natural stone by means of high repetition rate UV laser, *Appl. Surf. Sci.*, 278, pp. 268–272.

Fiorucci, M.P., Lamas, J., López, A.J., Rivas, T. and Ramil, A. (2011). Laser cleaning of graffiti in Rosa Porrino granite, *International Conference on Applications of Optics and Photonics. Book Series: Proc. of SPIE*, Vol. 8001, 80014A.

García, O. and Malaga, K. (2012). Definition of the procedure to determine the suitability and durability of an anti-graffiti product for application on cultural heritage porous materials, *J. Cul. Herit.*, 13, pp. 77–82.

Giacomucci, L., Toja, F., Sanmartín, P., Toniolo, L., Prieto, B., Villa, F. and Cappitelli, F. (2012). Degradation of nitrocellulose-based paint by *Desulfovibrio desulfuricans* ATCC 13541, *Biodegradation*, 23, pp. 705–716.

Goidanich, S., Toniolo, L., Jafarzadeh, S. and Odnevall Wallinder, I. (2010). Effects of wax-based anti-graffiti on copper patina composition and dissolution during four years of outdoor urban exposure, *J. Cult. Herit.*, 11, pp. 288–296.

Gómez, C., Costela, A., García-Moreno, I. and Sastre, R. (2010). Comparative study between IR and UV laser radiation applied to the removal of graffitis on urban buildings, *Appl. Surf. Sci.*, 252, pp. 2782–2793.

GRAFFITAGE: Development of a new anti-graffiti system, based on traditional concepts, preventing damage of architectural heritage materials. SSP (Policy Oriented Research) of the Sixth European Programme of the European Commission. FP6-2003-SSP3-513718 (August 2005 – July 2008).

Lettieri, M. and Masieri, M. (2014). Surface characterization and effectiveness evaluation of anti-graffiti coatings on highly porous stone materials, *Appl. Surf. Sci.*, 288, pp. 466–477.

Liu, K. and Garmire, E. (1995). Paint removal using lasers, *Appl. Opt.*, 34, pp. 4409–4415.

Marzolla, S.J. (2008). Modified fluorinated polymers for the protection of stone surface from mural writings, *Scient. Acta (Ser. 2)*, 2, pp. 12–16.

Ortiz, P., Antúnez, V., Ortiz, R., Martín, J.M., Gómez, M.A., Hortal, A.R. and Martínez-Haya, B. (2013). Comparative study of pulsed laser cleaning applied to weathered marble surfaces, *Appl. Surf. Sci.*, 283, pp. 193–201.

Oteo, J.L., Elvira, M.R., Alonso, L. and Rubio, J. (2006). Surface protection treatments of building materials by new organic-inorganic hybrid multifuncional materials, *Bol. Soc, Española de Cerámica y Vidrio*, 45, pp. 96–102.

Penide, J., Quintero, F., Riveiro, A., Sánchez-Castillo, A., Comesaña, R., del Val, J., Lusquiños, F. and Pou, J. (2013). Removal of graffiti from quarry stone by high power diode laser, *Opt. Las. in Eng.*, 51, pp. 364–370.

Rabea, A.M., Mohseni, M., Mirabedini, S.M. and Tabatabaei, M.H. (2012). Surface analysis and anti-graffiti behavior of a weathered polyurethane-based coating embedded with hydrophobic nano silica, *Appl. Surf. Sci.*, 258, pp. 4391–4396.

Rivas, T., Pozo, S., Fiorucci, M.P., López, A.J. and Ramil, A. (2012). Nd:YVO_4 laser removal of graffiti from granite: influence of paint and rock properties on cleaning efficacy, *Appl. Surf. Sci.*, 263, pp. 563–572.

Sanmartín, P. (2012). *Color Quantification in the Study of Biofilm Formation on Granite Stone in Historical and Artistic Heritage*, Ph.D. thesis, University of Santiago de Compostela, Santiago de Compostela, Spain.

Sanmartín, P., Cappitelli, F. and Mitchell, R. (2014a). Current methods of graffiti removal: a review, *Const. Build. Mat.*, 71, pp. 363–374.

Sanmartín, P., Chorro, E., Vázquez-Nion, D., Martínez-Verdú, F.M. and Prieto, B. (2014b). Conversion of a digital camera into a non-contact colorimeter for use in stone cultural heritage: the application case to Spanish granites, *Measurement*, 56, pp. 194–202.

Sanmartín, P., De Araujo, A., Vasanthakumar, A. and Mitchell, R. (2015). Feasibility study involving the search for natural strains of microorganisms capable of degrading graffiti from heritage materials, *Int. Biodeterior. Biodegrad.*, 103, pp. 186–190.

Siano, S., Agresti, J., Cacciari, I., Ciofini, D., Mascalchi, M., Osticioli, I. and Mencaglia, A.A. (2012). Laser cleaning in conservation of stone, metal, and painted artifacts: state of the art and new insights on the use of the Nd:YAG lasers, *Appl. Phys. A. Mater. Sci. Process.*, 106, pp. 419–446.

Troiano, F., Vicini, S., Gioventù, E., Lorenzi, P.F., Improta, C.M. and Cappitelli, F. (2014). A methodology to select bacteria able to remove synthetic polymers, *Polym. Degrad. Stab.*, 107, pp. 321–327.

Völz, H.G. (2001). *Industrial Color Testing*, Wiley-VCH, Weinheim.

Weaver, M.E. (1995). *Removing Graffiti from Historic Masonry*, Preservation Brief No. 38, National Park Service, Technical Preservation Services, Washington DC. Available at: http://www.nps.gov/tPS/How-TO-PREsERve/preservedocs/preservation-briefs/38Preserve-Brief-Graffiti.pdf

Wyszecki, G. and Stiles, W.S. (1982). *Color Science: Concepts and Methods, Quantitative Data and Formulae*, Wiley, New York.

Ye, Y., Qi, Y., Yuan, X., Xiang, X., Wang, H., Zhou, W., Cheng, X., Luan, X. and Zhang, R. (2014). Laser cleaning of contamination on sandstone surfaces in Yungang Grottoes, *Optik*, 125, pp. 3093–3097.

Index

www.ingramcontent.com/pod-product-compliance
Ingram Content Group UK Ltd.
Pitfield, Milton Keynes, MK11 3LW, UK
UKHW020349150926
8624IPUK00004B/76

9 781783 268856